D0271790

COPS & ROBBERS

THE STORY OF THE BRITISH POLICE CAR

COPS & ROBBERS

THE STORY OF THE BRITISH POLICE CAR

ANT ANSTEAD

WILLIAM
COLLINS

William Collins
An imprint of HarperCollins*Publishers*
1 London Bridge Street
London SE1 9GF

WilliamCollinsBooks.com

First published in the United Kingdom by William Collins in 2018

24 23 22 21 20 19 18
11 10 9 8 7 6 5 4 3 2

Text © Ant Anstead Limited, 2018

The author asserts his moral right to be identified as the author of this work.

All rights reserved. No parts of this publication may be reproduced, stored in a retrieval system or transmitted, in any form or by any means, electronic, mechanical, photocopying, recording or otherwise, without the prior permission of the publishers.

A catalogue record for this book is available from the British Library.

All reasonable efforts have been made by the author to trace the copyright owners of the material quoted in this book and of any images reproduced in this book. In the event that the author or publishers are notified of any mistakes or omissions by copyright owners after publication of this book, the author and the publisher will endeavour to rectify the position accordingly for any subsequent printing.

ISBN 978-0-00-824451-4

Typeset by Palimpsest Book Production Ltd, Falkirk, Stirlingshire
Printed and bound by CPI Group (UK) Ltd, Croydon CR0 4YY

MIX
Paper from
responsible sources
FSC **FSC™ C007454**
www.fsc.org

This book is produced from independently certified FSC™ paper to ensure responsible forest management.

For more information visit: www.harpercollins.co.uk/green

INTRODUCTION

I am hugely proud of my police career, and when I look back on the years that I spent as a police officer I do so with a great sense of pride and achievement, knowing that I played a small part in what I firmly believe to be the best police service in the world; an institution that puts the great into Great Britain.

Like any typical teenager, in my youth I never really knew what I wanted to do when I was older, except that I didn't want to follow my father into the hospitality sector. I was the second oldest of four boys; my two younger brothers were still at school, but my older brother was working as a scaffolder on building sites. I would often join him at weekends to earn extra pocket money, which I blew on car parts for the numerous classic cars I was restoring and selling from home. I had an auntie in the police, though, who was cool and I looked up to her, so I started to research joining the police. At 18 and a half I was just old enough to apply. I never told anyone what I was doing, and it was only in the week that I was packing my bags to go to police college that I told my parents of my chosen career. I was young and full of enthusiasm, and I developed a passion for the law. I loved being within the police family and still consider myself, somehow, part of that; I look back on that section of my life with great affection.

Being a police officer also taught me a lot, much of which I was able to use once I left the service, and that knowledge has given me a great sense of perspective. At the age of 23 I became one of the UK's youngest armed officers and a full-time member of the TFU (Tactical Firearms Unit). I hold two commendations for bravery and I can recall countless incidents that took my breath away with either fear, laughter, tears or amazement. I look back at those dangerous incidents now in a positive way, as moments of great luck, and I still have the scars from them, which remind me how lucky I am. I owe a lot to the police and I am grateful for the years I spent with them. It helped me to mature, and I have the utmost respect, admiration and affection for my fellow officers.

The police service in the UK is truly a *service*. I think of it as a service for the people of the nation, which was the original aim, unlike in many other countries where police work has a different ethos. The police taught me about self-discipline, gave me a work ethic and, I'd like to think, some empathy (after all, the best coppering is not always about arrests or box ticking). Also, it taught me, perhaps less usefully in civilian life, firearms training. I learned how to deal with people in all different circumstances. I became fascinated by human behaviour and I use all that learning today in a positive way in my own life. People often ask me if I get nervous hosting live television to millions of viewers, and I simply say 'No. I've had a gun pointed at my face, this is easy'.

Our framework of laws and the ability to enforce them is what makes modern civilised society function – without this we would have no hospitals, roads or even an economy in which to find useful work. The police are the bedrock of a civilised society; we need them in order to function and make our world bearable. We hope we never need them, or meet them, yet we feel reassured that they are there just in case. A great deal of police work goes under the radar because proper policing isn't shouted about, it mops up the bad in favour of the good. My tutor, Terry Walker, who was about to retire after 30 years' service when I joined, taught me about humanity and caring for the public that we aimed to protect. I joined in 1999, at a time when he had seen the worst of people during a changing police service that meant officers had to adapt too. Terry always wore his helmet, never carried a baton, never raised his voice and never asked for anything in return. He was an old-school gentleman. I remember him fondly.

I carried with me into the police a great affliction. I was a petrol-head, and the truth is I always have been. From my days pushing Corgi toys around the carpet as a four-year-old, I turned to Lego as a boy and at the age of 14 made my own go-kart from a lawnmower I bought at a flea market. I paid £2 and dragged it the three miles home before taking it apart and putting the engine into a wooden chassis. I never did get it working. I built my first car when I was 16 and bought a vermillion orange MG Midget the moment I passed my test at 17. Cars were where it was at for me. At an early age I was totally and utterly besotted with the looks, sounds, smells and shapes of cars. I had it bad! I could recognise every car on the road and I yearned to drive them. In fact, it was the cars used in police dramas on TV that also helped steer me into policing – who doesn't love Bodie and Doyle hand-brake-turning a Capri (see chapter fifteen on film/TV police cars) and who wouldn't want to be them when you are 12 years old? I certainly did.

The British police's relationship with the car and the motorist has at its very heart a central issue that is both contradictory and conflicting. When policing and motoring were both in their nascent stages, PCs were set to catch this new breed of people, the motorists, who were terrorising horses and the public with their frightening and, frankly, it seemed, death-defying 15mph machines. Horseless carriages (the term 'car' did not catch on until the early twentieth century) would be an increasingly common feature of UK life for some years before the police started thinking about using them as tools themselves.

Since then the, car has become essential to everyday life. However, it has also become a source of friction between the police and the public. Think about it; if you get burgled or attacked, the police action in catching the criminal and putting them behind bars is supported by 99.9 per cent of the public, but if a traffic cop pulls someone over for doing 83mph on the M1 at 3am, that person is quite likely to feel aggrieved, even though they have, pure and simple, broken the law, just as a shoplifter has. Society's attitudes still mean that the speeding driver is quite likely to moan about the police to their friends and family, or even on social media, in a way that would be unacceptable for a burglar or mugger, almost as if motoring offences are kind of accepted. The number of times I pulled a car over and I was met with 'Have you not got anything better to do?' was alarming. And I'll let you into a secret. The police feel some sympathy with that, too, because most cops love cars!

Motoring, whether it's dealing with driving offences or clearing up accidents, represents the vast majority of UK police interaction with the public and is the biggest cause of friction, too. Yet deep down everyone loves a police car, even when they have been done for speeding! Whether it's a Wolseley cornering on its door handles at 25mph, a Senator spearing down a motorway, or a Morris Minor Panda, the cars used by the police take on a special quality, so much so that countless enthusiasts collect model police cars in different liveries and particular legendary police cars have become part of the nation's shared consciousness.

That love for the police car and the men and women who drive them is exactly what this book investigates and celebrates.

And, like many people today, I share that love affair with the Great British police car.

So I joined the police . . .

There have been a number of stand-out moments during my time as a police officer. I joined in what I can only describe as a period of 'pre-red tape'. Lots of crazy stuff happened. While the police service brought me great enjoyment and I experienced some incredible events, I also saw great hurt and sadness. People can be damaging to each other, and I have witnessed at first hand the very worst of human behaviour. Researching this book and revisiting this period of my life has brought back many positive and negative memories, but I will share with you just a handful of those lighter memories, which whilst writing have made me laugh out loud and cry at the same time. The truth is, in the years I spent in the police, most of those stories are unrepeatable in a book celebrating police cars, and some events I have simply emotionally moved on from, blocking them from my mind. I was young, and revisiting this period in my life has made me reflect on the small impact that I made on public safety and that quite often the day-to-day role of the police officer needs to be secret. In truth, the public cannot know, and I think deep down they don't want to know, what it is that the police really do.

When I joined, a senior officer described being in the police as 95 per cent mundane boredom and 5 per cent total utter fear. And in some ways he was right. As an officer of Hertfordshire Constabulary I spent the early part of my career in the affluent, rural area of Bishop's Stortford. Staffing levels were so low I spent the majority of my two years there policing solo. The public perception was that there was a whole army of police officers patrolling the streets. Truth is, there were often only three of us. Three! Covering an area from the Essex border to the east, Cambridgeshire border to the north and as far south as Ware, the town in which I lived. This huge amount of space was known as part of 'A Division'.

A Division was so vast that to police it effectively we needed to drive – quickly! Hertfordshire Police allowed you to drive cars based on passing various levels of driving qualifications. The entry level driving qualification was known in-house as a 'G Ticket'. Pass this basic test and you can drive a marked police vehicle – but that's all: no blue lights, no rapid response and certainly no car chases. It was simply a marked police taxi to get from A to B. I spent many months policing Bishop's Stortford on a G ticket, and I remember the car of choice fondly, my first ever police car. It was a P-registered Vauxhall Astra – the rubbish one with rounded rear end and awful plastic bumpers. It had grey velour seats with that familiar generic car pattern, and the heater didn't work. It was the most basic of cars, with a single cone blue light on the roof. This car was slow – so slow – but it got me around my patch brilliantly, never missed a beat. Boy, that car could tell some stories . . .

There's something remarkable about driving a police car. It stops everyone. People stare, people behave differently, people certainly drive differently when near a marked police car. I became all too familiar with the varying degrees of public reaction and relied on that response when policing. The car, and of course the uniform, became my asset every day. The car itself became a tool for me in so many different ways; it carried a vast amount of items, it blocked roads when needed, and it became an ambulance when called upon, a refuge for victims, a safe place for informants, an escape when in danger and even a weapon when all else failed. It felt like an armoured vehicle because of everything it stood for, yet all it had was a badge on the door and a blue light on the roof – a blue light I couldn't even use! For me it was what the car represented that made it invincible. I drove that car across every inch of A Division. I got to know my patch so well, and got to know its residents, too. In the UK we are very lucky; we police with consent, with the majority of the public behaving like law-abiding citizens and crime being committed by a small minority. It meant I saw the same offenders over and over. I got to know them, and they got to know me, too.

Once I had established myself as a local police officer I pestered my sergeant on a daily basis to let me qualify to drive with blue lights. I cannot emphasise enough how frustrating it was for me to listen to the car radio for crime and public calls for assistance, knowing I had to drive there like another car in the daily traffic, to be literally sitting at the lights while a burglary was in progress was not what I joined the cops for! The 'Yankee' car was qualified to attend on 'blues and twos' and it would often pass my Astra, waving at me while weaving through traffic. I wasn't in the police long before I knew that I needed to drive a police car properly. I needed to get myself a Yankee ticket. And quick.

The Yankee qualification was a great course. I remember my instructor Vince really well. He was a very dry man who spoke in a monotone, and who, truth be told, really just wanted to ride police motorbikes and tell bad jokes. It took me two weeks to pass the course, which I did whilst driving up and down the country in an unmarked car with two other officers also itching to be allowed to be let loose in some county metal. After the initial days we all became quite familiar with each other, as you would when spending all day together trapped in a car. We covered various aspects of driving and I learned a lot from Vince, including how dark one man's sense of humour can be. Once Vince felt confident that we were competent, we would BLAT (blues and twos) in a police car to different locations across the UK. Vince would start each day declaring our intention, 'Today we are going to Southend to buy chips'. And that's what we did. I still look back and think I had the coolest job. The Yankee course was all about passing; the humiliation of not passing would have been career-ending, and my group would never have let me live it down. Plus they all knew I was a car guy! I had to pass. At the end of the course I was handed a certificate, a piece of paper that said: 'Anthony Anstead. Response Driver'. I still have it, in fact. And that was it, my ticket to get me to the front line of policing. My police career was about to change forever.

Once I was a response driver, A Division policing instantly became different. I could now attend all incidents as a first response. I could stop cars and chase vehicles. So far I had by default been somewhat protected from the public, but now the safety catch was off and I saw the real side of front-line policing. The really bad bits.

My first fatal car crash was horrific. It was on the A120 between Little Hadham and Bishop's Stortford. It was a night shift, around 3am, and three young lads had stolen a lorry, set fire to it and left it in the layby, making off in a second stolen vehicle. In their haste the driver lost control of the car within 100 yards of the dumped lorry and turned it upside down. Both front passengers left by the windscreen. The driver was killed instantly and was lying in the road when we arrived, while the other was alive but had serious head injuries. He was holding his face together like it was a latex mask split down the middle. The rear passenger somehow managed to crawl free and he ran off, nice chap.

As I got closer to the scene, I used my car to block the road at the Bishop's Stortford end and radioed for a roadblock the other side, but I knew assistance was a fair distance away. I could see the lorry on fire at the top of hill and assumed it was involved in the crash. I ran to see if there was anyone inside but the flames were so bad I couldn't get close. I passed the lorry and used cones from my car to block the road. I asked for fire brigade and ambulance while sitting in the road with the injured man, near his dead friend. I was bandaging his head with blood pouring everywhere. He was silent, and it must have been the shock that prevented him from feeling the pain from his injuries. And the sight of his friend's body. It was a strange moment. It felt like hours until assistance arrived, and once I was cleared from the scene I had a few moments to clean myself back at the station before I was then sent at 6am to a report of a broken window at the local Tesco. My Yankee ticket took me to the coal face of incidents and that night became the start of familiar relationship with RTAs, which I attended on an almost daily basis, as Hertfordshire has some fast open roads. I'm often asked if, as a car guy, I like motorbikes, and there's no doubt that, because of this period of my life, and attending numerous bike crashes, my answer is a resounding no.

I conducted numerous traffic stops, mostly mundane for small incidents of speeding or poor driving, and often I just had to have a peek at a car that didn't quite look or feel right. One stop that stands out was in South Street on a sunny afternoon. I was at the traffic lights and an estate car passed me going the other way. The boot was wide open, hinged upwards, and a young man was sitting on the tail of the car with his legs dangling over the edge towards the road and holding the front of a 15-foot rowing boat on his lap which had some wheelbarrow wheels on the rear. No tow bar, no trailer, just this kid holding a boat that was being pulled along by a car. It was one of those 'what did I just see?' moments. I quickly spun my car round and pulled the car over as it got to over 40mph. Just a bump in the road, a clip of the kerb would have dragged that kid out of the car. And they couldn't really see what the issue was – what was wrong with dragging a boat by hand out of the back of the car? There's no real obvious ticket for such an offence, but, needless to say, I didn't allow them to drive a moment longer.

One evening around midnight I was patrolling the edge of the A Division near the M11 junction when a blue BMW roared past me. I instantly gave chase with blue lights, giving the details of my pursuit on the radio. Past the industrial estate I was doing my best to keep contact and get sight of the number plate. I was struggling to keep up and knew the car would be long gone once it was on open roads, and the M11 was nearby. I called for assistance from our faster Traffic cars and requested further help from our neighbours in Essex. I was losing the car but still revving the nuts off my little Astra, trying my very best to keep up. The BMW entered the slip road for the M11, and I was several hundred yards behind. Then the car suddenly braked, screeching to a halt as I sped closer. The driver got out, waving his arms as I closed in on my target, I jumped out to be met by this furious man who shoved an ID in my face and screamed 'I'm fucking Special Branch you fucking prick, check the fucking number plate.' Then he ran back into his BMW and drove off. Yes, I got my ass handed to me by my sergeant for that. But hey, the thought was there. Whoops . . .

I had an appetite for policing, and soon I wanted to leave the rural scene, so I transferred to the very busy K Division, covering Cheshunt and Waltham Cross. There were no green fields and farms; it was a totally different style of policing. And it was busy!

I've had numerous memorable police chases, but one sticks out purely because I'm a car man. It was a normal afternoon on a normal weekday and I was patrolling alone in a slightly newer W-reg Astra. I passed a silver Porsche 911 convertible with the roof down, and instantly recognised the driver – a well-known local toerag. And I knew that there was no way he could afford a 911! I followed for a few hundred yards, requesting a PNC check on the plate and the status of the known villain's driving situation. The car came back normal, owned and registered locally to an address in Little Berkhamsted. Weird how I remember that little detail? Still, it didn't add up, and whatever the information, I was stopping that car. The moment I put my blue lights on he was off. It was now a Porsche 911 versus a 1.4 Astra – mmmm . . . Had he stuck to the A roads he would have left me for dust, but he didn't. He entered a housing estate, weaving in and out of the roads and losing his back end on almost every corner. His lack of car control meant I kept up easily, and we raced around the estate until we reached a dead end, where he jumped out of the car and ran off into a park. I gave chase, and as I was a pretty quick runner back then. He was arrested within a few minutes and we walked back to the car. Of course, he had a perfectly reasonable story: the car was his friend's, etc, etc. But I know cars pretty well, and having a simple look around it I found the poorly modified chassis number concealing that it was, in fact, a stolen car and he had just copied the number plate from another local 911 he saw to avoid suspicion. However, he couldn't resist the temptation of some roof-down cruising. Sure enough, I had caught red-handed one of our most notorious local offenders in a stolen Porsche 911, which was then reunited with its owner, and I'd had a pretty cool police chase, too. That was a good day.

Driving police cars is dangerous and I have had numerous scrapes along the way. I once parked my police car on the A10 to direct the flow of traffic down the very tight Theobalds Lane and to block a crash scene. While waving a lorry on, the back of his rig caught the front bumper of my car, dragging it for about 20 yards while I was frantically waving my arms to get him to stop. Then 'PING', he pulled my front bumper clean off. It was a light and funny moment, looking back. However, that same road was also the scene of my first serious POLAC (police accident). I was a passenger in a marked police car with a member of my team driving. We were on 'blues and twos', pulling onto the A10 when – BANG – we were smashed into by a silver BMW. We spun off and hit a fence and he went into the oncoming traffic. It was a heavy hit. My partner Sue was stunned. I turned the sirens off and ran to the car that had hit us, trying to get the man out of the car. He wouldn't leave the vehicle, even as I was pulling at him harder and harder to get him out. It was only when a member of the public came over and said 'he has his seatbelt on' that I realised what an idiot I was. Whoops. The next moment I was in an ambulance on the way to hospital. Shock is a weird thing.

K Division was crazy, a busy place to police, with varying crime and a melting pot of cultures. It was on the fringes of north London, making it a Metropolitan police area until Herts took over the patch. I was on the transition team for that takeover, sharing the patrols with the Met until they slowly thinned out to being all Herts officers. A memorable period for policing at this time was that of the petrol strikes, which crippled the nation. It was chaos! Only priority motorists were allowed access to fuel (which meant that, as a member of the emergency services, I was okay). Seeing the public descend into chaos and the queues and distress at the pumps made it clear how important the motor car is to the daily lives of so many people.

My time in K Division was great; I saw and did some crazy things in that period; there was a lot of crime and violence, but I knew that I wanted to lean towards more action, so I applied for probably the most specialist role in the police: the TFU, the Tactical Firearms Unit. That decision changed my life forever.

It's important to remain focused on the fact that this is a book about police cars, and although I have numerous stories from my time in the police, as I've already said, many are not fit for this publication. There was, however, for me, a changing relationship with the police car at the moment when I was handed a gun. I learned very quickly that the police car was now both an effective weapon and a place of safety. Before being armed, I had never really considered the ballistic properties of a car, such as, when the shots start coming which areas are effectively bulletproof? It's not like the movies; the door of a Volvo T5 won't stop a bullet! The engine bay, however, would, and that Volvo bonnet became a regular vantage point to lean over while armed with either an MP5 or a G36 weapon. It was the Volvo that became my favourite police car. It was amazing! We used Volvos and Mercedes as ARV (Armed Response Vehicles); both were estate versions, providing more space to carry the extra equipment. The only real performance upgrades were ceramic brake discs and fancy callipers, otherwise it was pretty much a standard car. Between the front seats there was a gun safe that carried larger weapons. First the MP5, a small, compact assault weapon that was really accurate, then some months later we changed to a G36, which was more of a rifle. The safe was locked and the weapons unloaded. We would have to 'self-arm' when en route to incidents, and of course for immediate issues we would use our side arm (in my case a Sig Sauer 9mm self-loading pistol) that we carried each day in a holster. Open the boot of the car and there was a huge locked pull-out tray that contained loads more goodies: baton guns, flash bangs (Stun grenades), shotguns with varying types of cartridges, CS rounds and a bunch of other cool non-firearm stuff. Under this tray we carried additional equipment like first-aid kits, a defibrillator, a stinger (a bed of nails for puncturing car tyres) and so on. This ARV was basically a mobile armoury, and therefore it was heavy! The Volvo seemed the perfect choice, and I could not believe that car's performance ability; it was rapid, but ever so reliable. I can't remember a single moment when that car let us down – and we put her through her paces, we did not hold back! As we patrolled with two ARVs at a time there was always a race to grab the Volvo ahead of the Mercedes, as all the TFU team knew the Volvo had the edge. Strangely, of the standard road-going cars the Mercedes would be superior, but I guess the Volvo just carried the extra weight better. It was a reliable Swede. The TFU had a host of other unmarked cars, and I spent many hours concealed in the back of a green unmarked Mercedes Sprinter van, in particular. It's amazing how dark the humour is when there are half a dozen armed coppers trapped in the back of a van. Now if that van could talk . . .!

The TFU also had an armoured Land Rover 90. I was told it was an ex RUC (Royal Ulster Constabulary) vehicle from Northern Ireland, and it certainly looked the part. Constructed with fully riveted flat steel panels and thick bulletproof glass, it weighed tons. It was also the slowest car I think I have ever driven, taking minutes to get up to 40mph, and it barely went round a corner. I drove it several times but only ever used it operationally once. It was in December one year when we received a report of a man at the Christmas-tree-sales place brandishing a gun. We entered the farm in the Land Rover using the loud-hailer to give the chap instructions. Turned out the gun was a toy.

In 2005 I left the police to follow my true passion of building and restoring cars. My television career kicked off with the Channel 4 car show *For the Love of Cars*, which ironically was hosted by a TV police legend Philip Glenister, who is well known for playing Detective Inspector Gene Hunt in *Ashes to Ashes* and *Life on Mars*. In *For the Love of Cars* I restored an ex-Scottish Police Rover SD1 known as 'The Beast', which went on to sell at auction for a new auction world record for a SD1. Working on that car brought back many police memories, and in the show we retraced the infamous 'Liver Run' in the car as a tribute to the original SD1's dash through London with an organ that was urgently required for a patient mid-surgery.

This book was written in order to share not only my passion for the Great British police car but as a nod back to those proud and amazing years I spent serving the British public.

This is a Hertfordshire Vauxhall Astra Mk3 just like the one I started my policing career driving, only mine didn't have a bonnet box. It wasn't fast, but it was a faithful servant and I loved it at the time.

CHAPTER ONE

CREATING THE BLUE LINE:
THE BIRTH OF THE BRITISH POLICE FORCE

In modern times we take the police for granted, don't we? Politicians wrangle about bobbies on the beat, budgets or efficiency savings (which even in my time in the force was a euphemism for cuts!), and if a serious crime happens we're used to seeing the police going about their business on the news. We all know we should pull over for a police car rushing to an emergency on 'blues and twos', and we take this sight for granted, because it is now expected that in a civilised democracy we will have a police service that will enforce the rule of law on the criminal few for the good of the law-abiding many.

However, it wasn't always like this, and although this book will concentrate on the service cars, some background to the police is both instructive and interesting, especially when you learn how the police first dealt with the early motorists on the roads! The story starts long before the famed 'Peelers'; although they are often seen as the origin of the British police, in fact local areas had, from the earliest days of society, utilised some form of law either by force or consent. The Roman conquest of Britain, which began in 43 AD and lasted until around 420 AD, brought with it a monetary system and a form of organised policing. The Anglo-Saxon kingdoms also had a police service of sorts, based on the regional system known as a hundred or (and I love this word) '*wapentake*'. The first Lord High Constable of England was established during the reign of King Stephen, between 1135 and 1154, and those who filled this position became the king's representatives in all matters dealing with military affairs of the realm.

The Statute of Winchester in 1285, which summed up and made permanent the basic obligations of government for the preservation of peace and the procedures by which to do this, really marks the beginning of the concept of nationwide policing in the UK. Two constables were appointed in every hundred, with the responsibility for suppressing riots, preventing violent crimes and apprehending offenders. They had the power to appoint 'Watches', often of up to a dozen men, or arm a militia to enable them to do this. The Statute Victatis London was passed in the same year to separately deal with the policing of the City of London, and, amazingly, today the City of London Police, dealing only with the 'square mile', remain a separate entity to the larger Metropolitan Police. The current City Police Headquarters is built on part of the site of a Roman fortress that probably housed some of the London's first 'police', and these 'square mile' officers can be distinguished out on the streets by the red in their uniform and by their red cars.

By the end of the thirteenth century the title 'Constable' had acquired two distinct characteristics: to be the executive agent of the parish and a Crown-recognised officer charged with keeping the king's peace. This system reached its height under the Tudors, then the oath of the Office of Constable was formerly published in 1630, although it had been enacted for many years previous to that. Sadly, this arrangement disintegrated somewhat during the seventeenth and eighteenth centuries, and nationally nothing replaced it fully until the Victorian era. However, all police officers still hold the Office of Constable, and when I was appointed as a Constable, in 1999, I too had to take a similar oath in a rather splendid passing-out parade.

London led the way in the development of policing because by 1700 one in ten of England's population lived in the metropolis, and the population of the area would rise to over a million before the century was out. Sir Thomas de Veil, a silk merchant-turned-soldier-turned-political lobbyist, became a Justice of the Peace in 1729 and worked out of his office in Scotland Yard. He conducted criminal hearings, moving to Bow Street in 1740, and as the first magistrate to set up court there effectively created the Bow Street Magistrates' Court. His ideas for curtailing crime were ambitious, and, following his death in 1746, the novelist Henry Fielding was appointed Principal Justice for the City of Westminster in 1748 together with his brother, Sir John Fielding. They appointed a group of 'thief takers' as they were initially called, who were paid a retainer to apprehend criminals on the orders of the magistrates. These men soon became known as the Bow Street Runners; by 1791 they had adopted a uniform, and by 1830 over 300 officers were based in Bow Street, some using horses in the battle against crime.

A similar process happened in Scotland, where the first body in Britain to actually be called a police force was set up by Glasgow magistrates in 1779. It was led by Inspector James Buchanan and consisted of eight officers, but by 1781 it had failed, through lack of finance. However, the force was revived in 1788 and each of its members was kitted out with uniforms with numbered 'police' badges on, in return for lodging £50 with magistrates to guarantee their good conduct. The force of eight provided 24-hour patrols (supplementing the police watchmen, who were on static points throughout the night) to prevent crime and detect offenders. Their list of duties, which would fit comfortably into the basic duties of policing today, included:

- Keeping record of all criminal information.
- Detecting crime and searching for stolen goods.
- Supervising public houses, especially those frequented by criminals.
- Apprehending vagabonds and disorderly persons.
- Suppressing riots and squabbles.
- Controlling carts and carriages.

Thus Glasgow had established the concept of preventative policing, and it's hard not to wonder just how much knowledge of this Robert Peel had or how much inspiration he perhaps took from Glasgow's pioneering force when he brought this idea into Parliament forty years later. On 30 June 1800, the Glasgow Police Act 1800 received Royal Assent, and John Stenhouse, a city merchant, was appointed Master of Police in September 1800. He immediately set about organising and recruiting a force, appointing three sergeants and six police officers and dividing them into sections of one sergeant and two police officers to each section – to some extent copying military practice while also pioneering a line of command that is not unfamiliar today.

However, in order to understand the history of policing in England we must also look at the most efficient freight vehicles of the eighteenth century – boats. The modern police service has its roots in a river security force that was set up by Patrick Colquhoun in 1798. He persuaded fellow merchants to pay into the scheme, which was designed to combat the epidemic of cargo thefts in the Pool of London (a stretch of the Thames), and which, in 1799 remember, amounted to over half a million pounds. The area was so congested it was said to be possible (most of the time) to walk across the Thames simply by stepping from ship to ship. The government absorbed this service in 1800 and renamed it the Thames River Police.

When reformist Robert Peel (whose Christian name is the reason why police are still commonly referred to as bobbies) became Home Secretary in 1822, his initial attempts to create a national police force failed. However, he succeeded in putting through the Metropolitan Police Act in 1829, creating a force with just over a thousand officers. Peel was determined to establish professional policing in the rest of England and Wales, and the Special Constables Act of 1831 allowed Justices of the Peace (JPs) to conscript men as special constables to deal with riots, and this was built on by the Municipal Corporations Act 1835, which established regular police forces under the control of new democratic boroughs, 178 in all. It was not a great success and was often ignored; however, the later County and Borough Police Act 1856 built on the idea, committed proper financing and decreed that a rural police force was to be created in all counties; it was one of these, the county force of Hertfordshire, that I joined.

At this point in history there was still a way to go to achieve the police force that we have today, but the beginnings of our modern policing system could be seen in the counties and in Glasgow. By the 1860s a nationwide force of separate bodies acting to one set of rules and connected by transport links (horse-drawn initially, but increasingly using the railway system, which was expanding rapidly by 1850) existed. The car would eventually be invented in 1885, but police forces all over the world simply weren't ready to manage this new and politically divisive mode of transport. Britain was no exception.

The Office of Constable: the basis of UK policing

On appointment, each police officer makes a declaration to 'faithfully discharge the duties of the Office of Constable', and all officers in England and Wales hold the Office of Constable, regardless of rank. They derive their powers from this and are servants of the Crown, not employees. The oath of allegiance that officers swear to the Crown is important as it means they are independent and cannot legally be instructed by anyone to arrest someone; they must make that decision themselves. The additional legal powers of arrest and control of the public that are given to them come from the sworn oath and warrant, which is as follows:

'I do solemnly and sincerely declare and affirm that I will well and truly serve the Queen in the office of constable, with fairness, integrity, diligence and impartiality, upholding fundamental human rights and according equal respect to all people; and that I will, to the best of my power, cause the peace to be kept and preserved and prevent all offences against people and property; and that while I continue to hold the said office I will to the best of my skill and knowledge discharge all the duties there of faithfully according to law.'

This is the oath that I myself took.

CHAPTER TWO

THESE DEVILISH MACHINES SCARE THE HORSES, YOU KNOW

On Monday 20 January 1896, Mr Walter Arnold, who worked for his family's agricultural machine-engineering concern in East Peckham, Kent, was driving along the Maidstone Road through the neighbouring village of Paddock Wood at an estimated 8mph, 4mph or so below the maximum speed of the 'Arnold Motor Carriage Sociable' (a Benz copy produced under licence) that he was driving. Local constable J.C. Heard, of the Kent County Constabulary, saw this outrageous behaviour from the front garden of his cottage and set off in hot pursuit on a single-speed bicycle. In that one moment the relationship between the UK police and the motorist was, partially at least, set.

The speed limit in the area was a mere 2mph, so Arnold was thus exceeding it fourfold – although how Constable Heard measured this accurately enough to definitely state this is, of course, unknown. Arnold was chased (on the bicycle) and eventually caught by the presumably quite fit local bobbie after a five-mile pursuit! It is recorded, with some understatement I suspect, that Constable Heard had to pedal at his hardest for quite some time (at 8mph that's just under 40 minutes' pedalling time) before catching the unfortunate Mr Arnold. Arnold was charged with four offences; three pertaining to the Locomotives Act and one offence of speeding. The case went to trial, and although Arnold's barrister, a Mr Cripps, argued that the law shouldn't apply to the new lightweight 'autocars' and that Arnold had a carriage licence (a system designed for horse-drawn carriages), Arnold was found guilty on all counts and ordered to pay a total fine of £4 6 shillings, of which only a 1 shilling fine and 9 shilling costs was for the speeding offence. When you take into account that his car sold for £130, this fine seems quite paltry, but remember that cars back then were a luxury, and the average weekly wage was less than £1 for a 56-hour working week.

Arnold was not only the first motorist fined for speeding in the UK, but it is believed that he was the first person in the world to claim this dubious honour. The car as a concept was young, and, much like laws surrounding the internet in the early twenty-first century, the legislators took some time to draft new laws that applied to these vehicles. Some eminent politicians of the time even felt it wasn't worth the bother, as they deemed that these horseless carriages were bound to be just a phase, the attitude being that they would be short-lived as they frightened the horses for goodness sake!

Arnold probably wasn't too unhappy with the publicity that his case generated, because he was in fact one of the country's first car manufacturers and dealers. His career began by selling new, imported Benz cars from Germany, then between 1896 and 1899 he manufactured a licensed copy of the Benz called the Arnold Motor Carriage. At this point demand for these newfangled machines was high and Arnold is quoted as saying in a newspaper report towards the end of 1896 that, 'if we had twenty in stock they could be disposed of in a week'. No documents exist hinting in any way that he deliberately drove through Kent intending to get caught, but you have to wonder, don't you? If he did do it deliberately he was a marketing genius and well ahead of his time. He apparently sped past the Constable's house, and as a local himself he would surely have known where the local bobbie lived. If he did get caught deliberately, he exhibited an understanding of the media and the gentle art of car marketing at least fifty years ahead of his time – secretly, I kind of hope he knew exactly what he was doing, and thus Arnold placed down a permanent marker for all car sales people to follow. Take a bow, my man.

Arnold kept a scrapbook of press cuttings about his cars and their exploits that still belongs to the owner of the very car that he was driving when he was apprehended, so it's a real possibility that this was a deliberate act. Having said that, that scrapbook also shows that he and other family members committed similar offences at least twice more, so perhaps they just considered this the price paid for the freedom to 'motor' at will. This was certainly the prevailing wisdom among motoring pioneers of the time, and the pages of the first editions of *The Autocar* (which began circulation in 1895) are full of motorists testing the judicial system in one way or another, using their new machines. Newspapers also picked up on this, carrying drawings of 'future townscapes' showing horseless carriages and stating that science had made this possible but the law was preventing it happening.

MOTOR "SOCIABLE."

WALTER ARNOLD,
EAST PECKHAM,
KENT.
Apply to—
Station—PADDOCK WOOD, S.E.R.

This Carriage is provided with a Motor of about 1½ Horse-power and Cycle Wheels with Ball bearings, and is intended to carry two persons.
PRICE—Delivered in London, complete with lamps, £130.   Weight about 5½ Cwt.
DIMENSIONS—Length, 7½ feet.   Width, 4 feet.   Height, 4 feet.

Arnold Motor Carriage Sociable, Chassis Number 1, MT20

Engine: 1200cc single cylinder with open crank case and separate water jacket.

Power: 1.5bhp@650rpm.

Range: 60–70 miles on one tank of 'Benzoline' oil, as early petrol was called. This cost 2s 6d per gallon in 1896.

Ignition: Electric, with enough current stored in two 2-volt accumulators to run the car for approximately 600 miles. The concept of a charging system was not initially included.

Max speed: Approximately 12mph, able to average 10mph comfortably.

Transmission: Fiat belt with movable bearings to take up slack.

Production: 12 examples made between 1896 and 1898, plus a small number of related vans.

After being development-tested in Ireland in order to avoid further persecutions by the local constabulary, Arnold Number 1 was sold to engineer H. J. Dowsing and became the first car ever fitted with an electric starter motor, called a Dynomotor. Amazingly, this was originally designed as an electric 'help motor' to give the car greater power when needed for hills. Yes, the concept of the petrol-electric hybrid so trumpeted by the Toyota Prius in recent years was in fact invented in 1897 by Dowsing, although its development was not pursued at that time.

Remarkably, this original car has still only had five owners from new: the Arnold Company, Dowsing, a Captain Edward de W. S. Colver, RN (Rtd), who partially restored the car after buying it from Dowsing in 1931 and whose family sold the car to the Arnold family at auction in 1970. In the mid-1990s it was sold to Tim Scott, who retains ownership today and uses it regularly on the London to Brighton run and other events.

With pollution from cars a regular topic in the TV news, and even a consideration in political campaigning, it seems counter-intuitive to say that the car was bound to catch on because it was so much cleaner than the alternative, but that's exactly what happened. The alternative, which had existed for many years, consisted of wading through a horse-manure-based, and extremely pungent, soup. Period dramas on TV skim over this, but it was definitely a factor in the growth of motor-car use, and compared to getting horse manure all over your skirt or trousers, car exhausts must have seemed amazingly clean and green!

However, UK legislation initially made the technical and physical progress of mechanised transportation difficult, whether powered by gasoline (the word petrol did not enter use until it was patented by Messrs Carless, Capel and Leonard of Hackney Wick in 1896), steam, electricity or even coal dust. In 1865 the Locomotive Act (Red Flag Act) had introduced a speed limit of 2mph in towns and villages and 4mph elsewhere. It stipulated that self-propelled vehicles should be accompanied by a crew of three: the driver, a stoker, and a man with a red flag walking 60 yards ahead of each vehicle. The man with a red flag or lantern (when dark) enforced a walking pace, and warned horse riders and horse-drawn traffic of the approach of a self-propelled machine. After centuries of animal training and use, both the politicians and general population trusted a well-trained horse far more than these newfangled steam-driven road locomotives!

Parliament thus effectively framed legislation that trusted the established system, horse-drawn wagons, and prevented the scaring of horses, which were then the backbone of commerce. The country's legislators put more trust in the behaviour of horses than in the safety of human-designed and built machinery ... This seems amusing to modern sensibilities, but perhaps they were right to create these laws, for these early machines weighed an enormous amount, had little in the way of brakes and made all sorts of scary clanking and hissing noises which could spook even the best-trained horse. However, the development of smaller petrol and electric carriages made this law outdated by 1890, and it was this law that was most often tested by pioneer motorists who claimed they could use an Autocar on a standard (horse-drawn) carriage licence rather than the much more restrictive Locomotive Licence. Arnold also claimed this and was convicted of this offence as well as speeding. This Arnold chap is my type of guy. I like to think we would have been friends.

Although the use of motor cars was growing in the 1890s, the horse was king and would effectively remain so (at least in volume terms) until after World War I, which was, like so many conflicts, the catalyst that drove change technically, economically and socially.

The world's first petrol-powered proper road trip. . .

Britain's legislation did not help our nascent car industry to grow, but it also didn't stop the rest of the world coming to a collective consensus about the fuel of the future. Karl Benz produced the first gasoline-engined car in 1885, and in August 1888, without his knowledge, his wife Bertha made the first ever car journey of significant length using an improved version of his design, to prove its worth to the less-than-confident Benz as well as the world.

Bertha and their two sons, Eugen (15) and Richard (14), travelled from Mannheim to Pforzheim, her place of birth, and although Bertha apparently used a hatpin to unblock a fuel line, she demonstrated the practicality of the gasoline motor vehicle to the entire world – and perhaps should have to put to bed any criticism of women drivers at the same time . . . Without her daring – and that of her sons – this book might have been about steam-powered police cars!

Left: Portrait of Bertha Benz. Right: Bertha Benz and her sons buy petrol during their long-distance journey in 1888.

Some UK pioneer motorists deliberately challenged the Red Flag Act, believing their new lightweight steam-, electric- or gasoline-powered vehicles should be classed as 'horseless carriages' and therefore be exempt from the need for a preceding pedestrian and could be operated using a carriage licence. A case was brought against motoring pioneer John Henry Knight in 1895; he lost and was convicted of using a locomotive without a licence and fined 2 shillings and 6d. However, he had neither a carriage nor a locomotive licence, so the law had two ways to turn, which meant that he was guilty of using a locomotive without a licence whichever decision was made!

Early editions of *The Autocar* (a weekly magazine, the first edition of which was published on 2 November 1895 and whose cover proclaimed: 'A journal published in the interest of the mechanically propelled road carriage') were full of lengthy reports of court proceedings being taken against motoring pioneers for a variety of reasons. Legislators caught up with the reality of mechanical development before the law created any real precedents, and the Red Flag Act was repealed on Saturday 14 November 1896, when the speed limit was raised to 14mph; a fact that was celebrated on the day by an Emancipation Day Run (in which our speeding friend Arnold took part in the very same car in which he had been done for speeding) and which today's London to Brighton Veteran Car Run (often incorrectly called a race) celebrates.

Sadly, the first recorded automotive road fatality in the UK was also in 1896, when Bridget Driscoll, 44, was killed in the grounds of the Crystal Palace on 17 August. She was walking with her 16-year-old daughter, May, and a friend when she was run over by a Roger-Benz car being driven by Arthur Edsall. At the inquest, Florence Ashmore, a domestic servant, gave evidence that the car went at a 'tremendous pace', like a fire engine – 'as fast as a good horse could gallop'.

The driver, working for the Anglo-French Motor Co., who were part of an automotive exhibition taking place at the Crystal Palace, said that he was doing 4mph when he killed Mrs Driscoll and that he had rung his bell and shouted 'Stand Back!' as he approached. However, it was suggested that Mrs Driscoll seemed bewildered in front of the car. She may not have interacted this closely with a motor vehicle before. Who knows? There were then fewer than twenty in the whole of the UK and the public were certainly not educated in how to cross roads safely. Edsall had been driving for only three weeks at the time, and as no licence was required, the fact that he apparently had not been told which side of the road he needed to travel on was overlooked . . . The case went to court, but with conflicting evidence about the speed and manner of Mr Edsall's driving, the jury returned an accidental death verdict. Interestingly, there was no outrage in the newspapers at the time, just a sense that it was an unfortunate accident. It would be some years before the press became powerful enough to express outrage in the way that we take for granted today. It's also worth noting that, although people had been killed by horses or carriages for many years, high-profile autocar accidents did start to bring up some public discussion of these dangerous new machines, which were loved by those who had them and often feared by those unfamiliar with them. For that reason accidents were reported perhaps a trifle more vociferously than they would have been if someone had simply fallen off a horse and broken their neck. It was, after all, the start of the unknown.

Perhaps ironically, the first death in an autocar accident was during an incident with an International Co. electric vehicle on 12 February 1898, when Mr Henry Lindfield of Brighton tried to stop his vehicle in order that his passenger, his 19-year-old son Bernard, could retrieve his bag, which had fallen off as they descended the hill off Purley Corner, in Croydon. When the brakes were applied the vehicle began to weave, before eventually hitting a post and then a tree, sandwiching poor Henry between the carriage and the tree. Son Bernard was thrown over his father and survived with relatively minor injuries, but, sadly, his father died in hospital the following day. Just over a year later the first fatal accident involving the driver of a petrol motor vehicle was recorded; 31-year-old engineer Edwin Sewell's 6HP Daimler Wagonette crashed into a wall after a rear wheel collapsed following a rim breakage when the brakes were applied too harshly while descending a steep hill at speed. He died almost instantaneously. At the time he was demonstrating the vehicle to three passengers who were assessing it for possible purchase by their company, the Army and Navy Stores, then a well-known department store. The front-seat passenger, 63-year-old Major James Stanley Richer, was thrown clear of the vehicle with Sewell and suffered such serious injuries that he died three days later in hospital. The other two passengers were injured but survived.

The roadside plaque (unveiled on 25 February 1969)
records the site of Britain's first fatal road accident,
on 25 February 1899.

As a result, the UK police started to take cars seriously, and the weight of public pressure – and in some cases outrage – meant that they had to. As far as research can prove, it was the Met that led the way, acquiring a Léon Bollée Voiturette in 1896, or possibly early 1897. The arrival of the Bollée, as it was usually called, is important because it represented the moment when the British police moved from enforcing legislation on motorists to becoming motorists themselves. Quite what those early Met police pioneer drivers would think of today's high-speed chases on the M25 we shall never know, but to modern eyes the pioneering car does seem almost laughably poor in both performance and engineering terms. At that time the template for what makes an acceptable usable car, as opposed to an experiment down a dead-end avenue of design, was yet to be set, and the Mercedes-badged Daimler that made every other autocar old-fashioned overnight, and so set the car on its course to dominate transportation in developed nations by the early 1920s, would not appear until 1901.

1896 Léon Bollée Voiturette

The birth of the Léon-Bollée Voiturette light autocar goes back to the 1870s, when Amédée Bollée, son of a foundry owner specialising in bells, started making steam vehicles in Le Mans, France, way before the town became part of motor-racing folklore after Renault's victory in the 1906 French Grand Prix there.

Although Amédée Bollée was a steam enthusiast, he could see the potential in gasoline-engined cars and helped both his sons, Amédée Junior and Léon, design and build them. Léon's design was immediately successful, although I suspect anyone who ever crashed into anything while sitting on that front-mounted chair may have disagreed! The vehicle was considered fast in its day; even in their early days the police bought the fastest car they could afford, kicking off the arms race between the good and the bad guys earlier than most of us had realised!

Bollée's design for a very short-wheelbased, 3-seater motor-tricycle, the Voiturette, weighed only 264lbs, used a tubular frame and featured a single back wheel, making it far more stable than having the single wheel at the front. The engine was an air-cooled, horizontal, single-cylinder unit of 650cc that was hung beside the back wheel on the near-side and used a large, heavy flywheel. It featured suction-opened inlet valves and mechanical exhaust valves with plain bronze bearings, hand-lubricated by grease for the main bearings and a splash-lubricated big-end. A tubular con-rod kept the weight down, while hot-tube ignition and a single-jet carburettor were so close to the float chamber that fires were common. The 3-speed transmission was by three virtually unlubricated gears on the end of the crank, which could be meshed with the three layshaft gears. The drive was by flat belt, a mechanism that was commonly used then to drive lathes and other machines in workshops. There was no actual clutch; the mechanical movement of the back wheel tightened or loosened the driving belt so that drive was achieved or neutral was obtained. Full-forward movement of the back wheel applied the belt rim to a stationary brake block. In some ways it was an elegantly simple solution; one fore and aft movement of the rear wheel achieved one of three outcomes. However, the Bollée was apparently quite tricky to actually drive, partly because it featured a small steering wheel on the driver's right with a handle familiar to traction-engine drivers, which actuated an early rack-and-pinion steering system. Meanwhile, the left hand had to ease the gear stick gingerly back and forth to engage the drive or free it. Turning the spade grip at the top of this lever engaged a gear, and, in order to stop, the lever had to be hastily pulled backwards. I can't imagine many of my former police colleagues (or myself, for that matter) taking it easily, but perhaps sheer fear would have engendered that on this device . . .

The Voiturette was powerful for its size and weight, and its low build made it stable, but the short wheelbase made spinning on slippery cobbled roads quite common, and comfort was not a priority because they had no suspension, relying on pneumatic tyres and a C-sprung seat to provide a modicum of relief. However, driven by what we must assume were men made of granite, they dominated their class in early motor races, famously occupying the top four places in the 1896 Paris-Mantes-Paris race, and winning their class in the 1897 Paris-Dieppe race with a time that was 0.6mph faster than the best four-wheeled car.

Legendary English racing driver and Le Mans winner S.C.H. Sammy Davis bought an example in 1929 to use in the London to Brighton Veteran Car Run (a Léon Bollée Voiturette was apparently the first car across the line on the original Emancipation Run in 1896) and was pleased to achieve a maximum speed of 19mph when testing it at Brooklands! This would have been impressive in 1896; the first Land Speed Record was recorded in 1898 at 39mph. Amazing to think now how far we have come.

After being developed in the family's Le Mans workshop, the car was produced in Paris, at 163 Avenue Victor Hugo, and at Le Harve by Diligent et Cie. It's believed that around 750 units were made; Michelin ordered 200 and the Hon Charles Rolls (we all know who he was) was a customer in 1897. By 1899 new four-wheeled Bollées were being produced, some with much larger 4.6-litre engines, but the marque was bought by UK maker Morris in 1924, who produced the Morris-Léon Bollée until 1931. It finally faded away for good in 1933.

What's in a name?

Léon Bollée's tricycle may have been short-lived, but the word he appears to have invented to name it, Voiturette, lived on and became the generally accepted term for a small lightweight car. It comes from the French word for automobile, *voiture*, and became so ubiquitous that it was used to name a specific Voiturette racing class for light-weight cars with engines of 1.5 litres or smaller. The supercharged Alfa Romeo 158/159 Alfetta that so dominated F1 on its inception was, in fact, originally designed as a Voiturette class racer.

CHAPTER THREE

ADAPT AND ARREST

Or, ignore and watch them flee . . .

"and that such signals are not understood . . ."

The police force had started policing roads prior to the advancement of the car, and complaints had been made about congestion in cities, especially London, in the first half of the nineteenth century, almost 100 years before cars caught on. Horses had a will of their own and were not always as well trained as they should have been . . . Indeed, the Great Exhibition of 1851 had taken traffic congestion at Hyde Park Corner as a theme, with various solutions being hypothesised. The passing of the Metropolitan Public Carriage Act 1869 vested many responsibilities into the hands of the police with regard to both Hackney and Public Carriages. By 1895, horseless carriages were appearing and major cities were getting motorised omnibuses or trams.

However, it's easy to forget that in the 1890s it was by no means clear what cars were going to be. They were obviously going to be *something*, and that something was going to be important, but what that was, no one was quite sure. Some saw them as a miracle cure for congestion (they took up less space than a horse-drawn wagon) and some saw them as evil machines likely to kill everyone who came into contact with them. Most people, of course, held opinions that were between these two extremes. Petrol eventually came to the fore as the way forward for powering cars, but prior to that, steam, electric (which it now seems was 120 years or so ahead of its time, because battery technology and electronic control have only recently made this viable and it is now very obviously the future) and even coal dust and various other strange ideas were explored.

However, by 1901 the Maybach/Daimler-designed Mercedes had built upon the 1891 'Système Panhard' (a mechanical layout which set the front-engine transmission then rear-wheel-drive layout on most cars until the 1960s) and basically set the template for the twentieth-century car: a petrol engine, a gearbox, and brakes on a separate pedal. Everything else that appeared – from disc brakes to fuel injection and rain-sensing wipers – is a refinement of that original concept.

So the government had to act, and in 1903 came the Motor Car Act, which was the first parliamentary legislation pertaining specifically to the internal combustion engine. The original intention appears to have been to abolish speed restrictions on open roads, but ultimately that did not happen and a blanket 20mph limit was imposed instead, which lasted, officially at least, until 1 January 1931. The fact that it displaced legislation known as the Locomotives on Highways Act tells us much about the rapid rate of technological change and the British government's typically measured approach to this. Only in the last ten years have the basic principles laid down by this been seriously challenged by technology, and as self-driving cars become a reality the legislators

will have an ever-greater role in managing traffic and perhaps they will
even control traffic flow through a central computer system.

In the same year, 1903, a Royal Commission on London's traffic was
set up, and after three years of investigation, having heard evidence
from all interested parties, it made the following recommendations to
the Home Secretary:

- Improved regulations for traffic.
- Street improvements.
- The regulation of street works.
- The regulation of costermongers.
- The removal of obstructions to traffic.

It also recommended that an advisory Traffic Board be set up for the
Greater London area. In making this recommendation, the committee
stated that it was of the opinion that the creation of such a board would
enable traffic matters to be considered as a whole, and not in the paro-
chial manner of existing local authorities. Apart from the phrase
'costermongers', whose meaning I confess I had to double-check (it's a
street trader, and I for one think we should try to bring it back into
regular use! It's so much more elegant than 'street-trader' or 'pop-up
shop'), that set of recommendations could have been written last week
. . . It was a serious business, though, as London had reported around
10,000 traffic accidents a year alone in the preceding few years, and
some other large cities had little better statistics.

However, as cars became more numerous on the roads, so motorists
devised ways of warning each other of speed traps, which, in the early
1900s before World War I, were numerous. As early as 1901 the importers
of Panhard et Levassor cars operated an official service for their
customers to notify them of the location of speed traps. That's the
equivalent, in today's terms, of Ferrari sending text messages to owners
saying,'Go on, we've checked the M54 is clear today, go and max out
your 458, you know you want to!' The Automobile Association (The
AA) also warned of speed traps, having advised its members to stop if

a patrolman failed to salute – the driver would then be given information about 'road conditions' ahead, which usually meant police speed traps in which officers used a stopwatch to calculate whether or not a car was going too fast over a measured distance. Oh, how times have changed . . . Amazingly, the AA continued this practice until 1961.

The police remained faithful to using foot patrols and horses during much of this period, especially in poorer and rural regions, and in some areas even bicycles were discouraged; for instance, Lancashire did not supply bicycles to their officers until 1908, although since 1903 they had paid a 'cycle allowance' to those constables wishing to use their own bike. Some forces' hierarchy regarded the ever-faster rise of the motor car as a bad thing for society as a whole; humans have not changed much in the last 200 years, save getting a touch bigger on average, but I think we have got better at adapting to change, haven't we? There is proof, too, that the police were often negative towards car owners simply because they had a car, and that may have as much to do with Britain's long-standing class war as it does with the car. This is to some extent proved by the Inland Revenue, who, until at least 1906 (and maybe beyond), levied a tax of 15 shillings a year on 'Automobilists who employ Male Servants as a motor-car driver' . . . I assume female chauffeurs were tax-free?

Even after the advent of the 1903 Motor Vehicle Act, driving was still very much seen as a recreational activity, and, as this 1904 advert makes clear, the dress required to go any distance tended to confirm this.

As cars became the norm and horses the exception, this changed. However, that unenforceable 1903 law had pitched the police into conflict with the motorist, and the pages of *The Autocar*, *The Motor* and other publications throughout the early twentieth century are filled with tales of pioneer motorists battling what they considered to be overzealous police. As speed cameras and average speed zones proliferate today, you could argue that that's never changed, but the British police like to police by consent and the media has now educated large sections of the public to the dangers of speeding, so it's no longer socially acceptable to boast of speeds attained on the road in the way that it was up until the 1980s.

The first occasion on which a motor car was used by the police to apprehend a criminal was, as far as we know, on 15 August 1900, the very beginning of the twentieth century, at Moor Edge, Newcastle upon Tyne. A rider of a horse, who was drunk and causing havoc, was chased for a mile and arrested by a constable who had commandeered a Gladiator Voiturette from its owner, one F.R. Goodwin, who, perhaps unsurprisingly in this era when cars were few and far between, later achieved some fame as a racing driver. There are many subsequent incidents of private cars being commandeered by the police, so that fantasy of Clint Eastwood getting in your car and shouting 'follow that car' (or horse) may well have been more likely in 1904! Oh, and for the record I have, of course, commandeered a number of cars belonging to members of the public, and yes, of course I said the very same words. The Moor Edge incident demonstrates initiative and a grasp of then-current technology on the part of the arresting constable; both elements that were in shorter supply among the higher echelons of the police at this point. Much was made of this in the press and it certainly helped further the cause of the motor car over the horse in the general population's mind, so it was a significant moment for the motoring 'movement' as a whole.

Gladiator Voiturette

Ironically, the Clement Gladiator 3.5hp Voiturette could not have been a more unlikely motorised hero, and it was far from the fire-breathing performance machine that its name suggested. It featured a 402cc single-cylinder engine and had a top speed of about 20mph, which it could obviously sustain far longer than a drunk man on a horse . . .

Bicycle magnate Adolphe Clément first made a fortune producing pneumatic tyres, then later whole bikes, cars, motorcycles, airships and even aeroplanes! He saw the potential of the motor industry early on and invested in or promoted several marques. His Clement Gladiator was basically a Benz copy made by one of these firms, Cycles et Automobiles Gladiator of Le Pre Saint-Gervais, France, which was imported then by Goliath Co. of Long Acre, London. It's thought they probably made a small number in the UK as well, although sources disagree on that.

Rather wonderfully, the first cars officially recorded as having been purchased by the police are two 1903 Wolseley 10hp wagonettes (remember this is RAC hp tax, *not* BHP), touring cars which were purchased for use by the Met's Commissioner of Police and the Receiver (a strange title to modern ears, which makes you wonder if the police have been placed in receivership, but it is actually the title of the officer responsible for the financing of the force). They were numbered A209 and A210; these number plates were still being used until very recently – A210 was on the Home Secretary's car in the 1970s, but I've not been able to find out where it is now. I'd not realised that the police loyalty to the Wolseley marque stretched back to what were really the first proper police cars in the UK. It gives me a rather warm glow to think of this.

1903 Wolseley, and why the police used this marque for so long

The last Wolseley-badged car, an ADO 71 'WedgPrincess', was produced in 1975, but even today, mainly because of the proliferation of period crime dramas from *Miss Marple* to *Buster*, the Wolseley brand is indelibly associated with police cars. There is actually a reason of sorts, but you'd never guess it in a million years because it is to do with, wait for it, submarines . . .

In 1900 the ailing car side of the Wolseley machine tool business, which had grown from sheep-shearing machines, had been for sale for some months and had not found a buyer. The board had decided that making cars was too complex and, after some exploits that had not been entirely profitable, decided to concentrate on what they knew. That company still exists, successfully, today. The car side of the company was offered with their chief designer, one Herbert Austin, who would of course go on to found a motor company of his own. After almost 12 months of being touted around, and just when the directors were losing faith that it could be sold, the company was purchased by Vickers, even then a much larger concern with interests in defence engineering. The Wolseley car-making operation was moved to an existing Vickers factory, at Adderley Park, in Birmingham, in 1901, and new models were planned.

Vickers was then working on Britain's top-secret new weapon, the Holland-class submarine, which was the first submarine built for the Royal Navy. This project was meant to be conducted in great secrecy initially, although unfortunately an Edinburgh newspaper leaked the details and left the Admiralty no alternative but to admit to its production. However, although they had been considering diversifying into car production for some time, in hindsight it's fairly clear that Vickers' main motivation for the purchase of Wolseley was to set up a discreet operation to work on developing and then building engines for the submarine engineering project; which they did. Doing this in a car factory in Birmingham, many miles away from the boat's development centre in Barrow-in-Furness or Vickers' facility in Sheffield, was also of some value when it came to keeping the details secret. Of course, this meant that Wolseley was effectively the car arm of major government defence contractor, Vickers, so what was more natural than ordering government cars from Wolseley? They were actually genuinely very good cars, with a reputation for reliability forged in long-distance trials and a customer list of the great and the good, including royalty both at home and abroad, plus celebrities of the time such as Sir Arthur Conan Doyle and politicians such as Cecil Rhodes. Wolseley continued to supply a variety of the War Office's needs (the name of this department was changed to the Ministry of Defence in 1964), including re-engineering the SE5 fighter aircraft's troublesome Hispano Suiza HS-8 liquid-cooled V-8 engine into the much more reliable higher-compression Wolseley Viper used in the SE5a.

Between 1902 and 1906 Wolseley supplied 27 vehicles to the War Office, including large heavy tractors and six 10hp cars, which were fitted with 'special tonneau' bodies for the Royal Engineers. The 10hp wagonette cars ordered by the Met featured a 2-cylinder horizontal engine (not opposed) of 2606cc, lying transversely across the car and allowing the driving of the rear axle by means of chains through a transmission of three gears, although later examples were of a 4-gear design. The engine only revved to 750rpm, which, proclaimed Austin, ensured a longer life than that of its rivals that revved faster. The cost was approximately £370, although that depends on specification since customers could, for instance, choose pneumatic or solid tyres. The first car delivered to the Met was number 284, a 10hp wagonette (meaning the seats in the rear were running along the length of the car rather than facing forward), delivered in May 1903, and the second, number 479, another 10hp wagonette, was delivered in June 1903.

A further car, number 480, yet another 10hp wagonette, was apparently also delivered in June 1903, but there is some doubt over whether this actually was supplied to the Metropolitan Police or to another government agency.

A 1903 Wolseley 10hp wagonette taken from a Wolseley company advert.

Herbert Austin left Wolseley to start his eponymous company in 1904, but the company carried on producing well-engineered cars and became one of the pioneers of mass-production OHC engines, with a design that was in some ways similar to that used in the submarine and aero engines . . . Unfortunately, Wolseley over-extended themselves in the early 1920s and went spectacularly bankrupt in late 1926. William Morris then bought the company, privately, and integrated it into his car business, which already contained Morris and MG and would, 12 years later, also include Riley. The distinctive illuminated radiator badge, or 'Ghost Light' as it was often called, was introduced in late 1932. The formation of BMC in 1952 was effectively a takeover of the Morris (or Nuffield) Group by Austin of Longbridge. Sadly, Herbert Austin had died in 1941 so he never saw his own firm merge with the one where he had built his initial reputation. However, that link and the police's loyalty to the Wolseley brand, forged in the early days of motoring, continued, especially in the Met, until the Wolseley 6/110s went out of service in the early 1970s.

If you wish to learn more about Wolseley's fascinating history, I heartily recommend Anders Clausager's excellent book, Wolseley, A Very British Car, *and the importance of his research in uncovering the story above must be noted.*

By 1906 the Met Police had six Siddeley vans, which were of a curious technical make-up to modern eyes and reflect the diversity of car design in this experimental era, being fitted with single-cylinder vertically mounted, 6hp 1173cc engines. To show that the car industry has not changed all that much, they were advertised as being 'Essentially British' despite the larger models being basically copied from Peugeots! The smaller ones used by the Met were actually re-badged Wolseleys because the Siddeley and Wolseley companies were intertwined at this time, which is no doubt why they were purchased.

At least two were used as despatch vans for carrying correspondence from the Commissioner's Office at Scotland Yard to the various divisions, and in 1907 they were joined by two larger, 10hp, Adams-Hewitt vans. The Adams-Hewitts were made in Bedford but designed in New York by the Hewitt Motor Co. and, to modern eyes, had quite the most ridiculous advertising slogan: 'Pedals to Push – That's All', because they used a 2-speed epicyclic gearbox. One is tempted to ask whether they steered themselves . . . By 1912 the Met had added a small number of De Dion-Bouton vans as well.

What's interesting at this point is that although other forces were following the Met's lead and buying vehicles, they were basically for use to move messages, equipment or, in some cases, men; they were not at this point actually being used for (what I would call proper) police *work*, which was still done by constables on foot patrols. Instead, they took the role of support vehicles, vans to move things, or cars to transport the higher echelons of the service between different police divisions. Chief Constables tended to be the first to get a proper car rather than a van (and perhaps hindsight could argue that they were the ones who least needed them . . .), and these were usually reasonably prestigious large cars, nearly always British machines, and often built locally to the regional force that used them. That was not always the case, though; the Chief Constable of Bedfordshire used a Scottish-built Arrol-Johnston, which was, like most of these cars, chauffeur-driven by a constable. Essex Police Chief Constable, Captain Showers, ordered a Belsize 10/12hp car in 1909 and Kent constabulary was quick to adopt this car too, as they had, at this time, the highest percentage of motorists in the UK, probably because Kent was among the richest areas in the country. It should also be remembered that most police officers could not drive, and paying to teach them was an expense that the police tried to avoid, at least initially.

That slow adaptation of mechanised transport and move away from the horse would probably have continued were it not for World War I. In 1914 the army was basically a horse-drawn service with around 120,000 men and fewer than 100 mechanised vehicles in total, including motorcycles. By the end of the war, in 1918, 6.7 million people had joined the British armed forces, and its quantity and variety of vehicles was huge. World War I really was the engine of invention and it's true to say that it was a major contributor to the advancement of the internal combustion engine; aviation in particular drove advances in engine design and ignition. At the start of the conflict most planes were lucky to make 50mph, but by the end both sides had fighters capable of nearly 150mph. In four years that is astonishing progress, by any measure.

The war also taught a huge number of people, well, mainly men but some women, to drive, so when the government, in debt and battling other issues such as the post-war flu pandemic, decided to sell off the military's massive vehicle roster there was no shortage of qualified (or otherwise) drivers. Suddenly war-surplus trucks were cheap and most tradesmen could afford to buy one and pension off their horse because the truck was substantially cheaper to run, more reliable and much faster. This was also seen as a move to clean up a society that had been literally awash with horse manure for many years, but in historical terms it was almost an overnight change – from the war ending in a still largely horse-drawn Britain to massive auctions of war surplus vehicles, such as Mac Bulldogs, being sold cheap to anyone who wanted one, in less than a year. This presented three very separate new issues for the police: vehicle theft became a problem, and in addition criminals could use this sudden availability of vehicles to get to and from crime scenes, and the roads suddenly became congested and needed policing – a wholly different problem that put the police into potential conflict with ordinary and normally law-abiding citizens. Policing motorists for speeding or parking wrongly was and remains a totally different problem than catching criminals.

The motor car had come of age, the public were falling in love with their cars, motor-racing heroes such as Sir Henry 'Tim' Birkin and Malcolm Campbell were celebrated for their achievements in the new cinema newsreels, and Brooklands was *the* place to be and to be seen. The post-war world was going to be a wholly different place and Britain's police service simply had no option but to become a properly mechanised service in order to keep up.

The police ambulances

As the police service became organised and matured through the nineteenth century it may surprise you to learn that a major part of their work was operating an ambulance service and sometimes even a fire service for the local area. This was usually with a hand ambulance – a cart that an officer pulled – while richer boroughs, or areas with a bigger distance to cover, had a horse-drawn ambulance. This continued into the early part of the twentieth century, when both tasks were gradually devolved to separate services in various ways. The government formally devolved the responsibility for ambulances to local authorities in 1930 and, of course, the creation of the NHS in 1948 launched a whole new service. For reasons of space I have focused on 'policing' duties and vehicles in this book, as the fantastic work done by ambulance and fire crews deserves a whole separate volume. On a personal note, I worked alongside both services during my time in the police force and I have huge admiration for their professionalism, competency and, it must be said, humour; they do a fantastically difficult job brilliantly well.

Constable (to motorist who has exceeded the speed limit): 'And I have my doubts about this being your first offence. Your face seems familiar to me.'

CHAPTER FOUR

POST-WAR PEACE;
LONDON SETS THE WAY

The Met police reorganised their management of traffic just after World War I started and placed it under one commissioner's auspices, which, in hindsight, can be seen as the beginnings of the first traffic department. At this time congestion from the mixture of horse-drawn and motor traffic was far worse in London than anywhere else in the UK, even during the war: a conflict which was initially seen as a short-term inconvenience likely to end by Christmas.

However, this scenario lasted five years and the practical experience gained by policing the city's traffic during the war years had shown the value of having a separate department to deal with these matters in London. The Commissioner, too, saw the wisdom of this and, in writing to the Home Secretary at the end of the war, stated that he proposed to make the temporary wartime arrangement a permanent one. He went on to state that, 'with the object of securing still further continuity and uniformity, it was desirable that a department should be formed to deal with traffic and to devote its resources to the study of the various and intricate problems that were constantly arising. This would act as an intelligence department by noting all points arising in other Public Departments, or in the press, and also the effect of the statistics prepared by the Executive Department and the Public Carriage Office.' The

Commissioner also said that he 'envisaged that the new department would also provide a member for the Committee for the examination of Parliamentary Bills and, when required, for the Cab Drivers and the (splendidly named!) Noise Committee (no, this is not a Monty Python sketch). It would also deal with all correspondence on the subject of traffic, including by-laws and licensing. In addition, it would also generally collect and collate all matters appertaining to every phase of the traffic question.'

The Commissioner's ideas on the need for a new department were accepted by the Home Secretary, and on 24 May 1919 the department officially came into being, having responsibility for all traffic matters, whether vehicular, aerial or pedestrian. The Commissioner, Sir Nevil Macready, in his annual report for the year, said of its formation that:

'Owing to the increasing complexity of matters concerning traffic, it has been necessary to form a separate department to deal with this matter. The traffic advisers are in close touch with the Ministry of Transport and are thereby enabled to issue and co-ordinate suggestions. There is no doubt that the congestion of the traffic in the Metropolis is, to a great extent, due to existing thoroughfares being out of date and unadapted, either to the volume or the nature of the present traffic conditions. Another factor is the length of time taken to repair streets and roads, no attempt being made to work at night, with the result that traffic becomes hopelessly congested in main thoroughfares for a period which could probably be very considerably curtailed if more up to date ideas prevailed.'

And it appears that almost a century later nothing much has changed!

Traffic policing was born and the new department was to be responsible to the Assistant Commissioner 'B', Mr Frank Elliott, and came under the immediate control of Mr Suffield Mylius, a civil staff member, and Superintendent Arthur Bassom, who was in charge of the Public Carriage Office, both men carrying the new title of 'Traffic Advisor'.

Arthur Ernest Bassom OBE KPM (14 June 1865– 17 January 1926), Director of Traffic Services

Bassom was, in many ways, the father of UK motorised traffic control; he thought far ahead of most of his contemporaries and was famed for his incredible memory and encyclopaedic knowledge of all aspects of the motorisation of society. Police Commissioner Sir Nevil Macready admitted publicly at the time that Bassom was the one man in the Metropolitan Police who was indispensable! When he reached the retirement age of 60 (for officers below the rank of Chief Officer), in 1925, he was promoted to Chief Constable and given the title of Director of Traffic Services in order to retain him. He died the following year, still in harness. He was also awarded the Road Transport (Passenger) Gold Medal by the Institute of Transport, just two months before he died. As a mark of respect, it was decided not to fill his position after his death and his duties were absorbed by Assistant Commissioner 'B', Frank Elliott.

Bassom joined the Royal Marine Artillery as a gunner at 17 and passed his gunnery examinations with flying colours, which meant he was therefore almost certain of promotion. However, in 1886, just before his twenty-first birthday (the minimum age at which he could be promoted), he joined the Metropolitan Police as a constable. He was posted to 'D' Division (Marylebone), and the following year was transferred to the Public Carriage Office at Scotland Yard, where he spent the rest of his career. In 1901 he was appointed Chief Inspector and given charge of the branch just at the time when it was being forced to regulate motor vehicles.

He had a detailed knowledge of motor mechanics and engineering, having qualified in the subject at Regent Street Polytechnic, and kept pace with developments in this area. He produced the Police Regulations for the Construction and Licensing of Hackney (Motor) Carriages, 1906 (The Conditions of Fitness for Motor Hackney Carriages). This included the requirement for a 25-ft turning circle, something that immediately influenced the design of London cabs, and still does, and has had echoes worldwide as well. In 1906 he was promoted to Superintendent on merit. He visited nearly every town in the United Kingdom and many in Europe to observe their traffic problems and learn from their errors, without being too proud to learn from their successes. He framed 'The Knowledge', the test undertaken by all London taxi drivers, and devised the Bassom Scheme of London bus route numbers.

He was awarded the King's Police Medal (KPM) in 1919 and appointed Officer of the Order of the British Empire (OBE) in the 1920 civilian war honours list.

I'd like to have met Arthur Bassom. His obituary in the *Commercial Motor* contained this paragraph, which rather amused me:

'As a matter of fact, however, he was in nearly every instance able to give better advice than he received. One of his attributes was imperturbability, and he always had the courage of his convictions. He will be sadly missed by the Commissioner of Police, the Assistant Commissioner, Mr Frank Elliott, with whom he came into daily personal contact, and his associates at the Yard.'

You could argue that he was one of the most important men in British history; his theories and actions certainly affected everyone's life, hugely, and still do today, although very few have heard of him.

Amazingly, as the war commenced, the Met's records show they still only had two cars – the 1903 10hp Wolseley wagonettes – so at least we can be sure the taxpayer received value from these vehicles, which did at least 11 years' service! However, the police did have a number of commercial vehicles and forces around the country that subscribed to this principle, with commercials being used to carry messages and equipment and one or two senior officers having cars, although the records of different forces are harder to pin down and some have even been lost. This is why the Met is discussed so much when talking about this early period, and also the Met's policing of the nation's capital tended to create policy that other forces followed in this era to a greater or lesser extent. It's interesting to note that in 1919 the Met actually started to keep a record of its vehicle fleet as a separate entity, acknowledging, at least clerically, that this was going to be a significant part of policing in the 1920s. Following the established principle, the first car purchased in 1919 was a Vauxhall Tourer for the use of senior officers. This model had been a popular military staff car during the war and had an established record of reliable service. It's worth noting that Vauxhall, in this period, some six years before GM bought them, were known as makers of high-quality, superbly engineered sporting and large luxury cars. GM sent the company in a different, more volume-led, direction.

By the time the traffic department was created in late 1919 they inherited a fleet of 35 vehicles together with workshop facilities for their servicing and repair, which consisted of:

- 1 Vauxhall Cabriolet, for the Commissioner.
- 1 Vauxhall Tourer, for Assistant Commissioner 'A'.
- 1 Austin 15hp Landaulet, for Assistant Commissioner 'B'.
- 1 Austin 15hp Landaulet, for the Receiver.
- 1 Siddeley 14hp Tourer, for Superintendent B.3.
- 4 two-seater Humbers, for District Chief Constables.
- 11 two-seater Fords, for Superintendents on outer divisions only.

- 5 Ford vans and 4 Warwick tricars as despatch vans.
- 2 Austins and 1 Arrol-Johnston as ambulances.
- 1 Austin 15hp Landaulet, 1 Renault 11·3hp and 1 Crossley 20hp Open Tourer, to be used as needed.

On 21 February 1920, *The Autocar* reported that 'the London Police are to be provided with motor cars for controlling traffic, the object being to speed up traffic generally and the confining of slow-moving vehicles to the kerb', where one has to wonder if in London's narrow streets they would get in the way and create bottlenecks . . . That statement, though, is important because it's the first example I've seen of the essential dichotomy of traffic policing in the UK. The police have a dual role, to keep traffic flowing and to ensure the safety of all, whether they be pedestrians or motorists. That statement, which was printed in the magazine but obviously comes from a government announcement, shows that commerce and the free movement of the population were still, at this point, being thought of as the pre-eminent problems, and that accidents were just a sad but inevitable consequence. The opposite is, correctly, very much the case today.

The car that the Met adopted for this role was the Bean 11.9hp, which had won out after extensive tests of over 10,000 miles, and chosen because it was the most suitable and affordable. The Met ordered at least four of these cars, probably the largest single order for police cars made in the UK at that time. The force remained loyal to Bean throughout the early 1920s and had at least three other larger examples, plus two or three vans. This was absolutely the beginning of mass motor-car use by the police, and statistically this matches the wider population's adoption of the car. In 1919 there were 109,705 private cars in the UK, and by 1924 this was 482,356. By 1930 it was 1,075,081 and by 1939 it had just been pushed to over two million. However, the war surplus commercials that were being sold off by the government at least doubled the early 1920s' figure, although reliable statistics are harder to find on that.

Bean 11.9hp: the UK's first Traffic car

The Bean 11.9hp was an adaptation of an Edwardian pre-war car (it seems even then that the police purchasers were conservative of mind and favoured tried-and-tested technology) called the Perry, which had been designed by Tom Conroy from Willys-Overland and launched in 1914. It's believed that around 300 were made before World War I broke out, which curtailed production at Perry's small facility in Birmingham. The design was purchased by Bean Cars from Dudley, just outside Birmingham, in 1919, apparently for around £15,000, which was quite a sum in 1919. They got a separate chassis of unusual strength for the period, in which was mounted an inline 4-cylinder side-valve engine of 1796cc, mated to a 3-speed gearbox. It cost around £400, nearly twice the price of a Ford Model T. However, Ford were not considered British, despite building 46,362 Model Ts at their plant at Trafford Park, Manchester, in 1920 alone.

The Bean cars' story is a fascinating tale of 'what might have been', and it *should* have succeeded. The tale begins in 1822, when the splendidly named Absolom Harper founded the iron foundry A. Harper & Sons, Dudley. George Bean married Absolom's granddaughter in

1901, became the company's principal shareholder and expanded the company. They entered the car market in 1919 with a new factory built on what was then a large-scale production model by a company with a reputation for being good engineers. They had, correctly, identified that the motor industry was a good way to make money and was about to boom, so they recruited a sales manager from Austin, Mr B.G. Banks. They planned 2,000 cars for 1919 (which was a lot then for a British manufacturer starting out), 10,000 by 1921, and expansion thereafter, and they had good financial and technical backing and a link with large sales group BMTC. The Harper Bean group was capitalised at an enormous, for the era, £6 million and equipped its former munitions shops in Tipton for assembly and its former shell building shop at Dudley for bodywork. They had moving track assembly lines and in July 1920 produced 505 cars. However, financial chaos ensued and the car soon became outdated in comparison to Morris's offerings, although Bean did produce new and larger models. They were rescued from bankruptcy in 1926 by Hadfields of Sheffield and eventually ceased making cars in 1929, their dream of being Britain's biggest car maker shattered by Morris, Austin and Ford of America, although the firm lived on as a component supplier. Bean did build one last car in 1936, the enormous 'Thunderbolt', which was powered by two Rolls-Royce V12, 36.5-litre engines, each delivering 2,350bhp, and weighing 7 tons. Captain George Eyston used it to break the land speed record three times; his final record of 357.5mph was set in September 1938.

Like most of the Midland's motor industry, Bean eventually became part of British Leyland and began to cast engine blocks, axle cases, etc. In 1988 Beans Engineering became freshly privatised and bought Reliant . . . which went bust in 1995, taking Beans with them. The Tipton factory was purchased by the German engineering group Eisenwerk Brühl, who invested heavily and at their height made 40,000 tons of cylinder blocks yearly. However, the business suffered from financial problems because the German group had not managed to recoup the investment and left the company in debt. For a second time, the management team purchased the business, which then became Ferrotech. By then one of the most modern and efficient foundries in Europe, the new business became a large supplier of castings to MG Rover, but when they went into administration in 2005, Ferrotech failed to find replacement work and the factory closed its doors for the last time in August 2005. Thus ended the story of heavy industry on this site.

The police both in and out of London benefited from those ex-war office commercials, too. In 1919, the Metropolitan Police was split up into four detective areas and each division was allocated two ex-Royal Flying Corps Crossley Tenders, a design that would now be called a pick-up. These are again significant because they were the first vehicles allocated to do real detection and arrest work, rather than just being used to transport people or equipment between different police facilities, although in my head Crossley Tender will always be the 'Lesney Model of Yesteryear' that I was fascinated by as a kid and is a model vehicle that I still own now.

Crossley Tender – the first proper police cars were pick-up trucks!

Crossley of Manchester are actually one of the oldest names in the British motor industry and were the first company to make a four-stroke internal combustion engine in the UK. Amazingly, they were making units under licence from Otto and Langen of Deutz in the early 1860s, some years before the 'car' was invented.

Crossley came to car manufacture in a slightly circuitous way. Having become well known for the quality of their engines and other engineering products, early pioneer car dealers Charles Jarrott and William Letts asked them to make a high-quality British car to address a perceived gap in the market. Ironically, the first Crossley, which was announced in 1904, actually used mainly Belgian and French components which Crossley assembled. It was well received, however, and the firm expanded their car-making while Jarrott publicised the marque by making a record-breaking drive from London to Monte Carlo in 37.5 hours, a record that was broken by C.S. Rolls only a month later using one of his own 20hp Rolls-Royce Tourist Trophy models. Wonderfully, Jarrott complained that Rolls had done this by breaking speed limits that he had obeyed!

The legendary 20hp series was launched in 1908 and remained in production until 1925, albeit gradually modified, rather in the way that Porsche evolved the design of the air-cooled 911, becoming the 20/25 in 1912 and the 25/30 in early 1919. Designed by A.W. Reeves, it was launched as a fully equipped short-chassis touring car for £495 (in an era when coachbuilding was still common on larger cars especially) and featured a side-valve 4-cylinder engine of 4531cc and a 4-speed gearbox.

The War Office purchased a batch of six 20hp Tourers and were impressed enough to think about ordering more with differing body styles. By the time World War I broke out, the nascent RFC had around 60 (figures vary slightly), and by the time the war finished, 6000! Any childhood fans of Biggles' Great War adventures will recognise this. In all, well over 10,000 Crossleys of all types were used by the military in World War I as staff cars, ambulances, RFC Tenders and vans. Most of these were sold off cheaply when hostilities ceased and, having contributed greatly to the war effort, then contributed to the commercial motorisation of Britain as traders abandoned horses for ex-War Office Crossleys of one sort or another. This of course meant that the examples used by the Met blended in, in a similar way to how a Ford Transit would today. So successful and highly regarded was the basic design that Crossley bought back a few, especially staff cars, and refurbished them to 25/30 specification. They continued this operation until 1925, alongside production of new examples which finished in the same year.

Unfortunately, Crossley never built a car that captured either the government's or the public's imagination in the same way again and eventually ceased making cars altogether in 1937. They continued to make trucks, though, and were acquired by the Associated Commercial Vehicles Group, better known as AEC, in 1948, who acquired Maudslay at the same time. The Crossley name faded away in 1956 and eventually became part of the British Leyland melting pot. It was a sad end for the marque that had been the backbone of the British military's move to motorisation during World War I and subsequently provided the UK's earliest police patrol and response vehicles. Rest in peace, Crossley.

The Crossley also played its part in the formation of the Flying Squad, formed in October 1919 after a post-war crime surge; it became known as 'Sweeney Todd' in rhyming slang quite quickly, then just Sweeney, and initially consisted of just 12 officers. Manchester City Police followed London's lead in forming what was originally called a 'Mobile Patrol Experiment' and patrols were made using a horse-drawn carriage that had been borrowed from a railway company. However, it was soon re-organised and issued with Crossley Tenders, the first vehicles used in Britain for actual police work rather than just moving things and people between police facilities. The Crossleys were heavy and fairly simple beasts that were apparently quite easy to skid in wet weather, but they did provide a good basis for patient underworld observational work and were used successfully in this role. They were much loved by the officers who used them because they were reliable and capable workhorses. Most were fitted with van-type backs, or, at the very least, a canvas tilt. They were actually liveried at times with false trade names to appear as delivery vehicles or furniture removal vans to aid their undercover work, the first time this was ever done in the UK and probably even around the world. These disguises were easy to believe because, as discussed, war-surplus vehicles were very common in the early 1920s, although apparently it was quite some while before most criminals cottoned on to the fact that the police also owned some of these vehicles. However, by the mid-1920s these vehicles had become outmoded and had begun to be replaced.

By May 1926, just as the TUC called a General Strike that plunged the country and the government into conflict with the unions, the Met's fleet consisted of 202 cars:

- 6 Austin saloons, used by the Commissioner and his assistants.
- 2 Austin ambulances.
- 6 Bean saloons, used by HQ personnel.
- 5 Bean saloons, used by CID.
- 4 Bean saloons, used by District Chief Constables.
- 12 Bean saloons, used by Superintendents in outer areas of the Met's jurisdiction.
- 4 Bean saloons, used as spares by whoever needed them.
- 18 Bean vans, used on inner and outer dispatch services.
- 4 Bean vans, used for accumulator service.
- 6 Bean vans, used as spares by whoever needed them.
- 17 Tilling Stevens, used as prison vans.
- 1 Dennis, used as a prison van.
- 52 Crossley Tenders for many and varied uses from Flying Squad to radio testing and everything in between
- 41 Triumph Solos and Combinations, used by sub-divisional and Detective Inspectors for patrolling.
- 24 Chater-Lea Combinations, used by sub-divisional Inspectors and Inspectors for patrolling.

And that is the last list of cars you will see in this book – you're welcome!

The General Strike is important in the history of the police car because it marked the very beginnings of police officers being moved around to deal with troubles, something I became all too familiar with in my time as a police (and especially PSU/Riot) officer. This particular role depended on motorised transport to get officers to scenes quickly, and the lack of this sort of capability within various UK police forces was soon exposed nationally to both the government and media. From this time on, policing fleets got bigger and bigger across the UK. At the early stages of researching this book we did look at trying to come up with a definitive list of cars that the police have used, but we gave up quite quickly, simply because it encompasses pretty much everything, with local forces sourcing cars individually and in some cases buying used as well as new, especially immediately after World War II. No records exist for many forces, and in the early days local philanthropic lords of the manor would sometimes even loan cars to the police as a gesture of public spirit, which meant that police officers were sometimes seen in the most unlikely of vehicles, from station wagons to Rolls-Royces. I can't imagine a member of the public today lending their car to the police to borrow. I wouldn't!

Car registration

The Roads Act 1920 had established the Road Fund licence and introduced tax discs to windscreens. This scheme was run by local councils, which meant it was fairly easy to register a car with a number plate that was also being used elsewhere because there was no central list of vehicles at this stage. The criminal fraternity exploited this regularly (as I discussed on the Channel 4 show *The Lost Lotus*, following my restoration of a mysterious Lotus Elite); a UK-wide system of checks was not properly introduced until the DVLA started to computerise information in the mid-seventies. This legislation, and consequent government revenue, was needed though, because by 1930 vehicle numbers had risen to a million (when the UK population was around 45 million, around 20 million less than it is as I write this book in 2018), but drivers were not educated in how to handle their cars in terms of etiquette, skill or plain common sense. Road deaths and casualties were rising alarmingly, and while the prevailing attitude was a feeling that 'one took that risk if one drove', the government felt they had to do something. Cars in this period were improving rapidly as well, getting cheaper and faster every year, with better handling and brakes. I've driven vintage cars (can I just say, vintage actually has an official meaning in the car world, referring to cars from post-1918 and up to 1930. Everything seems to be described as vintage these days, from Mk1 Escorts to 60s plastic handbags. Gosh, I sound like my dad!) and while they may seem, to modern eyes, very heavy to drive and possessing very poor brakes, they are much more usable than their pre-Great War cousins. It's quite possible to cruise at motorway speeds in the larger-engined vintage cars (remember vintage also encompasses Austin 7s and similar economy cars, but somehow the term 'vintage car' produces a picture of a Bentley in the mind's eye, or is that just me?) and they have 4-wheel brakes which, if properly adjusted and maintained, can actually stop a car reasonably well.

This increase in the performance of cars (and, perhaps perversely, the improved brakes as well, for humans will often go faster in a car that has good brakes! Indeed, legendary designer Alec Issigonis has been cited as saying 'putting a dagger on the steering column would lead to a great improvement in driving standards' and he was only half joking) led to the 20mph speed limit being so widely ignored as to be laughable, and the Road Traffic Act 1930 effectively removed speed limits, which caused much public debate. However, in short this was done because the speed limits then were just not enforceable and had become a joke. In fact, it was stated in Parliament in 1931 that, *'the reason why the speed limit was abolished was not that anybody thought the abolition would tend to the greater security of foot passengers, but that the existing speed limit was so universally disobeyed that its maintenance brought the law into contempt'.* So many people were breaking the speed limit that the police could not cope and the government dealt with this not by increasing the police's budget or improving the equipment at their disposal, but by getting rid of the law! The police had realised the situation was untenable and had by then been lobbying politicians to deal with this unenforceable law, and, certainly in the post World War I era, enforcing it with a light touch – code for often not enforcing it at all because they simply did not have the manpower to do so, although officers seeing driving they considered reckless or dangerous certainly did intervene, at their discretion . . .

What the government did was bring in the Road Traffic Act 1930, the first paragraph of which read:

> *'An Act to make provision for the regulation of traffic on roads and of motor vehicles and otherwise with respect to roads and vehicles thereon, to make provision for the protection of third parties against risks arising out of the use of motor vehicles and in connection with such protection to amend the Assurance Companies Act, 1909, to amend the law with respect to the powers of local authorities to provide public service vehicles, and for other purposes connected with the matters aforesaid.'*
>
> *[1 August 1930.]*

The Act required all police forces to institute motor patrols to improve driving behaviour by example, advice and ultimately legal sanctions and prosecution. There was a budget for this which was initially quite small, and because of this motorcycle combinations were often the chosen vehicle for such patrols. However, accidents proved this to be an unwise choice and they were quickly phased out of mass use in favour of cars. Some three-wheelers, especially the BSA, which, like the Morgan had two wheelers, at the front and was thus reasonably stable, were quite popular in various police forces for a short while after they were launched in 1929, until four-wheel cars came down to their price point and higher speeds made officers 'nervous'. These patrols naturally became engaged in other activities – preventing crime and dealing with emergency situations – and the line of the respon- sibilities became blurred. Judged on immediate results, the Act has to be looked at as a failure, for Britain's worst ever year for road casual- ties was 1934 when there were 7,343 deaths and 231,603 injuries recorded. However, in hindsight it was actually a very far-sighted and prescient piece of legislation that built on Britain's reputation for policing by consent and sought to educate the public into safer behaviour and only punished them if they really refused to come into line. I think it's important here to underline just *how* different the prevailing attitude to risk was in this era compared to what is, today, sometimes disparagingly called our health and safety nonsense. These modern regulations are a good thing, despite the press they sometimes get, because they actually make us think about risks and encourage us to take steps to minimise them – that can be anything from making sure your car has good brakes and tyres to putting your seatbelt on. In the 1930s these sensible steps evoked a society-wide mocking and were largely seen as not making any difference; but hey, smoking was good for you then as well . . .

Health and safety in racing

Oddly, motorsport has a very important role in changing this attitude and creating society-wide acceptance of the concept of risk management that gained social traction during the 1970s. Even in the 1960s, when Jackie Stewart started seriously to question why racetracks had trees on the side that would kill you if you skidded off and hit one, he was greeted by derisory calls of 'coward' from drivers and racing journalists alike, who, having recently fought in a war, even said things like 'this is safe, at least no one is shooting at us'. Jackie had thought it through, though; he was getting paid for his skill and courage, not to take what he quite rightly considered to be stupid pointless risks. After all, the racing and skill needed were the same whether the circuit had safe (ish) barriers or very dangerous trees. Jackie blew every other driver away on a wet Nürburgring and won three World Championships, so no one could credibly call him a coward, but it took a long time for that seemingly utterly sensible piece of thinking to become the societal norm. Jackie undoubtedly has a place in the history books for his racing achievements as one of the greats; however, it's his innovative thinking, which has became so mainstream as to seem obvious now but was revolutionary in the 1960s, and his tireless work on safety that are by far his biggest contributions to both motor sport and the world as a whole. His work had a massive impact on us all in one way or another. However, it's arguable that the government's determination to save us from ourselves when it came to motoring started with the Road Traffic Act 1930, and you could, thus, call it the very glimmer of health and safety culture.

First ever driving tests

Like so many other motoring firsts, the first ever driving test was taken in France, under the Paris Police Ordinance of 14 August 1893. It was introduced on a voluntary basis in Britain on 13 March 1935 but did not become official in Great Britain until 1 April 1935, and was not compulsory until 1 June 1935. The first driving test pass certificate in the UK was awarded on 16 March 1935 to the rather ironically named Mr R.E.L. Beere of Kensington.

As the Road Traffic Act came into effect, the UK police were becoming a motorised force and one major development was in its fledgling stages, which would totally change policing and the way in which the police used vehicles. The technology was called . . . radio.

Radio was first used by the British police in 1923. Again, the Crossleys enter our story here because the very first radio experiments were conducted using these cars with hilariously large bedstead-type aerials fitted. Nottingham and Lancashire Police were also at the forefront of this, only a year after the Met.

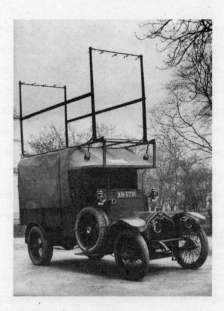

The introduction of police boxes had shown that communication was key to keeping pace with faster society, so it was logical that the next move should be radio. However, in the 1920s talking radio was not yet available; instead, police vehicles used Morse code radio telegraph. There were plenty of (mainly) men around in this era who understood this because it had been taught in the military, so that was the basis of the system fitted to cars. Its reflection of the change in health and safety culture is very evident. Today you are, quite rightly, not allowed to drive even an automatic car with a small mobile phone. These police drivers were expected to drive a car with a crash gearbox while tapping out a Morse code message, listening and translating while wearing a headset. The job of radio operator soon became the norm as the equipment was moved to the back seat.

World War I and the telegraph

World War I was the engine of much technological change, including wireless telegraphy, which used Morse code, and later what was then called radio telephony, which transmitted a voice. Both were refined and made more secure during the conflict, while radio communication took great leaps forward. Telegraphy was famously used by two-seater planes doing artillery observation (known as art-obs) above the trenches, where they tracked troop movements in order to improve the aim of their side's own gunners if they were firing beyond the line of sight, by giving signals to the aimers to shoot slightly to one side, further or less far. This was a fluid operation, as wind conditions changed almost minute by minute, so having instant updates was important. The early planes in question had a long aerial hanging underneath them which was wound in and out on a drum so they did not drag it along when flying to and from the observation area. The industrial production of radio valves made telephony a practical proposition, however, and in 1916 the Royal Flying Corps started developing this, which meant that planes could tell ground stations where shells were landing by voice rather than coded message. In this embryonic stage this was a one-way signal; those manning the ground station would raise a flag to show they could hear. By 1918, the British had mastered plane-to-plane radio communication, which would have been unthinkable a few years earlier. After the war, this first generation of radio operators became the roots of the amateur radio movement, something that would lead to broadcast radio.

The first major police operation that was assisted by using radio telegraphy was the 1923 Epsom Derby, where the force used aircraft as well as the Crossleys for traffic control, not crime prevention. The Met arranged for the use of a radio-equipped Vickers (Type 61) Vulcan fixed-wing passenger aircraft to cover the Derby traffic. This bi-plane was unwieldy and under-powered; its single 360hp Rolls-Royce Eagle engine gave it a maximum speed of about 105mph, but it had restricted banking capabilities and very restrictive small windows so it was unsuitable for observation duties, especially over the confines of the Epsom Racecourse. It was, wonderfully, nicknamed the 'Flying Pig'. The blue and silver Vulcan the police used, G-EBBL, was operated by Instone Airline Ltd and it carried the name 'City of Antwerp'. The head of the Met's Traffic Department, that man Arthur Bassom, was taken aloft along with two police wireless operators and their equipment by Donald Robins, the Instone pilot. The police team were in constant touch with Percy Laurie, who was in charge of the control room at Epsom, from where dispatch riders took instructions to traffic officers at the affected road junctions.

From these beginnings the radio technology developed quickly, and although we are not covering it in detail here, make no mistake, the organisation of police force cars into the categories of use discussed in the following chapters only happened because the police relied ever more on the improving radio technology.

CHAPTER FIVE

SEND THE AREA CAR

Operator: *'Police emergency, how can I help?'*
Caller: *'A bloke's going berserk with a knife down at the Rose and Crown.'*
Operator: *'OK, we'll get a unit to you straight away.'*
Operator: *'Foxtrot five-one.'*
Crew: *'Foxtrot five-one go ahead.'*
Operator: *'Make the Rose and Crown pub in the High Street, male going berserk with a knife.'*
Crew: *'Foxtrot five-one making.'*

A typical 'send the area car' call. All over the country, every hour of every day, divisional area cars and their crews will be dispatched to attend local emergency calls just like that one. This was something I was all too familiar with during my time in the police force. The area car is the emergency response unit for each shift, the backbone of the force – the real front-line cops.

The divisional area car, or response car as some forces call them these days, will generally be a mid-range saloon or estate with a fair degree of performance. The aim is to get a double-crewed unit to the scene of any local emergency as quickly as possible. It's worth noting at this point that your area-car officers are usually hand picked because they have a lot of front-line experience, are trusted to make life or death decisions, to take control of certain situations and have driving skills that are way above the average, having completed a four-week intensive driving course. Current-day models undertaking this role include the BMW 320d, Skoda Octavia vRS, Ford Mondeo 2.0d and the Vauxhall Insignia.

A Vauxhall Insignia area car deals with an RTA in Birmingham in 2016.

Unlike the panda car, whose introduction was down to one man and has a confirmed starting date, the origins of the area car are slightly more difficult to define. Nottingham Police may have been the first to have wireless cars in 1932. They used two-way wireless telegraphy (Morse), which was fully operational in vans, combos and single-manned cars. The method of operation was bizarre and would have gained them a £200 fixed penalty notice and six points on their licence today, whilst the health and safety people dived for cover to fill out their next claim form! Using his left hand, the driver would operate a

Morse key on a large box where the passenger seat would normally have been fitted. He listened via an earpiece attached to his hat, whilst continuing to drive using his right hand! The Chief Constable of Nottingham in his 1940s book *Mechanised Police Patrol* refers to Radio Motor Patrols in an area scheme and in specimen exercises speaks of No.1 Area Patrol, No. 2 Area Patrol, etc., so it is likely that the term was eventually shortened to just Area.

The other early form of system came from the Metropolitan Police, who, in 1934, introduced the 'Area Wireless Car Scheme', whereby the Metropolitan Police District was divided into 75 wireless areas, each patrolled 24 hours a day by a radio-equipped car, manned by two uniformed officers in direct radio contact with the Information Room at Scotland Yard. Their mandate was to deal with all matters and incidents that required immediate police attention. The Met's very first area wireless cars were Ford 14.9hp saloons, followed by Hillmans, Wolseleys and Humbers.

In 1937, Wolseley produced a special 'police tourer' version of the 14/56 chassis. It was not offered to civilian customers.

One of the earliest forms of area cars belonged to the Southampton Borough Police, who in 1956 devised a new concept in mobile policing that is still used today. Called Team Policing, it involved the use of Series 1 Hillman Minx estate cars, one being placed at each of the six sub-divisions within the Southampton Borough, and they were responsible for attending emergencies that fell outside of the remit of the Traffic Division. The cars were painted dark grey with a light grey roof and the only police markings consisted of a telltale roof-mounted radio aerial. To get through the traffic, crews would put the headlights on, and that would be enough in those days to ensure that the public would obligingly move out of their way! The team car experiment was a huge success; in later years the word 'team' was dropped and the term area car was adopted.

Unlike the panda car and indeed the Traffic Division cars, which were vehicles of a very specific type, the area car could be almost anything, depending on the force it belonged to and how it utilised the role. For example, in the 1960s the Cardiff City Police were using the Austin Cambridge A60 as an area car, whilst the Met Police used the much more powerful S-Type Jaguars, and by the late 1970s Kent County Police were sending their crews out in Morris Marina estates whilst the Met again gave their boys Rover 2.6 SD1s. We will concentrate on the most common cars, with perhaps an occasional oddity thrown in for good measure.

One of the big hitters in the 1960s was Austin and its Farina-bodied Cambridge. With its B-series 1622cc engine and 80mph top speed, it was used by a fair number of police forces as an area car. Although the Cambridge was an Austin design built and conceived at Longbridge, these cars were effectively the same model in five versions, because the Morris Oxford SeVI, Wolseley 16/60, MG Magnette MkIV and Riley 4/72 were basically the same cars with different lights, grilles, interior trim, or, in the case of the MG and Riley, more powerful 72bhp Twin Carburettor engine. The Austin was the cheapest of the range, which is one of the reasons why it was the most used by the police. BMC churned out around a million of these cars, in all badges, and they were used successfully. However, they were based on a floorplan and mechanical package, including all-round drum brakes, that went back to 1954 and by the mid-1960s were starting to be slow and under-geared for police work, which increasingly required more high-speed work, especially on the new urban dual carriageways that were starting to be built in this period.

The Austins did not catch on quite as much as had seemed likely, as the area car role was fulfilled by more modern designs such as the Mk2 Cortina. Why BMC never raided the corporate parts bin and updated this fundamentally sound if unexciting range by fitting the MGB's 1800cc version of the same B-series and its matching overdrive gearbox is one of many missed opportunities in the complex world of BMC/BL history. Many members of the Cambridge-Oxford Owners Club, which is a thriving group of enthusiasts restoring these cars, have fitted MGB engines and gear-boxes, which says it all really! (There was in fact a quasi-official attempt at this.)

It would be 15 years or so before another Austin product got the green light as a prospective candidate for area car use, with the Maestro and Montego both finding favour in several forces including the Met, Greater Manchester, Lancashire, Cumbria, Hertfordshire, Dorset and the West Midlands forces. They had adequate performance and a reasonable amount of room but suffered in the reliability department, like so many other British vehicles at this time. That said, let us not forget that the area car has an incredibly hard life; it's on the road

24 hours a day, seven days a week, and gets caned by a variety of different officers during its relatively short but high-mileage police career. I certainly would not want to own an area car that my team and I had used . . .!

Hertford area car, Austin Montego 1.6.

Morris only ever produced one car that was used as an area car, but that didn't stop it gaining legendary status; sadly, though, this was for all the wrong reasons. The Marina was virtually force-fed to the police by a government desperate to keep its ailing and by now state-owned car industry from total collapse, and by the mid-1970s several forces had been 'persuaded' to take them on, including Essex, Hertfordshire, South Wales, Cheshire, Hampshire, Thames Valley and the West Midlands. It wasn't the fact that bits would fall off or its tendency to boil over at regular intervals that gave it its bad reputation, but more its legendary handling – or, rather, its complete lack of stability whilst negotiating anything resembling a curve in the road! It was quite simply appalling. Official complaints were made to senior officers that the cars were unsafe, with some officers refusing to drive them at all. Those unfortunate enough to crash one were hailed as heroes by their colleagues for managing to get rid of the car! The photo shown here comes from the Hampshire Constabulary and is an official photograph taken to show how bad the car was. It is a 1.8TC saloon negotiating an average-sized roundabout at less than 20mph with massive understeer and a suspension set-up that induced motion sickness.

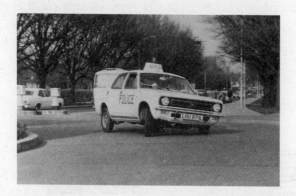

Ford has made more cars that are suited to the role of area car than just about all the other manufacturers put together, thanks largely to the Cortina. The original Mk1 wasn't particularly popular, with only Wakefield City, Surrey and Bedfordshire Police experimenting with them. But in Mk2 guise it received far more attention, especially in 1600 GT form. The car was good for 100mph, with decent handling and a fair degree of street cred to go with it. Essex Police, Devon and Cornwall, Cumbria, Merseyside and Hampshire all bought Mk2s for area car duties. The Mk3 Cortina was just as popular, although its 1970s build quality left much to be desired. As a police car it looked terrific, and several forces even used the estate variant as an area car, including the Devon and Cornwall Police and the Lancashire Constabulary, who painted theirs bright orange. Mk3 saloons were used by Kent, Surrey, Hampshire, Lancashire, Thames Valley and South Yorkshire Police, with most opting to use the 2000 GT model. But these were no ordinary versions, oh no – in order to keep the price to a minimum the plush GT seats with their integrated head rests were replaced with the standard 1300L seats with no head rests or back adjustment, the carpets were replaced with a rubber matting and most of the other interior GT fitments were also missing!

In 1976, with the launch of the Mk4, the Cortina had become a firm favourite as a company rep-mobile and found more police customers, including the City of London, Essex, North Wales, Cumbria, Dyfed-Powys, West Yorkshire, Hertfordshire, Hampshire and Thames Valley Police. In 2.0S trim it came with the excellent 2.0-litre Pinto engine married to a slick and very precise 4-speed manual gearbox. It also got an all-black interior and black treatment to the window frames and bumpers, giving it a rather mean look. By the early 1980s the Mk4 had got some slight cosmetic updates and became the Mk5, but few forces seemed interested as other cars were starting to make a dent on Ford's monopoly.

Along the way there were a few Ford oddities thrown in, such as the Mk3 Zephyr 4 from the mid-1960s, whilst in the 70s quite a few forces used the Consul GT and later the Mk1 Granada, including the fabulous 3.0S. The Granada was one of the few cars that managed to fulfil both traffic and area car roles. It was a large car with a good strong engine and excellent road holding for its size, although manoeuvring the thing at low speeds without power steering took some doing. Even the Ford Corsair got a look-in with a couple of forces, including Cheshire, Wiltshire, Somerset and Bath Constabulary and the Wolverhampton Borough Police. The Ford Orion saw service with Hertfordshire and Devon and Cornwall Police.

In the early 1980s Ford gave us the 'jelly mould', otherwise known as the Ford Sierra, in both hatchback, saloon and estate variants, and it was a very popular choice for area car services in virtually every force in the country. Here was a medium-sized car made available with a number of engine options from 1.6, 2.0 and 2.3 litre that the police could choose from, and they bought them by the bucket load. The car even starred in that new TV series of the era, *The Bill*, and was an integral part of the opening credits, with PC Alan Stamp, the Relief Area Car driver, behind the wheel.

In automotive terms, the Sierra was quite short-lived, being replaced by the first series Mondeo in 1990. Even though the cars were pretty similar in size and specification, the Mondeo was nowhere near as popular in police use. As the car morphed into the Mk2 Mondeo things did improve slightly, but by the mid-1990s there were huge changes taking place with the advent of common rail diesel engines being pioneered by BMW and Peugeot, which had Ford and every other manufacturer playing catch-up. It was over a decade before Ford could fight back with the Mk3 Focus that a lot of forces opted to take on in an era where costs were playing a significant role once again. At this stage, it's worth noting that just changing a car is only the tip of the police car iceberg. Before that decision is taken, the car has to be thoroughly tested to assess its suitability for the role, then the workshops have to be kitted out with all the latest computer software applicable to that make and model; specialist tools would often need to be purchased, and in some cases technicians were sent off to the manufacturers' training establishment to undertake courses to learn about the latest technology involved. However, at the end of the day, if the bean counters upstairs didn't like the figures there would be a sharp intake of breath through gritted teeth as they stated rather firmly that they didn't think the car was 'best value'. So in twenty-first-century Britain you'll find a huge number of Ford Focus area cars in your neighbourhood, although you might find it nigh on impossible to decipher which is the section car (current term for panda car!) and which is the area car, as they will both look identical, dressed in their modern Battenburg clothes.

Yorkshire Mk2 Cavalier after being involved in an RTA.

Vauxhall's area car contribution in the 1960s included the Victor FC with its 1600cc engine, which was similar in styling to the later Mk3 Cortina but not as popular with either the public or the police. Most of Vauxhall's other models were either the bigger-engined Traffic cars or the panda car range, with little in between. It wasn't until the advent of the front-wheel-drive Mk2 Cavalier that Vauxhall produced a car that would be able to cut it as an area car. The Cavalier was a good, strong workhorse made available as a saloon and a hatchback with a willing engine that produced reasonable performance, and, like the Cortina, it soon found its niche as a response vehicle. Merseyside, Derbyshire and Sussex Police took on the Cavalier, and, in later Mk3 and Mk4 versions, forces like the Met and Lancashire also brought them on board. But just like Ford in the mid-1990s, Vauxhall were caught napping and missed the move towards diesel power, and it was several years before we started to see cars like the Mk2 Vectra and the Mk5 Astra make a fight back towards the lucrative police market.

This picture of Woolwich Police Station yard in 1968 shows the eras changing, new Escort panda cars coming in to replace Anglias, but an S-type area car still being used along with a Hunter, Rover P4 and Mk2 Cortina.

Think Jaguar and most of us will think Traffic car; whether it be a 1960s Mk2 or the latter XJ6 range, we generally remember the big cats for cruising the country's motorway network. But the Metropolitan Police decided that the S-Type Jaguar would make an ideal area car – and who would possibly argue? Certainly not the boys tasked with driving them! Probably one of the best-looking saloon cars of all time, the S-Type Jaguar still stirs something in most of us, and in police trim it looks positively fabulous. The Met employed these cars as both traffic and area cars, with the former painted in white and the latter in black. These cars were produced on a separate production line – but only at the weekends – and were some 33 per cent cheaper to buy than the standard production car, but then there were some significant changes to facilitate that reduction that makes these cars unique. For starters, the 3.4-litre engine was changed in favour of the more powerful 3.8 unit, which was married to an automatic gearbox and a low-ratio Power-Lok limited-slip differential. There was no power-assisted steering and the tachometer was removed. But it was the interior where most of the savings were made, by removing the polished walnut veneer dash and door cappings so often associated with the marque and replacing them with very cheap wood painted matt black. Out came the deep-pile carpet and in went rubber floor matting. The door cards were plain black vinyl affairs with no door pockets or arm rests. In total there were more than 150 changes made to each car, and between 1966 and 1968 the Met Police bought 266 S-Type Jags, making it the single-biggest order for one model ever made. Of those 266 units, 183 of them were black area cars whilst the remaining 83 white cars were designated as Traffic cars. It must have been quite a thrill to drive one, on a shout with the Winkworth bell chiming away through the streets of 1960s London, although it's doubtful whether it was possible to get anywhere near the car's 120mph top end very often. A total of only five police-spec S-Type Jaguars have been saved and restored to their former glory, with three of them being area cars. As a restorer, personally I think it's a shame that more of these cars were not saved.

Prior to the Met buying S-Type Jags, their area car of choice was the Wolseley 6/110. The big Farina cars were powered by a 2.9-litre straight six engine mated to a 3-speed gearbox. They were all finished in black and carried a large PA speaker on the roof. The cars were good for just over 100mph and the officers who crewed them rated them as the best area cars they'd ever had. In most other forces the cars were used as Traffic cars, with the last ones still in service as late as 1972. Their subsequent success on the banger racing circuit (big BMC Farinas were the banger of choice in the 70s and 80s) oddly required much the same qualities as those needed in a police car. Shame so many got destroyed, though.

Rover's foray into the police-car market didn't really begin until the introduction of the P6, but with it came instant cult status. Prior to the P6's introduction a couple of forces had dabbled with the P4 – namely Cheshire, who had an entire fleet of them in green! Like the Wolseley 6/80 and the Jaguar Mk2, the Rover has become an iconic police vehicle revered by many, especially those officers who were tasked with driving them. Although many forces used the 3500 V8 as a Traffic car, with some using the smaller 2200 as an area car, only the Met sought to use the V8 as an area car, taking over the role from the outgoing S-Type Jags. They were all finished in Zircon blue, with a single blue light and a couple of Mickey Mouse spot lamps adorning the roof, and they will no doubt be remembered by Londoners of a certain age. You know who you are . . .

However, in 1975, when Rover replaced the P6 with the all-new SD1 3500 V8, the Met's area car crews were all reduced to the 2.6-litre model. Out went the spot lamps and blue paint, in came white cars with orange and gold stripes along the sides and little else to distinguish them from Traffic cars save the single blue light instead of twin beacons. Both the P6 and SD1 served the Met brilliantly over a period of 15 years, and that aforementioned cult status isn't a label applied lightly – although elsewhere in chapter eleven you can read more details about the SD1 and why the choice of the 2.6-litre, 6-cylinder over the 3.5-litre V8 made by accountants backfired spectacularly.

City of London Police Triumph 2500 Pis at speed.

Rover's biggest challenge came from Triumph, when it launched the 2000 range at about the same time that Rover introduced the P6 in 1963. Another big four-door saloon with a decent engine, it was seen by many forces as a Traffic car rather than an area car, and in later Mk2 Pi trim, even more so. However, some forces did utilise the Mk1 Triumph 2000 as an area car, including the Stoke-on-Trent City Police, Somerset and Bath Constabulary and the Nottingham City Police, who had their cars finished in black with the city coat of arms placed on the front doors. The Met Police used the Mk2 2500 TC as area cars about the same time as the Rover P6 models. They were painted in either Delft blue or Pageant blue and were said to be lighter and easier to drive than the Rover but did suffer from engine overheating when in heavy traffic and from a number of other niggling faults. The City of London Police used the Mk2 saloon as area cars, all of them finished in white to differentiate their cars from neighbouring Met vehicles.

A Rootes Group publicity image from 1970, taken to publicise their range of special police vehicles.

The Rootes Group, usually using their Hillman badge on smaller cars and Humber on larger cars, produced some ideal cars to fulfil the role, especially with its Hunter and Minx saloons. For a decade from 1966, the company produced this solid four-door saloon with a 1725cc motor capable of achieving 95mph. It had dependable handling and was a tough and solid car. It was everything an area car should be, and those officers who once crewed them remember them with a fondness not afforded to other cars. Many forces utilised the Hunter and its round-headlight twin the Minx, including the Sheffield and Rotherham Constabulary, Durham, Hampshire, Surrey, the British Transport Police and the Ministry of Defence Police. Developed under the internal code 'Arrow', the Hunter range is often forgotten now but it achieved an enormous amount in its quiet understated way, not just with the police. Not only did Andrew Cowan, Colin Malkin and Brian Coyle win the 1968 London–Sydney Marathon in one, but it also remained in production, latterly only as a pick-up, until 2015 in Iran where it was known as the Paykan, which is, apparently, Persian for Arrow.

The later Hillman Avenger was even more popular with forces such as the Avon and Somerset, West Yorkshire, Sussex, Dorset and Kent Police. Although it had a slightly smaller engine, at 1598cc, it was almost as quick as the Hunter/Minx and, being the younger of the two models, was obviously a more modern drive. Its styling wasn't to everyone's liking, though, due to its rather heavy-looking rear end. Sussex Police in particular bought lots of them and they remained the area car in that county for many years.

Volvo. Yes, Volvo. Many years before those legendary figures 'T5' hit the automotive headlines, Volvo had enjoyed a modicum of success with a few forces but none more so than in Hampshire, who had pioneered the Swedish brand in 1965 when it bought the 121 Amazon estates as Traffic cars, followed by the 144DL in 1968, again for traffic patrol, until about 1973 when the car got downgraded to area car role because more powerful cars like the Rover 3500 V8 and Mk2 Triumph 2500 Pi were now the favoured tools. But the force stayed loyal to the Volvo 144DL, which was followed in 1975 by the new 244DL with its 2.1-litre engine. Every station in the county had one as their area car, although a number of other cars like the Hillman Hunter, the Mk3 and Ford Cortinas were brought in, but in very small numbers along the way. Forget all the very unfair journalistic 'tank' labels given to it over the years, the Volvo was a big, solid, dependable car that was perfectly suited to the role. In 1979

Volvo upgraded the engine to a 2.3 litre, added fuel injection, a 5-speed gearbox, alloy wheels and a few other bits, and all of a sudden the 'tank' would top 115mph with a 0–60 time of just nine seconds. Those stats are identical to those of the Mk2 Ford Escort RS2000 that those same journalists all raved about! In 1985 Volvo introduced the 240 Police Special, built to Swedish Polis specification. It got the same 2.3i engine and performance figures, but Volvo built the car with frontline police work very much in mind, with steel wheels instead of alloys, half-cloth and half-vinyl front seats, all-vinyl rear seats to facilitate the mopping up of blood and vomit, no rear window winders and a rubber compound on the floor instead of carpet. The first 244s were released in 1975 and the last 240 version in 1990 – during which time the Hampshire Constabulary bought no fewer than 276 of them, all for area car use. There can be few other cars that have spanned 15 years and managed to outgun all the new arrivals, staying the course until it was no longer available. There were many police officers who genuinely mourned its passing – and you can't say that about many cars. Certain officers who were unfortunate enough to be driven into by other road users also credit the Volvo with saving their lives, at a time when few other manufacturers were prioritising occupant safety in accidents.

One of Hampshire's 1985 Volvo 240PS area cars, photographed at Brooklands' annual Emergency Services Day (which is well worth attending, by the way). It's been restored by ex-Hampshire officer Steve Woodward and is the only one preserved in police livery of the 276 244/240 models the force used. The light bar really was a roof-rack-type bar with two lights on, as this image clearly shows!

They were fitted with a switch panel for the police accessories which became almost legendary within the force, because the switches had a particularly tactile, satisfying action. And, okay, I'm admitting it here, I love a good switch action, so I totally get why the purple switches were so loved.

Left to right

S - Stop box on boot flashing stop
P - Police sign
B - Blue lights
H - Headlamp flasher
TT - Two-tone horns

In 1995 BMW entered the medium-sector police market like a bull in a china shop by giving us the 325 TDS. What? Diesel-powered patrol cars? Diesel? Really? No, this must be some kind of a wind-up, surely? Diesel is for tractors and buses! Coppers all over the country were now looking for the culprit who started this vicious rumour. But it wasn't a rumour; in fact, far from it. BMW had refined the diesel unit to such an extent that it gave the car a huge amount of useable, low-down torque, which was delivered in a very smooth manner without the clanking sounds of previous diesel-powered cars. It was an ideal power plant for area car use, with the added bonus of better fuel consumption and higher mileage possibilities before resale. And, of course, BMW had built its previous reputation on being a driver's car, so within seconds every copper had a smile on their face once more. Without a shadow of doubt the BMW 325 TDS was a huge success story, not just for BMW themselves but for the police service as a whole. The car had very few vices; it was fast, comfortable, handled well and had that all-important ingredient: presence. It looked the part. Forces all over the UK took them on, and over the next few years and into the twenty-first century the 3 Series has continued to dominate the area car sector because quite simply it's almost the perfect package.

Volvo story

The early Volvo 244 GLT cars were actually used as Traffic cars for a while alongside BMW 525 saloons. The Volvos were fitted with very nice-looking five-spoke alloy wheels; a first in its class. A traffic unit driving a Volvo 244 GLT picked up a Triumph Trident motorcycle one night that was suspected of doing a big drugs run into the city; the bike failed to stop and a very high-speed pursuit started, with the bike reaching speeds in excess of 100mph. The bike entered the city but its speed rarely dropped below 60mph. A second Volvo area car now joined the chase and positioned itself behind the Traffic car. As they sped towards a large crossroads the Triumph didn't slow down at all and went straight across it at 60-plus mph. The traffic Volvo hit the brakes hard and the rear nearside of the car slewed to the left. A huge arc of sparks, some 20 feet high, was seen coming from the rear before the car travelled across the same crossroads. Less than two miles later the biker fell off his machine and was arrested. A two-pound bag of heroin was found under the seat. After everything had calmed down a bit, one of the officers present called the traffic officers over to look at their car. We all stood in disbelief as we looked at the rear nearside alloy wheel, which now only had one of its five spokes left! A shudder went down the spine of the driver when he realised that after hitting that kerb he was still driving the Volvo at over 80mph.

Police Constable Gledhill GC

On the night of 25 August 1966, Police Constable Gledhill and his colleague PC McFall were on patrol in 'Papa 1', a Metropolitan Police area car, when they got involved in a high-speed chase with heavily armed robbers who fired fifteen shots at them during the pursuit.

For his bravery that day Tony Gledhill was awarded the George Cross; the highest civil award this country can bestow.

The citation for Constable (later Detective Sergeant) Gledhill's George Cross was published in *The London Gazette* (dated 19 May 1967), as follows:

Constable Gledhill was driving a Wolseley 6/110 police vehicle with wireless operator Constable McFall, when a message was received stating that the occupants of a motor car had been seen acting suspiciously at Creek Side, Deptford. As the officers reached the area they were looking for, a car (also a Wolseley 6/110) containing five men drove past them.

The officers immediately gave chase to the suspect vehicle which was being driven recklessly through the streets of South London, travelling on the wrong side of the road and against the one-way traffic system. In such conditions Constable Gledhill exercised considerable skill in following, at high speed, and in keeping up with, the vehicle.

During the chase, which covered a distance of 5 miles at speeds of up to 80mph, an attempt was made to ambush the police vehicle and no fewer than 15 shots were fired at it by the occupants of the suspect car, using a sawn-off shotgun and revolvers. Pellets from the shotgun struck the windscreen of the police car on three occasions.

Finally, at a road junction, the escaping car crashed into a lorry. The five men immediately left their vehicle and a group of three ran into the yard of a transport contractor.

The officers followed the group of three and as the police car reached the yard gates the men ran towards it; one of them was holding a pistol. He then proceeded to hold it to Constable Gledhill's head. He ordered the officers to get out of the car or be shot. Both officers left the car and the man with the pistol got into the driving seat with the obvious intention of using it to make a getaway.

Constable Gledhill, then backing away across the roadway, was targeted again as the man reversed away from the gates towards him, pointing the pistol at him as he did so.

However, when he stopped to engage first gear he momentarily turned his head away and Gledhill immediately grabbed hold of his gun hand. As the vehicle moved off, Gledhill managed to hold on to the car window with his left hand.

While this was happening, Constable McFall had run along the roadway to a group of men in order to get a lorry driven across it in the hope of blocking the hijacked police car, when he heard Constable Gledhill shout.

He ran back to the police car and saw him holding on to its window. He then saw the vehicle gather speed, dragging Gledhill along the road. At this point the front offside tyre burst, the car veered across the road, crashed into parked vehicles and Gledhill was thrown under one of them. McFall opened the front passenger door and as the driver was still holding the pistol, began hitting him about the legs and body with his truncheon.

Gledhill had then regained his feet and as he approached the driver's door it was flung open, knocking him to the ground.

The man got out of the car and backed away from the officers. He warned them not to move and at the same time fired a shot. The Constables then heard the gun click and both rushed at the man, and *as McFall struck at him with his truncheon Gledhill grabbed the man's right hand and took the gun from him. There was a violent struggle and the gunman fell to the ground, trying desperately to reach the inside of his jacket.*

At this stage other officers arrived. The man was subdued and another gun, an automatic pistol, was found in the pocket of his overalls.

Both Gledhill and McFall received injuries and had to receive hospital treatment. They had faced a sustained firearm attack and from the early stages knew the risks they ran of being killed or seriously injured.

For his conduct, Constable McFall was awarded the George Medal.

During an interview in late 2010, Gledhill was asked how accurate the citation was, to which he replied, 'It's actually a pretty good description of what happened'. On being asked why he and PC McFall got called to the scene, he answered, 'It was the school holidays and a school caretaker's son happened to look out of a window and saw five men putting on masks and dungarees and getting into a car. He told his dad and his dad called the police.'

Gledhill was asked what kind of car the robbers had. 'Actually, they had a car just like ours – another Wolseley – but in a two-tone blue colour. The only difference was, theirs was manual'. This actually makes the chase even more impressive, an automatic Wolseley, on cross-ply tyres, chasing a manual version, at speeds of up to 80mph, in the wet!

Of the chase he said, 'Once or twice I thought I would lose it. Not long after the start of the chase it started to go and I said, "I'm going to lose it". Terry, my wireless operator, replied, "You won't, you won't, you won't." And I didn't!'

Tony explained what Wolseleys were like as police cars: 'I used to really like them. They were not too bad at all. When I first passed my police driving course at Hendon, in 1963, I was based at Lee Green Police Station driving Wolseley 6/90s. They were much heavier. The 6/110s were nicer to drive, although around South London you could rarely get any speed up (but you could really get your foot down on the Sidcup Bypass!). I enjoyed getting up a good speed in them.'

By the time of the incident Tony was based at Lewisham Police Station.

'We had two Wolseleys at our station and there were three shifts. The cars were always used alternately unless there was a breakdown.' On being asked if he drove the Wolseleys all the time, he answered, 'No, as a policeman in those days you might find yourself on the beat for six or nine weeks, or as a police van driver on van duty, or driving the area cars (the Wolseleys).'

He was asked whether he was involved in the maintenance of them. 'No, but we were lucky as the Traffic Division workshops were on site at Lewisham so we didn't have to take them anywhere. The area car I was driving on the day was my favourite. Its call sign was "Papa 1" (registered CYK 360C).'

Tony was asked the question that is always asked of people who have done incredibly brave acts in the face of death or serious injury, 'What made you do it?' His answer? 'I have no idea.'

Burned-out Senator story

There's nothing an area car driver likes more than to borrow a Traffic car if theirs is in the workshops for some reason. And there's nothing a traffic cop hates more than having to lend one of 'their' cars out to a lowly area car driver, especially if they return it dirty or with no fuel in it.

Within my office, despite our best efforts to deny our area car colleagues the loan of our cars, we were told in no uncertain terms by the white shirts that the cars were in fact a force resource and not our own personal transport. So it was one wet wintry evening that the phone rang. It was the area car crew from Havant nick with the question that I didn't really want to answer: Could they borrow a car for the night shift as theirs was off the road? We ran a 50/50 fleet of BMW 525i saloons and Mk2 Vauxhall Senator 3.0i 24-valve cars at the time and I told them to take the oldest car on the fleet; a K-reg

Senator. After giving them the usual lecture about fuel and washing it down before returning it, I went home.

The next morning as I sat in the kitchen eating my breakfast, I switched on the local radio station to listen to the news. It was the main headline, 'Police car explodes into flames in Havant after crashing into a gas main but thankfully with no injuries!' A wry smile came over my face but it was immediately wiped off when a local resident was interviewed and stated that it was 'one of them motorway cars'.

It transpired that the area car crew were on a shout, in the wet, and as they negotiated a right-hand bend they lost it, over-corrected it, entered the forecourt of a local house and collided with a cast-iron box about two feet square that contained the mains gas pipe to the house. The Senator landed on top of the box and as the escaping gas met the exhaust manifold . . . boom, up it went. That same local resident went on to say that the 20-foot-high flame coming through the bonnet resembled one of those Bunsen burners we used at school! Needless to say, we didn't get the Senator back and the area car crew never asked us for another one!

CHAPTER SIX

PANDA CARS

'Panda Car; noun (plural, panda cars), a police patrol car; originally white with black stripes on the doors.'
Oxford Concise Dictionary

It's not every day you discover that one of Britain's best-loved institutions is wrong! But then it is somewhat of a surprise that the term is in the Oxford Dictionary at all. What is annoying but not a complete surprise is that their definition is wrong, because it states that the cars were white with black stripes on the doors. Over the years, the public's perception of what constitutes a 'panda car' has become blurred, so maybe now is the time to put to bed a few misconceptions, blow away a couple of myths and establish once and for all that not all police cars are in fact pandas.

So, let's go back to the Oxford definition. Where did the name 'panda' originate? And why has the term continued to be used into the twenty-first century when the last real panda cars became extinct way back in the early 1980s?

Mid-1960s Britain was facing a bit of a police recruiting crisis; there simply weren't enough officers to go around (nothing new there, then!), particularly on the big new housing estates that were being built all over post-war Britain. It must be remembered that most policemen during this period walked the beat or rode a bicycle and police cars were few and far between. Sure, every force had a Traffic Division, whose responsibilities consisted of attending road traffic accidents, escorting abnormal loads and, in one or two counties, policing the new motorways. Traffic didn't attend such routine calls as domestic disputes, pub fights or kids kicking footballs against people's windows.

Most stations would have had access to a van to transport prisoners, and there might have been an unmarked car that could be used by a supervisory officer of at least Sergeant in rank, to check up on the foot patrols at various 'points' throughout the day. As a member of the public requiring police attendance you could wait quite a long time for an officer to plod his way across his beat to get to you, and it has to be said that the beat bobby could be virtually invisible to the public, especially at night. The modern-day notion that in the 'good old days' there were 6-foot 5-inch coppers everywhere you looked, clipping young scrumpers around the ears, is little more than an urban legend, a myth that has been told over and over again and even appeared in *Just William* and other fiction. If your beat was just one mile square it would take you all day to patrol every street just once, therefore these legends of law enforcement were certainly not seen on every street corner, outside every school or lurking behind the walls of every orchard.

And therein lay the problem. There just weren't enough coppers to go around. And society in general was moving faster as cars became more accessible to the general public and reliable, too. The solution to this dilemma was brilliant, Unit Beat Policing, or UBP, changed the face of policing in Britain. In 1964 the Chief Constable of the Lancashire County Constabulary, Colonel Eric St Johnson, who was a frequent visitor to the USA, had noticed that in many cities their patrol cars were black with white doors and had the word 'police' on them, surmounted by the town's coat of arms. This made them stand out from the usual plain cars used by the public so that they were easily identifiable as police cars. He brought the idea home and had a couple of Mk3 Ford Zephyr 4s, which until now had been unmarked crime cars, dragged into the police workshops where they were resprayed into the blue and white colour scheme. Colonel St Johnson liked the idea that light blue be the chosen colour because he was a Cambridge University man! He worked with William 'Palf' Palfrey, the Chief Superintendent of Accrington, a division within the Lancashire County Constabulary that developed theories on how policing would change post war to cope with the needs of changing work patterns, mobility and mechanisation. Experiments with detailed versions of the basic idea were carried out in both Kirkby and Accrington; the Accrington idea being foot patrols backed up by teams of mobile detectives rather than bobbies moving between 'beats'.

Lancashire has long been recognised as an innovative force, and the concept of using blue and white cars developed very quickly during the Zephyrs' trial period in Kirkby – and of the two towns this was considered the very much more successful system.

Because the cars were readily identifiable as police cars, the force could use them to cover a greater area far better and quicker than a bobby on foot patrol. The cars were an instant success and there were calls to expand the idea. At the same time officers were being issued with new hand-held Pye pocket phones, a two-piece personal radio which meant they could now keep in touch with the station. The logic wasn't that cars became the patrolling medium; the theory of Unit Beat Policing was that the bobby on the beat drove to an area, did a foot patrol, then drove to another area and did another foot patrol or waited to speak to members of the public by his car. Of course, as radio response became ever more immediate, this blurred – but that's another story. The Government were very much keeping track of these experiments and eventually carried out further trials. The idea was even debated in Parliament under reforming Home Secretary Roy Jenkins at one point; if you are of a mind, you can read this on the Hansard Parliament website. Have a pot of strong coffee to hand.

Lancashire led the way, and in the wake of their success the Government issued a recommendation that forces follow their lead. Ironically, one of the last forces to adopt the system was the Met, who waited until 1970 and then, rather amazingly, introduced Morris Minors as pandas rather than more up-to-date cars, because Morris Minors were a known reliable car that was available and cheap, even if the basic design was from 1948 and the car was getting close to ceasing production – which it did in 1971. As explained elsewhere in the book, the Met had to buy cars that were built in Britain. Some forces, especially those with very different population densities and topography, never adopted the panda car system although they may have used the livery on their cars to a greater or lesser extent. These included the Stirling and Clackmannanshire Police, the Grampian Police and others.

In 1966 Lancashire ordered the first official panda cars. They opted to buy a huge number of Ford Anglia 105Es, 175 in fact, to be used as part of this new 'Unit Beat Policing' scheme. On 1 May 1966, local dignitaries and the press were invited to attend Lancashire's HQ at Hutton for the official launch of the scheme before a mass drive-off of all 175 brand-new blue and white Ford Anglias. The Anglia was chosen because it was cheap to buy, at just £500, and easy to maintain. It didn't need to be a performance car, as part of the scheme was for the officer to drive to part of his beat, park up in an area where the public could see him, put his helmet on and either walk that part of his beat for a while or wait for the public to approach him by his car. The car didn't need two-tone horns, just the distinctive blue and white paint to make it stand out.

So where did the term 'panda car' come from? None of us has ever seen a blue and white furry bear (well, I hope not for your sake) so it is likely that the name stems from the fact that press interest in this new policing initiative was huge and there were photos of the cars printed in the papers, which of course made them look black and white. The name was either penned by a journalist or more likely was coined by a Lancashire Police mechanic who, whilst reading one of said papers, passed comment that they now looked like pandas! Whichever one it was, the name has now become part of the English language and is a term still used today by police and public alike who are too young to have even been born when the first Anglias rolled out across Lancashire's sprawling housing estates.

It has also been suggested that the name originated from the term 'Pursuit and Arrest', but this isn't true because the cars were never intended for pursuit purposes. Just to add to the confusion, a couple of years after Lancashire launched the scheme the Home Office conducted an experiment along similar lines with a number of forces being required to paint some of their cars, both unit beat cars and traffic patrol cars, in black with white doors. Forces like Durham, West Riding, Salford City and Plymouth City Police painted cars like the Ford Zephyr, Hillman Hunter, Austin Westminster, Morris 1800S, Mk2 Jaguars and Mini vans in the distinctive colour scheme. They looked great and enhanced the visual impact of the cars, just as Colonel St Johnson had seen in the USA several years earlier. Somewhere along the way things may have blended together and become a bit confused, but whatever the case, the panda car name has stood the test of time even if the original concept hasn't.

Just about every force in the UK took the system on board, and from 1967 onwards blue and white panda cars were seen everywhere, making the thin blue line a little more pliable. But by 1979 the panda car's days were numbered, because the police service was being forced to make drastic budget cuts (again, it appears nothing changes . . .) by the new Conservative government. The fleet manager of the Hampshire Constabulary looked out of his office window at a yard full of blue and white Minis and Mk2 Ford Escorts and concluded that to order the cars in either Bermuda blue for Leyland products or Nordic blue from Ford cost him extra. To then paint the doors white and have to re-paint them blue again prior to selling them off at auction cost £80 per car. He dropped the blue cars, ordered them in white and at the stroke of a pen the idea was lost forever. Fleet managers talk to each other on a regular basis and it wasn't long before other forces followed suit, although one or two forces like Cheshire and North Wales Police perse-vered for several years and still had a number of Mk2 Vauxhall Astras and Austin Maestro vans in the familiar colours into the mid-1980s.

Enough of the history, let's talk cars! So, if Lancashire used the Anglia, did everyone else use the same car? No, thank goodness.

There was a huge variety in the models used, in the make-up of the colour scheme and in the equipment used in or on the cars. But the type of car remained the same; they were basic run-of-the-mill, bottom-of-the-range, small-engine vehicles that were cheap to buy and cheap to run.

The ubiquitous Morris Minor saloon will be forever remembered as the archetypal panda car – I've even used it on the cover of this book – but not every force used them; far from it in fact. The Metropolitan Police, who were one of the last forces to adopt the unit beat scheme, were avid users of the Moggy but also ran fleets of Anglias, Austin 1100s and later the Austin Allegro and the Talbot Sunbeam.

Other forces included Devon, Derbyshire, Bristol City, Sheffield City, Glamorgan, Dyfed-Powys, Cheshire and Edinburgh City, but no doubt the Minor found favour with several others, too. The marque was perfectly suited to this type of work – it was easy to drive, reliable, cheap and (back in the day) looked the part. It wasn't just saloons that were used either. Believe it or not, the Traveller was used by the Devon Constabulary, Edinburgh City, Leicestershire and Rutland, Wolverhampton Borough and the Ministry of Defence Police. The Minor vans, often referred to as 7cwt vans, had already been used for a number of years as Dog Section vehicles and in plain wrappers as

SOCO (Scene of Crime Officer) and Photographic Department units. But now some forces started to paint them in blue and white to utilise as panda cars; however, the Wiltshire Constabulary painted theirs in tan and mustard – apparently they thought that this scheme might blend in rather nicely in the countryside where these cars were used as rural beat units. What were they thinking? The colours made a mockery of the fact that the cars were to stand out from the crowd, not blend in with the hedgerows! Wiltshire stuck with their unusual colour scheme and in later years used it on the Austin A35 van and even the Mk1 Ford Escort. But the Morris Minor has stood the test of time as there are about 20 former police Morris Minor 1000s still in existence with enthusiasts – more than any other type of police car in history. I have even raced against one owned by Chris Rea. He and I had a great battle at Snetterton while I was in the seat of an Austin A35.

Unusually, the Lothian and Peebles Constabulary's Minors were in the darker Trafalgar Blue and White livery rather than the lighter Bermuda Blue used by most other forces. This was specified by the then Chief Constable, Willie Merilees, who wanted them to stand out from the lighter blue of the otherwise similar Edinburgh City cars.

Morris Minor 1000 'panda' racing car

Guitarist and singer Chris Rea is well known for being a petrol head and, partly because of his Italian ancestral heritage, having a real passion for Ferrari – so much so that he made a film about his F1 hero, Wolfgang 'Taffy' von Trips, and commissioned a replica of his famous 1961 Ferrari 156 'sharknose' F1 racing car. However, Rea has also raced in various series over the years, often but not exclusively in a selection of Lotus racers. He also worked occasionally as pit crew, anonymously, for the Jordan team in the early to mid-90s just to be involved (massive respect for that), and has owned a number of interesting road cars. However, as his song lyrics sometimes attest, he's a man with a dry sense of humour and one of his current racers demonstrates that perfectly: a racing 1959 Morris Minor 'panda'.

The car on the grid at Goodwood with other Historic Racing Drivers Club (HRDC) racers. The police sign was removed before the race actually commenced.

The car uses a BMC A-series engine as it did originally, but seriously modified and producing near three times its original output, as the 1.75-inch twin SUs and large oil-cooler attest.

Rea bought the car to enter historic racing, having raced as a guest in a couple of HRDC events, and decided he wanted to enter a Minor rather than one of the all-conquering A35s or A40s that dominate the smaller-engine class. HRDC organiser Julius Thurgood found this example rusting away in North Wales, removed it from that particular road to ruin and entrusted it to Alfa Romeo preparation experts Chris Snowdon Racing in November 2014 to be turned into a racing car. The project took the equivalent of one month's full-time work and, because it was so rusty, ten days of welding alone. Rea apparently briefly considered restoring it as a road car once he realised its police history but decided instead to follow the original plan and nod to the car's past by finishing it in a panda livery and entering it under the number 999. 'Pc Rea 6149' is painted on the doors in reference to Rea's 2005 song 'Somewhere Between Highway 61 & 49'.

Now I can hear every reader screaming, it can't be an ex-panda car if it's a 1959 as it would have been seven years old, at least, when the livery was introduced! And you're right, a little liberty has been taken here, but the car is a genuine ex-police car which now provides fun for competitors and spectators alike at events such as the Goodwood Members' Meeting, so I say good on him. It also shows just how iconic that panda car livery is, at least to a certain generation, plus it was kinda fun seeing Chris's police car chase me on the track in my rear-view mirror . . .

The Mini, without question one of Britain's greatest ever cars, was also very popular with the police in all its variants. Whether it was an Austin 850 saloon, a van, a Clubman estate, a Countryman estate, Leyland or Morris version, or even the Mini Moke, just about every model was used by a force somewhere in the UK. Bedfordshire and Luton Constabulary used large numbers of Minis before the locally produced Vauxhall Viva HB arrived, as did Durham, who chose to paint theirs black and white. Greater Manchester, Norfolk, Merthyr Tydfil, North Wales and Merseyside Police all had Minis, with their minuscule 850cc engine. From 1967 to 1979 the Hampshire Constabulary had no fewer than 900 Minis in service – a record for one make of car that still hasn't been equalled. The Mini Moke, incidentally, was used by the Devon Constabulary and by the Dyfed-Powys Police, which must have provoked some strange looks from the public as the TV series *The Prisoner* was on at about the same time and starred the same car in a distinctly more sinister role.

Driving police Mini panda cars could be a very dangerous experience for the unwary copper. Nothing to do with the vehicle's handling or its performance, but everything to do with the driver's colleagues. Police officers are amongst the worst wind-up merchants on the planet (together with the military and nurses!) and are constantly thinking up new ways of getting one over on their colleagues. The Mini gave them the ideal tool; it became imperative that at the start of your tour of duty, having got the keys to the Mini panda car, the very first thing you did on opening the doors was to check under the seats. You had to lift the hinged front seats very carefully to check if there was a glass phial stink bomb placed beneath the seat frame. If you failed to check it and merely jumped in the car and sat on it, you were guaranteed a very smelly eight hours!

Two police officers were sent to a domestic dispute at a house and arrived in their Mini panda car. It was such a regular occurrence at this house that only one of them bothered to get out of the car whilst the driver stayed put. But on this particular day it all went horribly wrong. The door to the house flew open and the male occupant came straight out and plunged a large knife into the officer's face before shutting the door again. The injured officer was bundled into the Mini by his colleague, who didn't wait for an ambulance and decided to drive straight to hospital. With the knife embedded just below his eye, all the officer could see was the handle bobbing up and down as his panicked colleague drove the Mini panda car with blue light flashing and two-tone horns blaring the four miles across the city to casualty. He drove it as hard as he could with no holds barred, the wrong way down a one-way street, along a pavement, on two wheels around a roundabout with cars and pedestrians leaping out of its way. After emergency surgery the officer later revealed that he was more frightened by the drive in the Mini than he was about the stabbing itself!

The Hillman Imp proved almost as popular with the likes of Kent, Glasgow City, Somerset and Bath Constabulary, Norfolk Joint Police, Newcastle Police and the Dunbartonshire Police. The Somerset cars were somewhat different in that they ordered theirs in the standard Rootes blue, which was a little darker than the light blue that everyone else had adopted. Kent County Police didn't bother painting the doors white on their early Hillman Imps, although they did eventually succumb to the full colour scheme. However, the prize for the 'best-looking Imp in the panda class' goes to the Dunbartonshire Police, who of course were not a million miles from where the Imp was manufactured in Linwood, Renfrewshire. Being canny Scots, they dreamt up a money saving idea to get themselves two pandas without the need for any special orders or additional paint jobs. The solution was to buy two cars – one white, the other one blue – then swap the doors, bonnet and boot lids over. Brilliant. But they did use some paint – bright yellow to be precise, on the roof. The roof colour was complemented by large 'Wide Load' signs placed front and rear. The cars were very high profile and gained the nicknames of Pinky and Perky; they were used to escort abnormal loads along the A82 Loch Lomondside Road from Dumbarton to Fort William during heavy construction work at Loch Awe and became tourist attractions in their own right. Although not strictly panda cars (they were actually Traffic cars), the livery alone makes their inclusion here a must. Incidentally, the diecast model manufacturer Corgi made a superb two-model set of Pinky and Perky, which you can occasionally pick up online.

Hillman Imp panda cars

In 1971 Northumberland Constabulary made a drastic error in buying a number of light-blue-coloured Hillman Imp saloon cars to be used in all areas as panda cars. Four burly giant police officers shoehorned into these tiny cars was always going to be a challenge, but the real challenge came whenever the police needed to make a stealthy approach to the scene of a crime. The Hillman Imp's aluminium, rear-mounted engine (which had its design roots from a Coventry Climax Fire Pump engine) made a distinctive high-pitched whine. Thieves could place lookouts keeping toot, who could hear police cars approaching from afar and make good their escape. The only thing that officers could do to overcome this was to try to approach the scene of crime from an uphill direction, so that they could free-wheel to the scene of crime to surprise offenders and catch them red-handed.

Birmingham City Police appear to have been the only force to have used the Austin A40 Farina for its panda car fleet, purchasing dozens of them in 1967 from the local Longbridge car plant. This was a move obviously designed to help with local community relations, as well as to clear some stock at Longbridge of what was by then an out-of-date car coming to the end of its life, thus it was probably available at a heavily discounted price. These cars were fitted with an illuminated roof box and a blue light together with two-tone horns. This was an unusual practice, although not unique. Some forces fitted their panda cars with emergency equipment to assist them in getting to an incident quicker, even though the drivers of such cars had only ever received basic driver training.

By the late 1960s and early 1970s a lot of our city and borough forces had amalgamated into the larger county forces, and at about the same time a variety of new panda cars came on stream, including the replacement for the Morris Minor: another Alec Issigonis masterpiece – the Austin 1100. The Met bought loads of them and they were just as popular with West Midlands, Gloucestershire and the new Devon and Cornwall Constabulary. The 1100 had plenty of interior space and its ride was very smooth thanks to its hydrolastic suspension, but it faced some serious opposition from Ford's replacement for the Anglia 105E: the all-new Mk1 Ford Escort 1100.

Introduced in 1968, the Mk1 Ford Escort 1100 was an instant success with the public and the police. It was cheap to buy and run, had decent performance and enough space inside. It proved itself to be extremely reliable and cost-effective and officers enjoyed driving them. Forces like Dorset, Merseyside, Hampshire, West Sussex, Lancashire, Suffolk, Wiltshire, Thames Valley, Stirling and Clackmannan Police all used this new Ford product, and most stuck with it when it was later replaced by the Mk2.

The case of the pink Ford Escorts

In 1978–79 Britain was hamstrung by the winter of discontent, so-named because of strikes in the public sector that took the unions head to head with Jim Callaghan's by then fatally crippled Labour government, which in turn led to Mrs Thatcher's first election victory.

However, the strikes also affected the car industry, and one of the longest-running disputes was at the massive Ford factory in Dagenham. At this time Northumbria Constabulary was wedded to Ford Escorts because it made sense for their maintenance work-shops to be set up to deal with only one type of vehicle. They also had been afforded a good deal in terms of the quantity of cars purchased each year. The problem arose when Ford, because of the long-running strike, could not furnish any white-coloured police cars, and in desperation the police, very reluctantly, accepted two pink-coloured panda cars from Ford. One of these cars was placed in Gosforth, a suburb of Newcastle upon Tyne, while the other one ended up at Prudhoe, near Gateshead. The end result was police officers drawing lots to avoid driving the pink panda, and lots of wolf whistles and jokes of dubious taste while they were on patrol if they lost. Perhaps this was a reflection of the attitudes of the period as well . . .?

Meanwhile, Vauxhall updated its lacklustre Viva HA model, which hadn't been looked at by the police, and gave us the Viva HB, which was a direct rival to the Escort. Again, it was cheap to buy and run and appeared to fit the bill (pun intended! You're welcome . . .) perfectly. It goes without saying that Bedfordshire Constabulary bought the Viva, as did the neighbouring Hertfordshire, together with Cheshire, Lancashire and the Ayr Burgh Police in Scotland, who did exactly the same as Dunbartonshire Police did with the Imps and bought vehicles in either white or blue then swapped all the opening bits over to obtain that panda car effect. Vauxhall knew they had a winner on their hands and produced a wonderful brochure selling the HB Viva's panda car potential.

Without doubt the blue and white scheme worked. It grabbed the public's attention – which was the whole idea, of course – and so a number of forces started to experiment by painting some other vehicles that were never intended to be panda cars in blue and white. Lancashire, for example, took several of its Mk1 Transit Section vans (station vans) and repainted them. The Met Police even utilised a couple of ageing Morris J Series vans in the early 1970s as mobile Careers and Recruitment Offices by painting them up as pandas simply to draw people towards them. Other forces started playing around with the scheme as well. Thames Valley Police had white Mk1 Escorts with dark-green doors, whilst Suffolk Police opted for light-blue doors on white-bodied Escorts. The Renfrew and Bute Constabulary couldn't quite make up its mind when it took a white Hillman Imp, painted the doors dark blue and then put orange stripes along its sides! Birmingham City Police repainted one of their Austin A60 area cars from black to panda livery with a curious breaking up of the colour on the C-pillars. However, they were used as supervisory units by sergeants in charge of local sub-divisions, who would use these oversized panda cars to check up on the PCs in their standard panda cars and to meet them at certain 'points' on their beat. This was an old-fashioned idea left over from the days of foot patrols, when sergeants would meet their officers at a 'point' on their beat at a certain time to ensure that they were actually doing their job in their respective area. The sergeant would then sign and date the officer's pocket notebook. Talk about policing the police!

As the 1970s progressed, other new cars entered the market. We saw the likes of the Vauxhall Chevette (my brother had a brown one) make the odd appearance in forces like Cheshire, Norfolk and Lincolnshire. Ford's new Fiesta got used by Northumbria, Hertfordshire and West Yorkshire Police. The Vauxhall Viva HC model was also reasonably successful but not quite as much as its predecessor. By the late 1970s one car stood head and shoulders above the opposition; Ford's new Mk2 Escort in 1.1 Popular trim made for a great panda car. It was light and easy to drive, with sound reliability – even if the driver's seat tended to collapse under the weight of an overweight policeman! Huge numbers of them were used by the Met, Cambridgeshire, Gloucestershire, Hampshire, Dorset, North Wales, Devon and Cornwall, Essex, Hertfordshire and the Lothian and Borders Police in Scotland, who incidentally still have one in their museum.

Some of the last cars to be adorned in the two-tone livery were the Austin Allegro and the Talbot Sunbeam. The Allegro was used extensively by the Met and in forces like West Midlands Police, who now ordered theirs in plain blue, but instead of painting the doors white they opted to place a large white sticker on the doors with the word 'police' on it. The Talbot Sunbeam wasn't as popular as it could have been, due to its cheap build quality and poor reliability, but the Met Police and the likes of Cumbria and Sussex did use it. Meanwhile, the Avon and Somerset Police used the Hillman Avenger 1500HL in blue and white as a response unit rather than an outright panda car. We were beginning to see a blending of roles, not just in the Bristol area but in most other counties as well.

By 1979 financial constraints were hurting the public-sector purse strings and police fleet managers had to look at ways to save cash. Thus the panda car died. Well, the livery did, but not the concept itself, which lives on to this day in most forces. Police still use run-of-the-mill, cheap, small-engine cars to do the basic running-around policing we now call anything from sector policing to community policing. Along the way we have seen cars like the Vauxhall Astra in all its variants, the Ford Fiesta and Focus, the Peugeot 306, 309 and the 208, the Rover 220 and several others, no doubt. In twenty-first-century Britain it's the Ford Fiesta and the Vauxhall Corsa that seem to be the current favourites. But thanks to the Home Office the entire concept has now been lost because of its insistence that all police vehicles carry the same livery nationwide, no matter what that vehicle's role might be. You can read the full story elsewhere in this book, but basically in the late 1990s the Labour government introduced the blue and yellow Battenburg livery to all emergency service vehicles. It wasn't well received and took some forces 20 years to comply. So now we have everything from a Ford Fiesta through to a BMW X5, from motorcycles to Mercedes-Benz

Sprinter vans all carrying the same graphics. In some ways the government's argument echoed that of Colonel St Johnson all those years ago, in that they wanted the cars to stand out and to be easily recognisable by the public – including foreign tourists – as a police car. But nowadays it seems others are using it, too; for example, Highways Traffic Officers use black and yellow Battenburg, and every security company and dog warden known to man decorates his van with Battenburg, as do your local recovery garages, emergency doctors and many others.

The original concept was brilliant and, more importantly, it worked very well indeed and without doubt changed the face of British policing forever. The only thing that hasn't changed is that Oxford Dictionary definition!

The Noddy Bike

The Unit Beat Policing, or UBP, scheme changed policing in the UK as well as providing a whole new livery and type of police car to talk about in this book! However, the very beginnings of the concept of mobile policing using a cheap economical vehicle so that a wider area could be covered was the quite wonderfully named 'Noddy Bike' scheme. From the early 1960s bobbies in selected areas were given Velocette LE motorcycles in an effort to achieve greater mobility. The name came about because those constables riding them were exempt from saluting senior officers, this being a potentially lethal practice while riding a bike at 40mph! They were allowed, instead, to merely nod their head, a massive change in protocol which, if we had the time and space, could be argued to be the beginning of the end of the old police services' quasi-military structure . . . The LE (which amazingly enough apparently stood for 'little engine') was introduced in 1948 and was a strong, tough little bike with a smooth and very quiet running transverse, water-cooled, side-valve, flat-twin, 4-stroke engine of, initially, 149cc, although later versions used a 250cc unit. It wasn't a bike for bikers but rather a machine for transport, and it added as much civilisation and comfort to the motorcycling experience as it was possible to do in that era, originally being fitted with a hand-operated starting lever, for instance, although this design was later dropped in favour of a kick-start. It started the police on a road towards mechanised movement for beat bobbies, but was used mainly in rural areas and by no means exclusively to move between beats in the way the UBP was developed. It's important to say it wasn't the start of panda cars, although people often speak of them as such on the web, but it's equally fair to say that they were a signpost on the road towards the panda car, and no discussion of this policing theory and practice is complete without at least a nod to the Noddy bike.

CHAPTER SEVEN

TRAFFIC CARS

What is it that stirs emotions in people when they refer to Traffic cars and the officers who drive them? Do they hold some mystical powers, or is it just because, no matter what the era, the cars used are top-of-the-range motors with the performance to outgun just about everything else on the road? Whatever it is, there is a definite fascination about all things traffic – both amongst the police and the general public – from the cars used to the type of work involved. Fast cars and car chases captivate us; it's certainly part of the job I loved doing – just look how popular those fly-on-the-wall TV documentaries are, such as *Traffic Cops* and *Police Interceptors*.

Before we look at the cars, though, here's a quick history lesson on the origins of the Traffic Division itself. In 1930, the government introduced the first Road Traffic Act, but it was Section 57(4) that was of particular interest to the police. It decreed that advances towards any expenses incurred by the police in the provision and maintenance of vehicles or equipment could be made out of the Road Fund (Vehicle Excise Duty). The Home Office proposed that annual grants would be paid in advance to cover vehicles running 12,000 miles a year: £60 per annum for solo motor bicycles; £80 for combination motorcycles; and £120 for motor cars. Up until this time each force would have purchased the occasional car as and when their own funds allowed in order to help combat the 'growing menace of the motor car', the numbers of which were expanding at an alarming rate. The Act now gave the police the necessary funds to buy a number of vehicles whose sole purpose was the enforcement of road traffic regulations, thus forming the very first Traffic Divisions. And so a new chapter in the history of British policing was born.

Many forces opted to use motorcycle combinations to begin with, but most of these were soon abandoned in favour of cars, which were seen as much more practical. For the first few years most forces only tinkered with the idea of using vehicles, until after the end of World War II. From then on the idea really became a necessity, as the growth in motorised transport started to have a big effect on the country. Police forces soon started to realise that they required the services of specially trained officers driving high-powered cars capable of getting them to a traffic accident or other serious incident as quickly and safely as possible. Officers were specially selected and sent on intensive advanced driving courses at Hendon for the Met and other regional driving schools like Preston and Maidstone.

Since those early days the role of the Traffic Division (today referred to as the Roads Policing Unit) has grown beyond all recognition and the diversity of cars and equipment used during the last 70 years or so would fill a decent-sized book in its own right. The role basically includes attending and dealing with all fatal and serious accidents, motorway policing, abnormal load escorts, VIP escorts, ambulance escorts, hazardous chemical transportation enforcement, overweight vehicles, HGV enforcement (including foreign goods vehicles), road-works supervision, enforcement of road traffic laws and driver education, armed response and to generally be the flagship face of the force.

So what makes a good traffic patrol car? It's not just top speed that is important, but handling, brakes, load-carrying capacity, comfort, safety, value for money, ease of maintenance, parts cost and availability, and in days gone by politics played a big part. Traffic cops are incredibly fussy about their cars; the vehicles have to be the best available at that time as they take an immense pride in what they do, in what they drive and how they drive it. Many of them consider driving to advanced level as some kind of art form and, trust me, they are deadly serious about the subject. Therefore the car they have to drive needs to match the ego of the driver! Put a traffic cop in anything less than a top-of-the-range car and they will bleat like a lost sheep for hours, weeks even, until they get their own way!

WOLSELEY

USED BY POLICE FORCES AND CONSTABULARIES THROUGHOUT THE UNITED KINGDOM

The Wolseley 6/110, a capacious 2·9 litre 6 cylinder car currently being produced for police work.   Price to Police Specification.

TWELVE MONTHS' WARRANTY and backed by B.M.C. Service.

BMC

WOLSELEY MOTORS LIMITED, (SALES DIVISION), LONGBRIDGE, BIRMINGHAM

Most of the early Traffic Division cars were Wolseleys. They were seen as tough, reliable and capable vehicles, whose legendary status lives on today. The 12/48, 14/60 and 18/85 models were all used between 1937 and 1948, with the 18/85 probably being the first universally accepted standard patrol car, with a good number of forces opting to use them as their first official Traffic cars. That is, until that most famous of police cars arrived: the Wolseley 6/80. It was first introduced in 1948 but it wasn't until the early 1950s that it found fame as a police car and was used by just about every force in the UK – but in particular by the Metropolitan Police, who used Wolseleys above anything else. The 6/80 had a top speed of 81mph and took almost 28 seconds to hit 60mph, which in comparison to some other cars of its era was actually quite slow. The new 6/90 took over the reins in 1954; by then performance had improved significantly, with a top end approaching 96mph with 10 seconds knocked off that 0–60 time. The big Farina-styled 6/99 and 6/110 saloons arrived in 1959, and although production of these cars ceased in 1968, some could still be seen patrolling our streets as late as 1972. The last Wolseleys to be seen in police livery were the BMC Landcrab-based 18/85 S Mk2 models used by the City of London Police in 1970; they were quite possibly the only force to do so. Despite all these models, it is probably the 6/80 that is remembered by people of that era as *the* police car, even though it's likely many will confuse these cars with other similar makes and models of the time. That illuminated Wolseley badge on the grille (called a 'Ghost Light') looming large in your rearview mirror still induces an unsurpassed sense of dread amongst a generation.

One of the cars that some less car-savvy members of the public might have confused with the Wolseley 6/80 was its Nuffield group stable mate, the Riley RM series (now famous to *Archers'* fans as the car that retired history professor Jim Lloyd apparently drives), whose long flowing lines must make them one of the best-looking cars of the immediate post-war period. The William Morris-led Nuffield Group added the Riley to its portfolio of marques in 1938, buying the company and assets from the receivers personally then selling it at a substantial loss into the Nuffield Organisation. Thus Riley joined Wolseley and MG, plus some other companies such as SU Carburettors. Riley had produced a plethora of models in too low a volume to actually make money pre-war, and this, combined with spending money on racing, had led to substantial losses and the eventual sale. However, they were a great engineering-led company known for making class-leading engines and cars with real sporting pedigree. Mike Hawthorn famously started his career racing in a Riley, and the English Racing Automobiles

(ERA) race cars can trace their roots to this marque, using a modified Riley engine. Morris were cash-rich at this time, partly because the Morris 8 had been so successful, but the war intervened before they could rationalise Riley, which was based in Coventry a little way from the group's headquarters in Cowley, Oxfordshire, and which, like every other firm, had turned to war work from 1939.

The result was that Riley developed a new car, the RM Series (with RM apparently standing for Riley Motors), almost autonomously using some Nuffield components. The old Riley-Coventry design team, led by Harry Rush, continued their sporting tradition and used their family of twin-high-camshaft, 4-cylinder engines (which owed much to Alfa Romeo racing-car design and were made as either a 1.5- or 2.5-litre form) in a new box-section chassis on which sat an ash-framed, steel-skinned bodyshell. The Citroën-esque torsion-bar front suspension and accurate rack-and-pinion steering, combined with half-elliptic springs, an anti-roll bar and traditional Riley torque-tube transmission (although this became a conventional propshaft on later models), was a winning recipe and was well received by sporting motorists and the police alike. At 3135 pounds unladen, it was perhaps a little heavy by the standards of the time, so acceleration was not quite as brisk as had been hoped by some, but even so enthusiasts loved what turned out to be the last 'real Riley'. However, in 2.5-litre RMB to RMF form they could nudge 100mph with a 0–60 time of 16 seconds and do an American standing quarter in 20.8 seconds – which in comparison with the 6/80 of the same period made it the cheetah amongst a herd of hippos. They were initially produced at Riley's in Coventry from 1946, but as Nuffield and then BMC began to rationalise, production was moved to MG's works in Abingdon and 8,959 2.5-litre RMs were produced – a fair number for a car that succinctly spanned the change from vintage-era separate chassis to modern 1950s monocoque and remained competitive in handling and performance until its demise in 1953. As you have probably gathered, I have a real soft spot for the Riley RM, but plenty of forces outside of London did too, and even though it was expensive at £1125 in 1946 (a Wolseley 6/80 was only

£767 when launched in 1948), plenty opted to use it, including Gloucestershire, Kent, Cambridgeshire, Portsmouth City, West Sussex and Lanarkshire, to name but a few.

The Chief Constable of Portsmouth City Police, Arthur Charles West (he was always referred to by his full name), was a real petrol head. In September 1946 he purchased a number of brand-new traffic patrol cars. Four officers were dispatched to the MG factory in Abingdon where they collected their Riley 2.5-litre RMB patrol cars. These wonderful-looking cars came in black, of course, and were covered in a coat of protective grease. They were driven back to a local garage in Portsmouth for de-greasing and servicing and were then fitted with a chrome bell, black loudspeaker and a rather unique item to the front. All Riley 2.5-litre cars were fitted with twin Butler spot lamps as standard, but on these police vehicles the offside lamp was removed and the standard lens was replaced with a blue one, which had the letter 'P' stencilled through it to reveal the white light when it was illuminated. Was this the first ever blue light fitted to a patrol car? If not, it was certainly one of the earliest examples. The light would be illuminated when the car was in a hurry or when it was in the process of stopping another car. A small, hinged police sign was also mounted on top of the nearside section of the dashboard, and it is likely that this was also raised into position during such operations. The cars cost £1,342 each and once fully prepared they were handed over to the Chief Constable at a formal ceremony.

The cars were greatly revered by the officers who crewed them, and they are still remembered today by the people of Portsmouth. With their 3-speed manual gearbox, they could reach 95mph and had handling to match. Once they were kitted out and ready to work, all four were driven to the Home Office in London to be inspected for their suitability for police work. One can only assume that perhaps these were the first ever Riley RMBs to enter police service and needed government-type approval before other forces could purchase the same cars. Certainly, other police forces did buy them in large numbers and it is unlikely that they all had to travel to London to be inspected. Unfortunately, one of the four cars didn't make it to the inspection, as it crashed en route!

A total of seven Riley RMs were bought altogether, the later ones being RMFs. For the record, the registration numbers were DBK 854 (this was the first one registered), DBK 956, FTP 322, FTP 541, ERV 847, JBK 227 and JBK 244. The RMF models were purchased around 1952.

Arthur Charles West was the president of the Southsea Motor Club, and on 8 August 1950 he entered drivers from his Traffic Division into the annual motor gymkhana on Southsea Common, to compete against other club members for the Faulkner Cup. In their gleaming Rileys they took part in three driving tests, watched by thousands of onlookers. The tests included the 'wiggle-waggle', a slalom course of barrels; the 'triple-garage' test; and the 'rallye-soleil' test, in which the cars had to negotiate a slippery grass course, against the clock, without spinning the wheels or losing control. The police team consisted of PCs S. Booth, W.H. Moore, J.B. Wilkes and L. Hoad, and they won the competition, much to the delight of the Chief!

The following year, though, saw the Southsea Motor Club win back the trophy. Entering this type of event wasn't so unusual then. Police patrol cars from the Southampton Borough Police and even West Sussex Police would come to this event to have their cars judged in the Concours d'Elegance competitions and usually went away with several of the prizes! Such was the pride in the appearance of one's patrol car then that very little extra work was required to enter. It worked both ways, of course, with Portsmouth City Police patrol cars attending similar events in Southampton and elsewhere.

Arthur Charles West, when questioned by the press about the cost of the new Rileys, would boast that they were quick enough to get anywhere in the city within four minutes. It was not unusual for him to telephone the control room at the force headquarters in Southsea *from his home on Portsdown Hill, some five miles away, and demand the immediate appearance of a patrol car. Having raced through the city, with bell ringing, the Riley would screech to a halt outside the Chief Constable's house and the driver would see him turn to his dinner guests and proclaim, 'There you are, under four minutes!'*

An original photo of the sole surviving Riley RMF, JBK 244, was used by Classic Car Weekly *newspaper in the 1980s as part of their advertising campaign. The photo was taken around 1954 at Byculla House, the headquarters of Portsmouth City Police, with Sergeant Sidney Booth standing beside the car.*

In 1953 Riley launched its replacement for the RM series. The Pathfinder used basically the same 2443cc Riley engine in a new chassis and should have been an advance over the old car. It had been designed as part of the Nuffield group's new range of cars by Gerald Palmer, who had moved down to work at Cowley having designed the revolutionary Jowett Javelin in Bradford. The new family of Nuffield group Palmer cars were basically two pairs: the MG Magnette Z-type and Wolseley 4/44 (which became 15/50 when fitted with a new engine) were the smaller 4-cylinder offerings, while the larger car was also built as the Wolseley 6/90 and aforementioned Pathfinder. However, although they looked superficially similar, the Riley was quite different in a number of ways, including its whole body being slightly lower, and, of course, the original Riley engine – the 6/90 used the BMC new C-series corporate straight six in 2.6-litre guise. However, it proved to be anything but an improvement over the old and much-loved RM. Being more aerodynamic, it was marginally faster, but on top end it was a full second slower from 0–60 and handled like it was running on four flat tyres! In fact, things were so bad that it quickly gained the nickname 'Riley Ditch Finder'. Sales to the police of this marque were poor as a result and very few saw service as Traffic cars – the Wolseley 6/90 was by far the better product and also actually cheaper. Sadly, that marked the end of Riley in the large car market, although some badge engi-

neering was applied to smaller cars based on the Austin A55/A60, the
1100 Morris Minor and the Mini. The last car to bear the Riley name
was the Riley Elf, a booted Mini. The Riley badge is currently owned
by BMW, who have made noises about bringing it back, probably on
the Mini, but thus far have never done so.

Just about every large Humber produced post-war was taken on as
a Traffic car, and, of course, they had a glorious history in World War
II as a staff car – did anyone else build that Airfix model of 'Monty's
Humber', as it was known? So government orders were very much part
of the corporate plan of Humber's parent company, the Rootes Group,
which may explain to some extent why the larger Humber Pullmans
were also the BBC's first camera cars. From the Mk1 Hawk of 1945 to
the last Super Snipe in 1967, the big Humbers were a legend among
traffic cops, but sadly Chrysler UK (as the Rootes Group had become)
chose not to replace them. It's worth noting also that the smaller
Humber-badged versions of the Hillman Super Minx 'Audax' and
'Arrow' Hunter did not find favour as Traffic cars, although both were
used in other roles, usually in their cheaper Hillman form. The larger
range all possessed an ideal blend of performance, handling and
load-carrying ability that is so necessary to cut it as a front-line patrol
car. The separate chassis Mk1 Hawk and Snipe made brief appearances
in a handful of forces, especially in Lancashire County where the dominant car at this time was the Hawk. Humber had announced this range
of cars in August 1945, the same month as VJ Day and only three
months after VE day. These side-valve engine cars were basically warmed
over pre-war designs, as many cars of this era were, and shared a body
that was offered with three different engines. The 2-litre 4-cylinder
Hawk, the 2.7-litre 6-cylinder Snipe, or the most upright English hot
rod of its era, the 4.1-litre Super Snipe: a car that offered an 80mph
top speed and 0–60 in 24 seconds, which in 1945 was impressive, especially at a price of only £889.

Humber's first truly post-war car was the MkIII Hawk of autumn
1948, which looked quite Studebaker like, which was not really
surprising because Rootes had consulted the famed Raymond Loewy

styling studio as well! It originally carried over the side-valve engine, and continued to do so through ever more flamboyant redesigns, but it then gained a new overhead head-valve unit when face-lifted (again) into the MkVI in 1954. However, the large 6-cylinder side-valve engines were not fitted to this new design and the older Snipe and Super Snipe continued until they were replaced by new models bearing those names in 1952. The new cars were based on the 1948 4-cylinder Hawk but featured a new OHV engine called, rather wonderfully, the 'Blue Riband', and a longer wheelbase longnose body.

However, it was the advent of the later unitary construction cars, launched in 1957 in 4-cylinder Hawk form and in 1958 as 6-cylinder Super Snipe, that really built their reputation as some of the toughest and most practical cars of their era. Although they had a transatlantic look, they were actually designed in-house at Rootes and the bodyshells were made at British Light Steel Pressings in Acton, West London. At the time they were apparently the largest mass-production bodyshell being made in Britain, being 15 foot 8 inches long from over-rider to over-rider, and having a front track of 4 foot 9 inches – at a time when cars were still generally incredibly narrow; just park an Austin A60 Cambridge next to a Ford Focus – it's the same length but a foot narrower. The Super Snipe's wheelbase was 9 foot 2 inches, which to put it in context is 11 inches longer than a Se1 Range Rover's! The Hawk had a 2267cc 4-cylinder OHV engine carried over from the previous generation, but the 2651cc straight-six OHV engine in the Super Snipe was a new design which was increased to 2965cc in 1959 when disc brakes and power steering were added to the Series 2. Amazingly, the Series 3 came out only a year later, which was distinguished by a handsome new (almost Edsel-like) four headlight nose which stayed with the 6-cylinder car until its demise in 1967. Both cars came with acres of polished wood and an entire herd of leather to give them a sort of Gentleman's Club feel on the inside, even on the police spec models, which were not trimmed down to price – unlike Wolseleys – Rootes taking the view that interrupting the line would cost more than would be saved by de-trimming. Despite what appeared to be barn-door aerodynamics and an unladen weight of 3415 pounds, the later model 129bhp Super Snipe could be coaxed to 106mph, not bad for a 60s barge, and its 0–60 time of 16.2 seconds was respectable as well.

The Met Police was probably the most famous user of the Series 3 Super Snipe estate when the car became the first of its new SETAC (Specially Equipped Traffic Accident Car) units; resplendent in their gleaming black coachwork, these cars were the forerunners of the Range Rover-esque vehicles we are used to seeing in more modern times. The SETAC units were heralded as a breakthrough in traffic policing as the estates were capable of carrying the huge amounts of equipment necessary at the scene of any accident, and they spawned an entirely new approach to keeping the accident scene safe for all to work at, with additional signs, cones and lighting. Humber was partly instrumental in this new concept as it offered the police a number of factory fit options including three different gearboxes; a 3-speed column change, a 3-speed automatic or a 4-speed floor-mounted manual. For the load compartment, they would build in a special signs box that replaced the rear seats and could be easily accessed by opening either of the rear doors. With the split rear tailgate open, a slide-out tray containing other equipment made light work of accessing everything that was needed quickly and efficiently, and it was certainly better than rooting around in the boot of a saloon car!

Other forces like Kent County, Hertfordshire, Hampshire, Sussex and Northamptonshire also used the Super Snipe estate for the same reasons. The Hertfordshire and Northamptonshire cars took over the role of policing the M1 from the first motorway cars of the era, the Ford Zephyr Mk2 Farnham estates. The Chief Constable of Northamptonshire, John Gott, was similar in many ways to Arthur Charles West and loved his cars. He was a regular rally car driver and captain of the BMC team at one point, and he was particularly associated with MGAs. Other forces such as Glasgow City and Cardiff City utilised the saloon as their Traffic cars.

The Austin A99/A110 Westminster was hugely popular and perhaps one of the great Traffic cars of all time. And it's easy to see why; a big 3-litre straight-six engine, a 105mph top speed, a 0–60 time of 16 seconds, enough space to swallow a large amount of kit, plenty of interior room for two big hairy coppers and a solid build quality that is second to none. Only the Vauxhall Senator and Volvo T5 have come close in recent times to being as dominant as the Westminster was in its day. Forces like the Met, Durham, Hertfordshire, Hampshire County, Portsmouth City, Cardiff City, Surrey, Halifax Borough, West Riding, Blackpool Borough, Staffordshire, Warwickshire and Worcestershire, to name but a few, all took the big Austin onto their rapidly expanding traffic fleets and, of course, it was basically the same 'badge engineered' car as the Wolseley 6/99 and 6/110, which were legends as well.

Austin and Morris also had a fair degree of success with a couple of its other products over the years. Of some surprise was the popularity of the 'Landcrab'. The 1800S, to give it its official title, was a four-door saloon with the same type of hydrolastic suspension used on other Leyland cars of the time, like the Mini and Austin 1100, so it was smooth over the bumpy bits and offered huge amounts of interior space for its time; a friend of mine remembers travelling in a Wolseley Landcrab with 11 other boys to represent the school football team when the minibus was being used elsewhere! Carefree days, at least until the car had an accident and all those unbelted occupants were thrown on top of each other. Like many other cars of the early 1970s, its reliability often let it down. That being said, a good number of forces did use it as their Traffic car for a while, including Staffordshire, Cheshire, West Yorkshire and West Mercia. The other car from this stable was the huge Austin 3 litre, which was very similar in design but with a large protruding boot and an engine almost twice the size of that of the 1800S. As far as is known, only a couple of forces used the 3-litre car: West Yorkshire and West Mercia Police.

I've often thought the Austin 3 litre was a great Traffic car in the making, so I was surprised that so few had been used when I started looking into it; it was basically a good car that never found its feet in BMC/BL or the police. It was announced at the 1967 London Motor Show but didn't actually go into production until the summer of 1968, by which time BL had been formed and its proposed rivals from Triumph and Rover were suddenly stablemates. It used the extremely strong (the most torsionally rigid body shell in the world at the time) basic hull of the 1800 but was rear-wheel-drive, although it also featured hydrolastic all-independent suspension, which, combined with its long wheelbase, gave it a superb ride, and the self-levelling aspect of this system would have been great, for all the kit police cars were starting to carry at this time. The new 7-bearing, crank, 3-litre, straight-six OHV engine was shared with the MGC (not the old Westminster or Healey 3000; it was a new unit) and was undoubtedly very smooth, but it was also very heavy, very thirsty and very under-carburetted, making the engines' breathing terribly inefficient. Experiments with MGC engines by people such as MG Motorsport show that this can be sorted fairly easily with some simple gas flowing and triple rather than twin carbs. However, once BL had been created it was always done for because the Triumph/Rover/Jaguar contingent could not see the point in having the internal competition, probably wisely, although BL Chief Engineer Harry Webster apparently ran one with a Rover V8 engine fitted and rated it very highly. With that much lighter engine you have to think its handling would have been much improved and its other virtues would have remained. Sadly, only 9,992 were made in the four-year production run and most of them ended up banger racing, so they are now quite rare.

Ford. Where do we start? Ford has surpassed itself over the years and provided the UK's Traffic police with a wealth of cars. In 1947 it gave us the Ford V8 Pilot, a monster of a car with the famous 3622cc V8 'flathead' engine that gave 85bhp@3500rpm – serious power in that era and a real taste of what was happening in America, although again it was basically a pre-war design. The Pilot could do 82mph, an almost unheard-of speed for a normal mass-market car then, and at £748 was incredible value for money, even though, bizarrely, it had hydraulic front brakes and cable rear brakes! The Lancashire County, West Riding and Cornwall constabularies all took on the services of the V8 Pilot but had to make some modifications. The 6-volt electrics were changed to 12-volt, which was done by adding a second dynamo and leaving the first in place as a pulley! Twin 6-volt batteries were fitted in the boot and the cars came from the factory with a zip hole in the head-lining for the radio aerial. Amazingly, the police didn't replace Ford's famously comedic windscreen wipers, which were driven by the engine's vacuum. Old Ford enthusiasts will know this, but for our younger readers, Ford persevered for some years with a frankly daft system in which the windscreen wipers were powered by the inlet manifold's vacuum pressure. In some ways this was a very clever use of the engine's natural characteristics, but with one flaw: when ticking over, the wipers went like the clappers, but under hard acceleration they stopped because there was no pressure to drive them! *'Sorry, sarg, I lost the bad guys because I had to back off so my wipers started working again'* is top line as far as police excuses go.

In 1950 Ford announced their first UK-made unitary construction car, the Mk1 Zephyr 6, a large four-door saloon with a 2262cc 6-cylinder engine capable of 82mph. It grabbed the attention of several forces, including Staffordshire and Bedfordshire, who used them extensively on their Traffic fleets. Of course, the Mk1 Zephyr, driven by Maurice Gatsonides, also famously won the Monte Carlo Rally in 1953. Gatsonides later invented the speed camera we all know as 'Gatsos'.

However, it was the Mk2 Zephyr, released in 1956 in both saloon and modified estate variants, that really gave Ford a firm grip on the police market. Its 85bhp, 2553cc straight-six was a gem, powerful, smooth and reliable, giving it a top speed of 88mph and a 17-second 0–60 time. Its ride and handling were class-leading at the time, as proven by Jeff Uren when he won the second over British Saloon Car Championship driving a Mk2 in 1959. It was capacious, reliable, strong and comfortable, everything a motorway cop needs. The model was a massive success for Ford UK, who made 682,400 of the 4- and 6-cylinder cars between 1956 and 1962. It would appear that the Lancashire County Constabulary was the first force to operate the new saloon and almost certainly the first to take on the modified estate version. Ford didn't make an estate car in the late 1950s and early 1960s, so they turned to E.D. Abbott Ltd of Farnham, in Surrey, who were specialist coach builders to the likes of Jowett, Bentley and Lanchester. Under licence to Ford, Abbotts built dozens of the new estates and were later commissioned to undertake the same conversion on the Mk3 Zephyr before Ford realised that there was a lucrative market out there for their own estate cars! The Mk2 Zephyr had ample cargo-carrying capacity via a rear-hinged barn door that opened from left to right.

Lancashire's first saloons and Farnham estates came in the traditional black, of course, but in 1958 the force was faced with patrolling a new type of road – the Preston bypass (later to become part of the M6) –

and a decision was taken to paint the cars that would police that road in white to make them more visible. They utilised the Farnham estate and placed a single blue rotating light on the roof with an illuminated police sign to the front and rear of the roof. A PA speaker, bell and VHF radio set plus all the necessary emergency equipment to deal with any high-speed accidents were stored inside its generous load area. Lancashire were renowned for their innovative approach to automotive policing and led the way on many occasions, much of which has stood the test of time.

But perhaps the most famous of these Mk2 Zephyrs belonged to two neighbouring forces that formed a unique partnership, and under the direction of the Home Office purchased near-identical cars to fulfil the same role on a new road running through their respective counties. The forces were Hertfordshire and Bedfordshire, the year was 1959, the road was the new M1 motorway and the cars were Mk2 Zephyr Farnham estates. And in a move replicating the Lancashire initiative, the cars were painted in white and fitted with new-style blue lights, and on the two Bedfordshire cars the leading edge of the bonnet sported the word 'police' in blue, which was the only difference between the two cars. The Bedfordshire version was modelled by Corgi in 1960, such was its significance.

The first section of the M1 was opened on 2 November 1959. There was no speed limit, no central crash barriers, no lighting and few or no restrictions, which meant that people would stop on the hard shoulder to have their picnics, do U-turns across six lanes of fast-moving traffic, with even pedestrians and cyclists thinking it was okay to use! As a result, the accident rate soared and it was a dangerous place for traffic cops to work on. To reinforce this point we have the story of one of the Hertfordshire cars that was involved in a serious crash on the M1 during those early days and its quite remarkable life ever since, which I learned about when I was lucky enough to see the car and meet Paul from Ford heritage at Gaydon Museum in early 2017.

The Hertfordshire cars were very well equipped for the day and carried a Lucas Acorn blue light, a VHF radio set and an illuminated 'stop' sign placed inside the rear window, and whilst the early cars had Winkworth bells fitted, these were later replaced with the new two-tone horns. A PA speaker adorned the front grille whilst a full-width mud flap was placed under the rear bumper to help suppress spray coming up onto the rear screen (no rear wipers in those days!). The rear seats were removed to help spread the weight of heavy metal police accident signs and all the other necessary equipment needed to police the new motorway. The signs were placed onto a metal racking system built and welded together by Sergeant Arthur Leach. The weight of all that heavy kit proved to be too much for the cars' standard rear rims, which had a tendency to split, so some heavy-duty ones were fitted. The car was originally fitted with over-drive, but this proved rather troublesome and was later disconnected.

3111 RO was in the second batch of Zephyrs that Hertfordshire purchased and it was based at Hemel Hempstead with the call sign VH-35. There has been some debate over that all-important call sign because once Hertfordshire formed its Traffic Division in 1962 all Traffic cars were designated with a T-Tango prefix. However, the letters VH are the force-designated code letters and it is believed that this car was in service just prior to that Tango system being implemented. Where did we discover the call sign? We were told that it had been painted on the inside of the glove-box lid, and despite the passage of time it is still there.

So that's the preamble over with, but what on earth happened to the car one snowy winter's morning on the M1? PC Ken Surridge was the driver and PC Frank Fuller was the observer when they got sent to an RTA on the northbound carriageway just south of Junction 8, the exit for Hemel Hempstead. VH-35 was travelling south and the crew stopped on the southbound hard shoulder then walked across both carriageways to the accident scene!

After ascertaining that there were no injuries, Ken stayed with the stricken vehicle to await the arrival of the recovery truck from the nearby Mount Garage. Meanwhile, Frank walked back across the M1 to the patrol car in order to write up the accident report in the warm. He was sitting in the front passenger seat when he looked in the observer's mirror and saw a car approaching from the rear, followed by a J. Lyons articulated lorry. Only the nearside lane was usable because of the snow. Frank saw the car slow down, probably because the driver had seen the police car, and the lorry driver braked hard but was unable to stop in time because of the conditions. The lorry driver tried to avoid the car but jack-knifed, and the back end of the lorry swung around and hit the rear of the Zephyr. Frank had seen what was happening and threw himself across the front bench seat before the huge impact. Thankfully he wasn't injured, but the Zephyr sustained major damage and was written off.

Another recovery truck was sent from Statham's of Redbourn to recover the Zephyr. When it arrived, Frank went to talk to the truck driver, whom the lads all knew as 'Wimpey'. At the same time a Sergeant and Superintendent arrived at the scene.

The Superintendent asked, 'Who was in the car?' and was told that it was Frank.

'Where is he and how is he?'

'He's fine, sir – he's just down the road talking to Wimpey.'

The Supt then asked, 'Is he smoking his pipe, because if he is, he is okay.'

Frank was smoking his pipe!

In 1962 Ford released the Mk3 Zephyr in both 4- and 6-cylinder configurations. The bigger 2.5-litre Zephyr 6 will be forever associated with the TV programme *Z-Cars*; albeit the first few episodes actually used the Mk2 Zephyr. It was also a very popular Traffic car in its own right. The new Zephyr 6 had it all; performance, handling, space, comfort and load-carrying ability. It was just like its predecessor (whose basic engine it carried over), only better in every way. It also had the looks and presence required and no doubt there was a certain kudos about driving a Z-Car when the television series was at its height. Numerous forces like Lancashire (obviously), Leicestershire, Exeter City, Warwickshire, Devon County, West Riding and Nottinghamshire all found the new razor-edged Zephyr to their liking. Some forces, such as Hertfordshire, Cardiff City, Lancashire and Manchester City Police, all went back to Abbotts and ordered the Mk3 Farnham estates again to be used on the growing network of motorways across the country.

However, the 1966 Mk4 Zephyr 6 had put on a large amount of weight, and despite getting a new compact 'Essex' V6, had a bonnet long enough to land a plane on. It was a huge car by UK standards and it was nowhere near as popular as its predecessor. Its build quality was also dubious and it quickly gained the nickname of the Dagenham Dustbin! Partly because of that new engine it now topped 100mph and had a 0–60 time of 11 seconds, so it was no slouch despite its ungainly size and middle-aged paunch. But it didn't make too many friends amongst the traffic crews, and so Ford's dominance stuttered for a while. It saw limited service with forces like Lancashire, Durham, West Riding, Cumbria and Nottinghamshire, and the Northamptonshire Police even bought a couple of Ford's own new estate version and attached one of the first ever elevated stem-light systems onto the roof.

There were a few left-field Fords that found their way onto the traffic fleets, but not in any great numbers. These included the Mk1 Lotus Cortina, a batch of eight being used by West Sussex Police, and the same force later continued the trend by taking on 30 of the Mk2 Cortina Lotus saloons whilst Hampshire and Thames Valley Police also opted to use them. The Mk1 Ford Capri 3000 GT was used extensively by Lancashire County, Greater Manchester Police and the Thames Valley Police, with rumours that at least one of the TVP cars was fitted with Ferguson four-wheel drive! By the time the Mk2 Capri 3.0S arrived, the likes of Hertfordshire, Dorset, Sussex and Greater Manchester Police took them on, and the Mk3 2.8i model was a real favourite in Manchester, and who could blame them for loving that particular car? To date at least four of the 2.8i cars have survived as preserved cars. Other cult Fords included the Mk1 Ford Escort Mexico and the Mk2 RS2000, which were used by Merseyside Police as urban Traffic cars.

Mention the words Mk1 Ford Granada and one of the first things an automotive journalist is likely to say is 'Sweeney car', or something equally inspiring/predictable like 'Your round, Guvnor'. Just like the Mk3 Zephyr 6, which got lumbered with the *Z-Cars* label, the Mk1 Granada will always have to wear the tag associated with the *Flying Squad* TV series. During the 1970s many of us were brought up on a TV diet consisting of *Z-Cars*, *Dixon of Dock Green*, *The Sweeney* and *Starsky and Hutch*, and we loved them all. But *The Sweeney* was always the favourite; it was brash, over the top, good-humoured and fun, but not afraid to be realistically tragic . . . Just don't look at the flares and the huge collars, which have dated far more than the scripts, which were impressive and still are in a way! But it was the car that we really fell for as it thundered towards us during the opening credits every Monday night at 9pm. The early bronze Consul GT they used had beige vinyl upholstery to match Regan's beige velour jacket. From our vantage point on the settee the car looked impressively fast, and it was obviously tough because George Carter was always yelling at the driver to slow down and to stop bumping over the pavement in case it messed with his hair. But it is sad that we only seem to remember it for its starring role on TV; surely it deserves a little more credit than that? Was it a popular patrol car? Did it fulfil the role it was required to undertake? Should it be remembered as a classic Traffic car? The answer to all those questions is an emphatic yes. It was brilliant, and the officers who drove it loved it.

It arrived in April 1972, taking over the reins from the Mk4 Zephyr 6 as Ford's new flagship. To begin with, the lower end of the range was badged as Consul L or GT, with only the range-topping model getting the Granada GXL label. Then in 1975 Ford revised the range slightly and dropped the Consul name and gave all the cars the Granada badge. It also introduced the 3.0S model as a replacement for the Consul GT. All the GT and S cars were fitted with Ford's acclaimed 3-litre V6 Essex engine, which produced 138bhp@5000rpm and gave it a top speed of 115mph and a 0–60 time of just nine seconds. There was also an estate version, but police use of these was limited to the base L model, as there was never a 3.0S version. These were also used on traffic patrol as load luggers, just as the Zephyrs had been previously.

The later S-series saloons were fabulous and were generally only ever used by the police, although they were made available to the public. The cars were all fitted with manual transmission and had no power steering, unlike most of the Ghia saloons, which were generally automatics with power assistance, something the public found much more to their liking. But as a police vehicle the 3.0S was excellent: very quick, sure-footed, predictable, with power-assisted brakes and with a real presence about it. It was, and still is, a very good-looking car, and in police livery it just looked the part. As a driver's car it felt secure and safe at speed, certainly not an edge-of-the-seat experience. It was as strong as an ox, with bags of room inside, and those seats were a model in design and comfort that one or two manufacturers could do with copying today.

The interior was definitely 'in vogue', as it was finished entirely in black; the seats, door trim, carpet and the whole of the dash were all one colour, which made it a bit dark inside. Although the lack of power steering made slow-speed manoeuvres a real pain in the shoulder area, it was something that police drivers were more than used to at that time, as Range Rovers and early SD1s were even heavier!

The Granada was one of those rare cars that managed to fulfil two roles – both as traffic and area car. So which forces used them? Well, plenty is the answer, including Avon and Somerset, the City of London Police, Cambridgeshire, Dorset, Essex, Gloucestershire, Greater Manchester, Lancashire, Leicestershire, Northamptonshire, Northumbria, Sussex, Suffolk, Thames Valley, Warwickshire, Glasgow City and then Strathclyde Police. No doubt there were a few others, so it has to rate as highly successful. The estate models were used by Greater Manchester, Lancashire and the City of London Police.

So it really was more than just a TV star; it was a truly classic Traffic car and deserves to be remembered as such. Your round, Guvnor!

By the time the Mk2 Granada was launched in 1977 it had a big reputation to uphold, but it became an icon in its own right and seemed to stay with us forever. Early cars came with the new 2.8-litre V6 Cologne engine capable of 117mph and with improved road manners, much better NVH control and improved aerodynamics, which made them more stable as well as more economical; a great public servant that officers loved but they never looked quite as cool in police livery as a Mk1, though, did they? Just my opinion; maybe it's my age. Later models gained fuel injection for the first time, which helped push the car past the 125mph barrier – a fact that meant it found itself competing head to head with the Rover SD1. It's fascinating to see which force used which car, and it was rare to find any force using both. The Met, for example, were loyal to the Rover brand because they had to be. Their funding mandated British-built cars, and the Mk2 Granada was built in Germany, so the engineering development and purchasing teams within the Met were told in no uncertain terms that it wasn't an option, in spite of Ford's large UK presence and manufacture of other cars and commercials here. After much development, which is covered elsewhere in this book, the Met were happy with SD1 V8s as Traffic cars but would have operated Granadas as area cars if they had been allowed to do so, because the 2.6-litre SD1 was so unreliable.

Of the forces that were not bound by political patriotism, Greater Manchester opted for the Granada whilst Hampshire chose the SD1. Surrey Police took the Ford in both saloon and estate, whilst it made sense for the West Midlands Police to stay loyal to the Rover brand, given that they were built on their patch. Surrey's use of the estate on its section of the newly opened M25 caused a mixture of hilarity and anxiety amongst the crews tasked with driving them. The cars were fitted with a large amount of equipment, including a large glass-fibre boot spoiler mounted on the rear of the roof, two additional roof-mounted spotlights and a huge stem light. These items tended to make the car very top-heavy and induced a large amount of body roll, even at slow speeds. Apparently this body roll was so severe that the cars gained the rather politically incorrect nickname the 'General Belgrano', after the Argentine warship of the same name!

The 1985 launch of the Mk3 Granada heralded a new style of car for Ford, which had been previewed with the amazing-looking Probe III concept car in 1981 and then seen commercially first on the 1982 Sierra, which was developed in tandem with its similar-looking but larger stablemate and actually shared some parts and sections of platform under the skin. The rest of the automotive industry soon followed suit. Although the 'jelly mould' body design was considered quite radical in its day, the same cannot be said for the engines, which were starting to lag behind the competition. The early Mk3s got the same 2.8 fuel-injected power plant as its predecessor, and this was later uprated to 2.9 litres. In truth, it was way past its sell-by date and the opposition were running rings around it. Ford tried all sorts of tricks to convince the police to continue using it, giving the car four-wheel drive, anti-lock brakes and the knowledge that it had been played about with by Cosworth, no less. But foreign imports from the likes of BMW meant that Ford's days of dominating the Police traffic fleets were rapidly drawing to a close.

The launch of the Ford Scorpio in 1994 was met with such derision that it wasn't long before the car got dropped completely. Its styling was curious, to say the least, and it was labelled as a 'wide-mouthed tree frog' by *Top Gear*'s Jeremy Clarkson. It was offered to the UK's Traffic police in 2.9i Cosworth trim with four-wheel drive, and although it was a slight improvement over the old Mk3 Granada it still wasn't as good as the opposition, even if Ford made every attempt to sell them to the police as cheaply as possible.

But Ford did have one last throw of the dice and threw a double six when it launched the performance-related Sierra range in the late 1980s. Although the basic model had been around since the early 1980s it was the bigger cars that got the nod from the plod. The XR4i was basically a revamped two-door model with the 2.8i V6 dropped into it. Only one force tried them, but they were made a bit more famous by their inclusion on the opening credits to a new TV programme that arrived on our screens about the same time; the car was used by the Devon and Cornwall Constabulary and the programme was *Crimewatch UK*. The next variant was the XR four-wheel-drive hatchback and estate, again with the 2.8i motor, giving it a top speed of 127mph. The combination of four-wheel drive and reasonable power won it many admirers on Traffic, and plenty of forces like Surrey and Nottinghamshire took them on. In 1990 Ford gave the public the Cosworth Sierra (albeit in limited numbers with the three-door to begin with), but then it put the 2-litre turbo-charged rocket ship engine into the four-door Sapphire bodyshell, added four-wheel drive and delivered an almost perfect police package. Top speed was a blistering 143mph with a 0–60 time of just 6.6 seconds! It was 'almost perfect' – it did suffer from a couple of weak points, namely its lack of interior space and the catalytic converter had a habit of filling up with oil when the cars were left standing at the scene of an incident, ticking over for long periods of time. In one particular case in Sussex, the car got slower and slower and it was some considerable time before they discovered the exhaust problem and managed to rectify it. After that it was known as a problem; police forces do communicate with each other about faults of this nature. Nonetheless, the Cosworth Sierras will be remembered for all the right reasons and many forces found them to be one of the best Traffic cars of all time, including Avon and Somerset, Bedfordshire, Cumbria, Devon and Cornwall, Greater Manchester, Humberside, Nottinghamshire, Northumbria, Sussex and Suffolk.

Following the success of the Sierra Cosworth came the even quicker and more agile Escort Cosworth, and a handful of forces used them not as mainstream Traffic cars but more as pursuit vehicles to help combat the scourge of 1990s Britain: the ram raider and the car thief. Only four forces – Cumbria, Northumbria, Humberside and West Yorkshire – had the decency to buy its traffic officers the awesome Cosworth Escort. The rest should all hang their heads in shame!

One of Ford's final efforts was the second-generation Mondeo-based ST220. The first-generation Mondeo had come out in 1992 and been widely criticised as lacklustre – not bad, but not good, *just average*. After this, Ford started really addressing the dynamic driving qualities of their cars, and chassis guru Richard Parry-Jones was really making his presence felt producing lovely accurate steering and all the 'feel-good' feedback and handling that enthusiasts want in cars; the first-generation Focus being a prime example. The Mondeo was face-lifted heavily for 1996, acquiring bigger lights and larger bumpers for visual effect, but most importantly a much-improved handling. However, in 2000 the second-generation Mondeo (which is sometimes confusingly called the Mk3, as people incorrectly call the Mk1 facelift the Mk2!) was enlarged enough to offer possibilities as a Traffic car and the 3-litre V6 ST220 form had the performance as well – after all, its V6 was effectively half an Aston Martin V12, as Ford owned Aston at this point and used the V6 as a starting point for the Aston V12 engine. Although much improved, this new Mondeo was up against incredibly stiff opposition from the likes of Volvo and BMW, who between them now held the monopoly on the Traffic fleets. The car never really caught on, which was a shame.

The Ford Focus ST, however, was similar in concept to the old RS2000s, in that Ford took a smaller-bodied car and dumped a big-performance engine into it, this time the 2.5-litre turbo-charged, 5-cylinder unit that Ford nicked from Volvo's T5 range (they also owned Volvo at this time). It was a very popular move, and several forces, including Dorset, Surrey and Northamptonshire Police, used them as urban Traffic cars and the officers involved loved driving them, although the petrol consumption was not liked by the accounts departments . . .

Jaguar. Just the name conjures up an image of police Traffic cars cruising the motorway or chasing the bad guys (also in a Jag!) onto a piece of urban wasteland and forcing them to crash into a stack of old oil drums, rendering them unable to continue their escape. For their looks alone they deserve every plaudit, but as a work tool in police hands they had a lot of problems – especially during the 1970s and early 1980s, when the reliability of Browns Lane's products sank to an all-time low, and in truth police orders for these vehicles never really recovered.

The earliest police use of Jaguars goes back before the war years and included the SS Jaguar and the 2-litre model of 1937. By the 1950s, some forces were using the MkV saloon, and even the huge MkVII got a look-in with the old Worcestershire force. Driving a car the size of a canal barge on a shout must have been an interesting experience, although of course they were raced and rallied with some success at that time, so they can't be as 'supertanker' as they look. By the late 1950s several forces had bought the sleek-looking (retrospectively named) Mk1 3.4-litre saloon, which came out in 1955 and used the legendary straight-six XK engine in a compact unitary construction body. These included Somerset and Bath Constabulary, Portsmouth City Police, Bedfordshire and West Sussex Police. Throughout the 1950s and 1960s, the police service became a significant customer for Jaguar cars, with one advertisement for the company quoting 33 forces using Jaguars. The company appreciated this custom, being at heart still a small independent company with a racing mindset as far as engineering was concerned, and because of this they were able to quickly adapt cars to an agreed police specification. They were keen to get the business, too, having invested in what was then, for them, a huge sum of money in bodyshell tooling for their new compact saloon. The expense was justified, though, because no other company in the world was making a compact 4-seater with the performance of the 3.4-litre version that did 0–60 in 9.1 seconds and could run on to over 120mph, given a long enough road. The fact that the engine had proved itself at Le Mans in C- and D-Types showed that it was basically reliable, and the Mk1s soon began to clean up in saloon racing as well. By the time Jaguar announced the Mk2 in 1959, over 35,000 Mk1s had been produced and Jaguar had gone from a niche luxury or sports-car maker to a volume producer with the image of niche luxury sports-car maker; a marketing trick that BMW pulled off a decade or so later.

On all the evidence...

...the finest police car of all

Over 50 Police Forces throughout Britain and the Continent testify to the efficient performance of Jaguar by its ever increasing employment for both routine and emergency duty on motorways and minor roads everywhere. The Mark 2, available in 2·4, 3·4 and 3·8 models, and the 3·4 and 3·8 'S' models, provide the unrivalled performance of the world famous XK engine together with exceptional manoeuvrability, outstanding roadholding under all conditions and the safety of four wheel disc brakes. For the more formal requirements, the new 420 and 420 "G" models offer dignity and luxury for all occasions.

JAGUAR

xiii

The Mk2 Jaguar was basically a heavily facelifted version of the Mk1, which rectified its commonly known failings: it had a wider rear track to address the massive difference front to rear on the Mk1 that had caused skittish handling, thinner screen pillars for improved visibility and improved handling and brakes. It was also available with the new largest ever XK engine of 3.8 litres, which produced 220bhp@5500rpm, giving a fantastic 0–60 of 8.5 seconds and a top speed of well over 120mph. It was, for the time, the ideal patrol car for many forces, while its position as a proud British firm flying the flag in international motorsport and beating Johnny Foreigner made it a politically correct choice, too. The Mk2 really established the marque as a car that the Police could make very good use of. From the south coast to the tip of Scotland, dozens of police forces opted for the fabulous-looking Mk2 saloon as its Traffic car of choice, often in top-of-the-line 3-litre form. However, many forces took the 3.4-litre car as well, which, with a top speed of 120mph, was not much slower, was cheaper to buy and had a reputation, at least initially, of being a more robust engine than the 3.8 and was certainly a sweeter, less gruff unit to drive.

By 1967 the Mk2 had been on sale for eight years. The XJ6 was due to be released soon, but Jaguar had by then spun two other cars out of the Mk2's basic hull – the 1963 S-Type and the 1966 flat-fronted 420. Both featured a longer tail and the new E-Type-derived independent rear suspension unit, and both were used by the police as Traffic cars, although the 420 was used in much smaller numbers.

Jaguar had decided to produce a new car, the XJ6, which replaced the whole four-car range with one vehicle! It was, of course, brilliant as a piece of engineering and styling, and in V12 form it was probably the most exciting car ever used by Traffic on a regular basis, but with the XJ6 delayed Traffic needed to refresh the Mk2 in 1966 to enable it to remain competitive for a few more years and not steal sales away from its IRS brothers. Thus they replaced the Mk2 with the 240 and 340, a cheaper, lower-spec version of the same car, designed to appeal to the police and other fleet users that used the straight-port cylinder head. The 2.4 went over to SU carburettors instead of Solex and the 3.8 engine was no longer offered, having been supplanted in their range structure by the 3.8 S-Type or 4.2 420. Are you with me? Because at this point I think even Sir Williams got a little confused . . .

The 240 and 340 had no Jaguar leaper, used 'Ambla' vinyl seats, thinner bumpers (which I must admit I quite like) and a lot less of the lavish walnut trim and picnic tables. The police versions featured matt-black-stained cheaper wood dash and door cappings instead of the polished walnut veneer, rubber compound on the floor instead of the plush carpets, no arm rests and plain door pockets among the 150 or so changes made to the standard car. It was available with the 2.4- or 3.4-litre engines and the cars were known as the 240 and 340. As part of further rationalisation, the 340 was deleted in 1968 and only the 240 was made available to the public. However, the 340 was available to special order for police forces and the Staffordshire and Stoke-on-Trent City Police was one force that ordered seven such cars, at a bargain price, to replace the previous motorway fleet of Mk2 saloons in 1968. Which explains why you could still find Mk2 Jaguars patrolling Britain's motorways into the early 1970s. A fitting tribute to one of the great all-time Traffic cars.

Staffordshire was one force that remained loyal to the brand for its motorway patrol cars from the 1960s through to the 1980s. Staffordshire's original southern border was close to the Jaguar factory, so this loyalty was entirely natural.

Our motorways at that stage were beset with fog, and probably none more so than the M6. In February 1967 Staffordshire Police had seen the death of an officer, PC Clive Blackburn, who was struck by another car when assisting at the scene of a breakdown in thick fog on the M6. Staffordshire then became pioneers in trying to make their vehicles more visible, so these cars were fitted with larger roof signs, rear reflectors and, for the very first time, a red livery on the boot, the rear face of the roof sign and the insides of the rear doors, together with a larger blue light that had twice the output of the standard blue lamp. The livery and all-new kit made these cars quite distinctive and was much heralded at the time.

To assist with the car's handling, the rear seats were removed and plywood boards bolted in place so that items of emergency equipment could be clipped to the wooden panels (shovel, broom, crow bar, etc.) with the other heavy items (12 cones, lighting equipment, etc.) stowed on the seated part of the boards to help spread the weight to make the cars more stable at higher patrol speeds.

Whilst most police forces favoured the Mk2, the Metropolitan Police bought 266 of the S-Type, making it the largest ever Met Police order up to that date – or since! Of those 266 cars, 83 were delivered in white and designated as Traffic cars, the first white cars that the Met ever had, whilst the remaining 183 cars came in black and were destined to become area cars. Although similar to the standard model, the 3.4-litre S-Type 'Police Special' had over 150 modifications, which included such items as a low ratio differential, uprated suspension and brakes, no power steering and a much-changed interior in line with the aforementioned 340 models. All the S-Type Traffic cars came with a 4-speed manual gearbox, whilst the area cars were all automatics.

The cars were fitted out with a Pye Vanguard VHF radio set, a Winkworth bell and a set of two-tone horns, whilst a calibrated AT speedo replaced the standard-fit Jaguar item. A Heller blue light and two Butler spot lamps adorned the roof whilst the Jaguar leaper was removed from the bonnet for safety reasons.

The S-Types were eventually replaced with Triumph 2000 and Rover P6s in the early 1970s. Most of the Jaguar cars had either SUU or WGK registrations. The last Jaguar S-Type in service was WGK 498G. The only other force to use them was the Ayrshire Constabulary in Scotland.

One of the cars' claims to fame came in 1969 when no fewer than eight Met Police S-Type traffic Jaguars escorted Moors murderer Ian Brady from the Old Bailey on the long journey north to Durham Prison.

In 1968 Jaguar launched the XJ6, a big four-door saloon with a punchy 4.2-litre straight-six motor capable of 120mph with a 0–60 time of 8.7 seconds. Like all Jaguars it looked fabulous; it was very comfortable, had plenty of boot space, performed well and was seen as an ideal motorway patrol car. It should have been the traffic officers' dream car but this didn't happen for a variety of reasons. First, there were two big rivals from Triumph and Rover to contend with, which definitely stole sales away in both the civilian and police markets, as these competitors were cheaper and almost as fast in real-world conditions. Then there was the reliability factor, which, let's face it, could have been a lot better than it was and no doubt cost the company dear. By 1973 Jaguar had revised the car slightly, making it the Series 2 XJ6, which in turn was replaced by the Series 3 in 1979. Only a handful of forces used the Series 1 and 2 cars, including Avon and Somerset, Ayrshire, Thames Valley, Dunbartonshire and Staffordshire Police – the latter were quite possibly the only force to have stayed loyal to the brand right through to the 1990s.

The later Series 3 XJ6 was a vast improvement on the previous two models when it was launched in 1979. Jaguar produced a Police Special and even badged it as such, with the letters 4.2 PS placed on the boot lid. It was also finally offered with BL's new LT77 5-speed gearbox (as used in the SD1), whereas previous manual XJ6s had used a 4-speed plus overdrive gearbox. All the luxury items were removed and replaced with hard-wearing plastic, as well as basic fittings like steel wheels, sports suspension, limited-slip diff, uprated alternator and additional wiring for the roof lights. The new car was faster, with a top speed of 130mph, handled better and was generally more refined. And Jaguar made a concerted effort to win over the police and promote the car as its best ever vehicle in terms of being suited to traffic patrol. It provided a test car for forces to evaluate and it was on that basis that orders were won.

This was at a time when the Rover SD1 and the Mk2 Ford Granada were the kings of the strip whilst BMW were attracting more and more interest. This is where Jaguar shot itself in the foot; many of those orders were late arriving or weren't the specification ordered. The test car was an automatic with sports suspension, and its performance was said to be excellent. At least one force ordered 16 new XJ6s based on that test and were dismayed to find that, when delivered, Jaguar had fitted them with manual gearboxes and limousine suspension! Legend has it that one of them didn't even get off the low loader on delivery because the gearbox had seized. Driving them on the motorway was said to be a joy, but handling on anything resembling a curve was best described as 'difficult' – or 'soggy', as one officer told me – especially when fully loaded. And when doing a U-turn on the motorway (in the days when police cars could actually drive through an ECP, or Emergency Crossover Point, to facilitate a quick turn around on the motorway!), the Se3 XJ6 was the only car that traffic cops ever drove whereby the turning circle was so bad that even on full lock the car wasn't capable of undertaking this manoeuvre in one go. In fact, it would only make it to the hard shoulder, where the car was then put in reverse and had to move back out into one lane of a live motorway before shuffling forward to complete the turn. I bet that raised a few heart rates among traffic officers! But despite the problems the Series 3 XJ6 was a big success with forces like Hertfordshire, Hampshire, Sussex, Northamptonshire, West Midlands, West Mercia, West Yorkshire, Derbyshire, Staffordshire, Thames Valley and Leicestershire Police, who all had the car as their mainstream Traffic car. Traffic cars tend to acquire nicknames or phrases for a reason, and at least two of the officers interviewed when asked about the Series 3 came up with 'Jaguar trait, great in a straight' as their immediate reaction.

In September 1986, the new XJ40 was announced and expectations were high as it was Jaguar's first all-new car since 1968, having been in development since 1972, on and off! The Project's Chief Engineer, Jim Randle, has steered it through several changes of ownership and corporate reconstructions since, which resulted in Jaguar being floated as a separate entity in 1984, only 14 years after they had joined BMC to ensure continuity of body supply. The result was a car that looked underwhelming, partly because its styling had been signed off five years or more before the final launch, and its rectangular lights and very plain styling reminded many of a 1970s' big Vauxhall. Journalists didn't like the cramped interior either, and they thought the new all-aluminium 2.9-litre AJ6 engine was underpowered. The 3.6 litre was better liked, and although its poor packaging (kind of opposite to the TARDIS – smaller inside than it looked like it should be on the outside!) and build quality were criticised, it won group tests because its ride and handling were actually very good. It sold well and was the reason why Ford and GM eventually got into a bidding war to buy Jaguar, which ended with a Ford victory in 1990. Despite its characterful traits and a quickly acquired reputation for electronic niggles, the police, who are extremely fussy about the cars used on Traffic, used it in surprisingly high numbers, and this included some forces who had never used Jaguars before.

There may have been political and financial incentives at work here. Jaguar was being trumpeted as Margaret Thatcher's shining star of privatisation, and some political pressure may have been exerted behind the scenes. That, combined with the square XJ40 actually not leaping out of the showrooms quite as fast they'd hoped, meant the police were offered the cars on the cheap during a time of austerity in the late 1980s. Whatever the case, forces like Derbyshire, Cumbria, Cambridgeshire, Durham, Northumbria, Northamptonshire, Sussex, Warwickshire, West Midlands, South Wales and, of course, Staffordshire all took on the XJ40. Most of these were PS models that included the fitment of an additional calibrated speedo into the glovebox lid, and I've spoken to at least two policemen who rated it as one of the best cars they ever used in 25 years of motorway policing. The model did suffer from one or two odd issues, though, including a number of engine management glitches which usually resulted in a holed number 6 piston. However, the weirdest was a portent of electronic issues to come. The Jaguar demonstrator had been built to a good standard, with the modern radio equipment well shielded. However, when the cars started entering service, the fitment of the elderly (and thus electronically very 'dirty') 28-watt VHF radio transmitters used by West Midlands Traffic on the M6 actually set off the ABS! Many XJ40s had to go back into the police workshop to be shielded or updated. Officers all say the later 4-litre XJ40s were noticeably poorer in quality than the 3.6, although this improved on the later facelifted (and much nicer looking) X300 because Ford had got control by then. They still didn't handle as well as rival cars, though; one West Mids officer who regularly drove an XJ40 remembers classic Range Rovers keeping up with Jags on twisty roads, and 24V Senators, which were less powerful, leaving them for dead as soon as the road had any bends.

By the 1990s Jaguar had reverted the styling back to the quad head-lights loved by all and given us something to lust over for its looks alone. Jaguar made a fairly low-level approach at promoting the all-new X300 4.0-litre automatic onto the police market and built two Police Specials, but the cars were now up against the likes of the Volvo T5, Vauxhall Omega MV6, Rover 827i and the BMW 5 series. The X300 PS differed from standard production models in that it had uprated suspension, a calibrated speedo and a rear screen wiper blade.

The two cars were first registered by Jaguar Cars in May 1996 as N610 SWK and N601 SWK. The first car spent the next three years being sent out to various forces on test. In early 1997 the Lancashire Constabulary trialled it on their Traffic Division at Skelmersdale for two months and in a written report concluded that its handling and road holding in the dry was outstanding but great care was needed in the wet. It described the interior as cramped and claustrophobic and stated that the boot was too shallow to house a sufficient amount of emergency equipment. The public loved seeing it, although many questioned police use of such luxurious cars; a problem Jaguar and the police have always had to put up with. At a cost of £31,000, though, it was double the price of a Rover 827i. Warwickshire Police borrowed the car a couple of times to assist during royal visits to the county, and in April 2000 the car was used by the South Wales Police on its Traffic Department until November 2005.

When the car was returned it then spent some time at the Metropolitan Police Traffic Museum at Hampton before being returned *to Jaguar Heritage Trust until August 2009, when the Trust decided to sell the X300 to Police Car UK on the understanding that the car would not be sold on outside of PC-UK.*

The second X300 spent several years with the No.6 Regional Police Driving School at Devizes, in Wiltshire, where it was used on Advanced Driver Training. The graphics used on both cars were unique to Jaguar Cars.

In 1997 Jaguar re-engineered the XJ40 structure yet again to take their new V8 engine, including the awesome supercharged 370bhp XJR, although sadly the police versions had normally aspirated 4.0-litre V8s. On paper, once again, this should have been a winner, but the competition and pricing were major factors. However, West Yorkshire and Staffordshire Police (of course!) had two each to use as motorway cars. They looked fabulous and reliability had improved dramatically; Jaguar at last had delivered the car it had been promising since the demise of the Mk2.

Two years later, Jaguar tried again with the introduction of the retro-styled S-Type with a 2.5-litre V6 motor. Two more demo cars were prepared and sent out to various forces for evaluation. A large number of forces gave them a fair trial but in truth they really weren't big enough or fast enough for Traffic patrol and were more suited to area car work, but there they were up against the BMW 3 series with its diesel engines, and the Germans had cornered that market already. Nevertheless, West Yorkshire Police were fans of the model and bought a couple. One of the demo cars was written off in an accident, whilst the second one went back to Jaguar Heritage Trust, where it remained for a few years before being sold to Police Car UK.

During Ford's brief ownership of Jaguar, the X-Type saloon and estate were released. Merseyside Police felt obliged to take on several of the saloons, given that the cars were built on its patch at Halewood, whilst the Grampian Police had a couple of the X-Type estates in their Traffic departments.

Then it all went very quiet. Other cars dominated the scene until 2007 when Jaguar introduced the XF to huge worldwide acclaim. In 2009, a police-spec XF-S with a 4.0-litre V8 motor was pushed towards the Traffic fleets. Plenty of forces trialled it, but in the end the Central Motorway Police Group (consisting of West Midlands, West Mercia and Staffordshire Police), who had eight of them, and Surrey Police, who had two cars, were the only customers. The car looked superb and its performance and reliability were up there with its rivals, although the boot was very shallow with a very small aperture. In 2013 Jaguar unveiled a police-spec Sport Brake (estate) model, but there were no takers.

It's unlikely we will ever see large numbers of big cats cruising the strip in the future or forcing the bad guys to crash into that stack of oil drums ever again, simply because of the way the police are tied to national framework purchasing agreements and large-scale procure-ment in twenty-first-century Britain, but Jaguar's history as a maker of legendary Traffic cars will never be forgotten.

Think Rover and most of us will think police car. The name Rover, it seems, is intrinsically linked to the police but in truth the two only really came together with the launch of the P6, or Rover 2000 to give its proper name, in October 1963 (it was voted Car of the Year in 1964), and in fact most police forces didn't start using them in numbers until the P6B V8 model came out in 1968. Prior to the P6 it was rare to see a police Rover. The Met Police used one or two models like the 75 and 105 as staff cars, and in 1936 they had a couple of Rover 10/12s as patrol cars. In 1956, the Cheshire Constabulary used a large number of Rover P4 saloons, all of them finished in green, as marked Traffic cars. Officially designated as the Rover 90, it came with a 2.6-litre, straight-six engine that pumped out 90bhp and gave the car a top speed of 90mph on a good day. Having the cars in green wasn't that unusual at the time, as the Somerset and Bath Constabulary had their Series 1 Jags in green, whilst a few other forces bought the occasional green car or van. The Cheshire Rovers were fitted with an illuminated police box on the roof and a rotating blue beacon, making it one of the earliest of its kind in the UK.

But it was the P6B V8 that really brokered the marriage between Rover and the UK police. The Met led the way and bought these cars almost from day one to be its mainstream Traffic car to replace the ageing S-Type Jaguars. As we have already learned, the new Rover was given two roles within the Met; blue cars were designated as area cars, whilst the white model was assigned to traffic patrol. By 1976 the last batch of 3500S saloons were the first Met cars to get reflective stripes along the sides.

The big V8s became a London icon and are remembered today with a great deal of affection. The car itself was brilliant for its time and its powerful 3.5-litre V8 could top 125mph with a 0–60 time of 9.1 seconds. It had good handling and road holding, it was very comfortable and was easy to manoeuvre in city streets. Outside of London it was an ideal motorway patrol car. It had the power and top end to keep up with everything, and Innovative technology also meant that the old chrome bells were now making way for the new two-tone air horns, and with that big lump of a V8 under the bonnet the cars sounded as good as they looked. Just about every force in the country was snapping them up whether they had motorways to patrol or not, and it was from here that the words 'Rover' and 'police' forged their relationship.

In the early 1970s, Leyland combined the design teams of the Rover and Triumph brands into one group called the Specialist Division to work on the single replacement for the Rover P6 and the Triumph 2000/2500. Their project became known as 'SD1', the first project for the Specialist Division, and in 1976 they delivered the first SD1 models. As a piece of design work it was truly outstanding; it looks fantastic and very futuristic. It must rank among the best car designs of all time. But looks aren't everything, as they say. The British car industry in the mid-1970s was a joke, as was the build quality of the SD1 – like several other cars of that era – and it suffered enormously. What should have been one of the best products that this country has ever produced fell from grace quicker than the inflation rate rose in the 1970s. But that didn't stop Britain's police forces buying the car by the bucket load, and why not? Top speed was still 125mph from the same Buick V8 used in the P6, handling was excellent, and the space both in the cabin and the boot was class-leading. It was an excellent package let down by poor build quality and an even worse parts availability. Bits would just fall off them, they weren't seen as police-man-proof (few things are!) and the complaints soon started to roll in. The quartic steering wheel on the early models wasn't seen as in keeping with the 'Police system of driving' and many forces changed the square-style wheel for a standard round one, including the Met who led the development of this. Interestingly, I learned recently that when the SD1 was launched in the North American market it did so with a conventional round steering wheel, no reference to quartic wheels anywhere! The manual gear lever was positioned too far to the left to make it comfortable to use (it was rumoured at the time among police officers that this was done for the American market, but I don't think this was the reason), and the position and location of the switch gear wasn't to everyone's liking either. Nonetheless, it still proved very popular as a Traffic car, and in later Series 2, 3500SE versions from 1982 to 1986 it was improved significantly. The Met even managed to obtain some Vitesse models and a couple of twin-plenum versions capable of 133mph with the 0–60 time down to 7.1 seconds!

The police were basically split over the SD1 or the Ford Granada, with your local force usually having one or the other in this era. The

Met were clearly in the Rover camp, as were Hampshire, West Midlands, West Mercia, Derbyshire, South Wales, Cheshire, West Yorkshire, South Yorkshire, Durham, Cleveland and Humberside. Incidentally, some of the early Met cars were painted in the same Zircon blue that their P6 saloons came in, even though they were experimenting with stripes at this time, but it was short-lived as the remainder were finished in white and got their stripes.

The SD1's replacement was the 800 series, and a strange amalgamation between Rover and Honda saw the legendary V8 lump banished to the history books, at least as far as non four-wheel-drive cars go, and Honda's 2.7-litre 6-cylinder engine was put in its place. The cars were made available as saloons or five-door fastbacks. It was a quick car, though, at 130mph, but its front-wheel-drive system (previous Rovers were all rear-wheel-drive) found that power difficult to cope with, especially as traction control was still an expensive option at this time. Preventing front wheel spin, especially in the wet, took some police drivers a long time to master, and of course it goes without saying that there were all the usual moans and groans! Although the engine on the 827 was considered almost bulletproof, the same cannot be said for the rest of the car. It was flimsily built and the door handles, of all things, would just snap off on a regular basis. That said, plenty of forces bought them, including the Met, Avon and Somerset, Cheshire, Dorset, Greater Manchester, Hampshire, Staffordshire, West Mercia, Strathclyde and Grampian Police.

In later Series 2 models the whole body and interior were revamped and the cars were dramatically improved, especially with the introduction of traction control to the front wheels. The fastback 827i version became the number one choice for the Met's traffic boys and the cars were used extensively throughout London until it went out of production. They still have one in their museum collection. Most other forces also used them on their Traffic fleets but not as the sole car. Many forces were now using a variety of cars so as not to put all their eggs in one basket, and this proved to be a wise decision following the discovery of a major fault on the newer 800 series cars. Cracks were found to have developed along the boot floor between the rear suspension turrets, causing the cars to flex and have a severe effect on the handling. At first it was thought to be a one-off, but checks revealed that most of them were suffering from the same fatigue. Rover accused the forces of overloading the cars with equipment but then conceded that they would rectify the fault immediately as the police had always been upfront about how much kit the cars were expected to carry and Rover had said they were up to the job.

The 800 series replacement was the Rover 75, but it didn't get a look-in and even its beefed-up brother, the MG ZT-T range, struggled to make any kind of impact, with Sussex Police being the only customer. Mind you, theirs had the Mustang V8 engine! And, as we all know, Rover went to the wall and the police had to look elsewhere.

Rover's big rival in the 1960s and 1970s was of course Triumph, and its only real Traffic car contender was the 2000/2500 range, which was made available in saloon and estate variants. Although used by a good number of forces as a Traffic car, it was also one of those rare cars that crossed the divide and was operated as an area car in some forces. Launched in 1963, the Mk1 2000 got off to a slow start, with only a handful of forces taking them on. Top speed from the 1998cc straight-six engine was a respectable 98mph, but it was the car's space and interior comfort, together with the excellent handling and road manners, that won it much acclaim. Nottingham City Police (whose cars were finished in gloss black with the city's coat of arms emblazoned on the front doors), Dorset, Somerset and Bath Constabulary, Stoke-on-Trent City Police, West Yorkshire and the Lancashire County Constabulary all had the saloon, whilst Hampshire operated two estates in an effort to stem the wave of political pressure they were suffering for daring to buy foreign.

As in civvy street, the Triumph was battling hard against the Rover P6, and when it launched the Mk2 2500 model it really gave Rover a fright. The new car became the first family saloon car in the country to use fuel injection, and the police saw this 150bhp Pi as an opportunity to make time on the motorways that were springing up all over the country at this time. The new engine pushed the big Triumph up to 108mph, which was still some way behind the Rover, but it was lighter and much more agile. Many consider it to be a better driver's car than the Rover, and no doubt the arguments continue to this day. Whatever the pros and cons of each, it remains a fact that the Triumph 2500 Pi and its twin-carb aspirated 2500 TC stablemate were very popular Traffic cars with many forces. The Met used loads of them as both traffic and area cars, whilst the City of London Police would only use the Triumph and not the Rover.

Other forces, like Avon and Somerset, Devon and Cornwall, Dorset, Hampshire, Lancashire, Nottinghamshire (who used estates), Surrey, West Yorkshire and the West Midlands Police, all used the Triumph as a Traffic car. There is a preserved West Midlands Police Triumph 2500 TC in the Coventry Transport Museum.

Has there ever been a more useful workhorse in police hands than the Range Rover? As British as a Sunday roast, it must rate as the most versatile car the Old Bill has ever used. As a motorway car it could do it all; from blasting down the strip to the next RTA at 100mph to dragging a fully laden stricken artic over to the hard shoulder, from undertaking VIP escorts to acting as an ARV gun ship, and from being a serious off-roader to being the flagship motor of the force, the Range Rover has done it all.

When the Range Rover was launched in 1970 it was an instant hit with the public, who couldn't get them quick enough, and the police were also quick to see the car's potential. The nearest thing to it at this time had been the standard Land-Rover (it still had the hyphen then), but that marque was little more than a utility vehicle, and what made the Range Rover so different was the addition of that famous Buick-derived Rover V8 engine, the superbly comfortable long travel coil springs and its car-like cabin. The 3.5-litre V8 could top 100mph and was capable of carrying large amounts of additional equipment to the scene of any RTA, whether on the motorway or elsewhere. The cars were introduced at a time when the police service in general was equipping its cars with more and more officer safety kit, with the addition of better blue lights, large roof boxes, reflective markings, better protective clothing and much-improved VHF radio systems. One of the innovative ideas was the roof-mounted stem light. Basically the blue light was mounted on top of a box, but at the push of a button the light was raised on a telescopic mast to about 10 or 12 feet above the car. This helped to alert drivers who were approaching the scene of an RTA much quicker than if the blue light had been kept at roof level. Later versions of the stem light had floodlights positioned beneath the blue light to help illuminate a darkened scene. However, the system wasn't always policeman-proof. It had to be raised and lowered whilst the car was completely stationary, and invariably some officers forgot to lower it at the close of an incident and the mast would bend or break completely once the car started to move!

The original Range Rover (often now referred to as the Range Rover Classic) was a two-door car with a 3.5-litre V8 that pushed out 130bhp and permanent four-wheel drive via an additional diff lock lever. It had no power steering and a four-speed manual gearbox. The Lincolnshire Constabulary was the first force to use the Range Rover, but it wasn't long before most others followed suit.

Prior to that a company called Radio Telecommunications from Blackpool developed the Range Rover Vigilant especially for the emergency services, for which the rear side windows were replaced with roller shutters which, when opened, gave access to a range of properly stored emergency equipment in lockers measuring 12 inches by 36 inches. On the roof, they added a huge fibreglass pod that incorporated faired-in illuminated POLICE and ACCIDENT signs, a blue light and two 11-inch spotlights that could be elevated to a height of 15 feet to help illuminate a darkened scene at night. It proved too expensive for the police, and the only real success it had was as a diecast model from the toy manufacturers Corgi.

Those force areas that had a lot of motorway to cover or which suffered from adverse weather conditions in winter, or both, took the Range Rover on in big numbers, with Lancashire and Greater Manchester Police being the biggest customers. For a brief period Rover made a

Range Rover Commercial, which was basically a van. It was similar in concept to the Vigilant model in that the rear side windows were removed and simply replaced with metal whilst the interior lost the rear seats to extend the load area to enable more emergency kit to be carried and to help spread the weight more evenly, which should have helped with the car's handling. If you have ever driven a Range Rover Classic you'll know just how much it rolls from side to side on bends and roundabouts. At least two forces – Greater Manchester and Cumbria Police – had a small fleet of Range Rover vans.

In the 1980s the Range Rover gained two significant features: rear doors, which made it much more versatile, and fuel injection, which upped the power to 155bhp. In 1990 they increased the engine size to 3.9 litres and it was referred to as the 3.9 EFi. This was probably the Police Range Rover's high point, with just about every UK force now employing the big car for a variety of roles never envisaged by anyone when it was first introduced. Some 25 years after its birth, the Range Rover got its first major redesign in 1995 with the P38A model. The body was completely different but somehow the car still looked like a Range Rover, and the engine was now a 4.0-litre EFi V8 pumping out 190bhp. In 2002 the third-generation Range Rover was launched, and during its development period Land Rover was taken over by BMW, so early versions were fitted with the engine and transmission from the BMW 7 series whilst later cars got a Jaguar 4.2-litre V8 or a 3.0-litre BMW diesel, all of them mated to an automatic gearbox as the manual was no longer available. The car had now reached superstar status with the public with a price tag to match, and as the luxury specification went up and up so the demand from the police just nose-dived, in part also because by this time the BMW X5 was quite literally taking over the baton from the Range Rover and by around 2010 the German brand had complete dominance. No police versions of the fourth-generation Range Rover that was introduced to the market in 2012 were ever made.

TJA 972X arrived at Greater Manchester Police's Openshaw Workshops in October 1981, one of a batch of ten new Range Rovers destined for the Motorway Group. However, the Technical Communications Branch (Tech Comms) was looking for a new command vehicle and they obtained 972X. To make the car stand out from Traffic Range Rovers, individual black squares were painted along the waistline of the car in a chequerboard effect. The rear of the car was fitted with multi-channel, state-of-the-art radio equipment, including a radio telephone. The car also towed a 40-foot hydraulic mast, which would be erected on site when used as a command vehicle.

The Range Rover was based at GMP's Chester House Headquarters in Old Trafford and was in use for ten years, covering 47,000 miles in that time. In 1991 the car was no longer required as a command vehicle but was kept on the fleet due to its low mileage. Openshaw Workshops removed all the radio equipment and refurbished the car with bits of three-year-old, high-mileage, motorway Range Rovers that were being sold off. The car was resprayed and fitted with a Woodway strobe light bar, then given back to the radio branch as a runabout. Few people would drive it, as it had no power steering.

In 1991 Inspector Geoff Taylor of the Motorway Group set up a Road Safety Display, which was taken into police station open days, classic car shows and the like, so that motorway officers could talk to people about driving on the motorway. TJA 972X was borrowed from the radio branch, equipped as a 1970s motorway patrol car and fitted with a 'blow through' roof sign. The use of the car as a display item was so successful that 972X was transferred to the motorway fleet, kept purely for PR work and road safety displays, and continued in this role until 1997. Having now completed 55,000 miles from new, the car entered GMP's Museum as a static exhibit.

In 2005 a review of the Range Rover found it was again in need of refurbishment, and as the museum didn't possess the necessary funds, a decision was taken to sell the car. Having retired from GMP,

Geoff Taylor asked if he could buy the car direct from the museum, but that permission was refused and he was told it would have to go to auction. The museum removed all the police equipment from the car prior to it going to BCA Auctions at Preston. Geoff Taylor bought the Range Rover at auction, refurbished it and then returned the car to the GMP Museum, where it was reunited with all its original equipment. The Range Rover was loaned to the museum as a display item, until a visit from the Health and Safety 'police' raised concerns about an emergency evacuation of the building and this big police car being in the way. So 972X was evicted from the museum, but it attends events whenever requested.

TJA 972X has featured in British Bus Publishing's The Police Range Rover Handbook, *and in* James Taylor's *book* Original Range Rover, *and has appeared in the Channel 4 trilogy* Red Riding. *BBC North West Tonight also utilised the car for three days, touring the North West during the original elections for the Police and Crime Commissioner. When in the Police Museum the model maker IXO made a 1/43 scale model of TJA 972X which is now highly sought after. The tooling for this model appears to have been bought by a Hong Kong model manufacturer, and an identical model bearing number plates GMP 999 is now also available.*

TJA 972X is the sole survivor of some 300 Range Rover Classics that have been used by Greater Manchester Police between 1975 and 1998, and it is one of only two two-door Classics still in police livery. It is an important piece of GMP's vehicle heritage and a unique survivor of these often hard-worked police vehicles.

I was lucky enough to drive the car at Gaydon during a Police Car UK club rally, and what a privilege! I loved it. It gave a real feeling of being King of the Road, that mellifluous V8 purring away in a brilliant piece of history that I'm really glad has been preserved.

Some police cars become a little famous during their careers, usually on a local level if they have appeared in the local paper or TV attending an incident of some kind, or others might appear in books or magazine articles, but only one has reached a worldwide audience, and that was a Range Rover.

M751 CVC started its life as a white Vogue SE, second-generation P38 and was one of six such cars turned into police demonstrator models by Land Rover in 1994. As such they were the first cars from Land Rover to appear in the new Battenburg livery that the government was attempting to make a national livery at the time. The car was used in Scotland and then by Warwickshire Police before being transferred down to London in March 1997 to be used by the Metropolitan Police. The Met weren't that keen on the new graphics and so replaced it with their own standard livery. The car was decommissioned in December 2001 and sent off to auction, then had a number of owners before the current owner found it on eBay in April 2011.

The car needed a full respray before being put back on the road. The owner knew it was an ex-police car, and whilst researching its history he got in touch with Stuart Exelby at Police Car UK, who found out that the car had been used by the Special Escort Group and sent him a couple of photos of it when it had been in service. He also sent him a YouTube link but wouldn't tell him why. The owner clicked on the video link and was shocked to see that M751 CVC was the escort car for Princess Diana's funeral, escorting the hearse from Hyde Park Corner to Althorp on 6 September 1997 in front of one of the biggest TV audiences of all time. Since then the owner, Richard Hopkins, has been in contact with John Swain who was the officer controlling the funeral cortège from the rear of the Range Rover, and he has assisted Richard with a lot of the detail to make sure that this historically important car is restored back to the exact specification it was on that very sad day.

In 1989 Land Rover introduced a new car to its stable; the Discovery. Although a slightly smaller SUV than the Range Rover, it looked likely that, given its cheaper price and the fact that it could be offered with a 2.5-litre diesel engine at a time when the police were heading in that direction, all the signs were that Land Rover had another winner. But the police didn't really take it on board until about 1993; its early diesel motor was a noisy and very slow unit and the cars quickly gained the name 'The Tractor' from Traffic cops, who would find any excuse not to drive it. And in truth you can perhaps understand why. When fully laden with all its kit it was just about on the manufacturer's recommended weight limit. Now place it on the hard shoulder of a motorway, facing an uphill section, and ask its driver to pull away quickly and rejoin the main carriageway! It wasn't exactly safe. Later Series 2 Discoverys got a better 300 TDi motor, and there was an option for the faithful V8 lump to be installed, but almost all police versions were diesels. The Series 3 Disco (as it is often called) was a lot more sophisticated and more like a Range Rover than anything else, but the car was now much improved, with better common rail diesels that gave far superior performance. The Series 3 and 4 models, launched in 2004 and 2009 respectively, now had a 2.7-litre diesel and the car's popularity grew massively. Although a few forces took them on, the BMW X5 was now firmly entrenched at the police dining table and it wasn't going anywhere. As good as the later Discoverys were, they were more expensive to buy and run than the X5, so its appeal waned just as that of its bigger brothers had done.

Vauxhall's overall contribution to the police has been fairly substantial, but it did struggle to gain a major foothold on our traffic fleets until the late 1980s and early 1990s. One of its earliest efforts was the 1950s Velox and Cresta models most famously used by the Stockport Borough Police, whilst the earlier model Velox was used by the Lancashire County Constabulary. Vauxhall upped its game in the 1970s and tried to push the big 3.3-litre Cresta onto the police with a couple of trial cars and a serious promotional campaign with glossy literature.

As far as can be ascertained, only the West Yorkshire Police had the Cresta and for some inexplicable reason got them in light yellow! The 1972 Victor FE was good for just over 100mph with a 0–60 time of 12.4 seconds. It was roomy enough with bags of torque but it suffered somewhat in the road-holding department, especially in the wet. Several forces had the uprated Ventora model, which was good for 107mph, but again its road holding was said to be awful.

Things took an upward turn in 1984 when Vauxhall launched the Senator A and its V-Class platform sisters – Carlton, Monza, Royale, etc.! It was big, comfortable, fast, reliable and gained almost instant success for the Luton-based company, although it has to be said that this was, of course, the end of Vauxhall as a design arm of General Motors Europe. The Senator and all its siblings was a German car developed in Germany by Opel using Opel's famous CIH (Cam-in-Head) range of engines, which were known to be fantastically strong and reliable and in 3-litre injected form produced a lovely smooth 180bhp@5800rpm. However, here was a car able to compete head-on with Ford and Rover, and with a top end close to 120mph it made a good motorway car. But the competition was tough at this time; the mid-1980s saw Ford and Rover go head to head and there was little room for the Griffin to push in, but it did steal sales away from them, not in any great numbers but enough to grant it a modicum of success, with forces like Hertfordshire, Bedfordshire, Durham, Leicestershire, Cleveland, North Yorkshire and Humberside all taking the big saloon on board.

The early Senator laid the foundations for what has become one of the most successful and popular Traffic cars of its era; the Senator B of 1987. The Senator B, which used the same 3.0-litre, straight-six motor, featured a distinctive cheesecutter grille and an overall shape that was inspired by a concept car designed by Pininfarina, the Ferrari Pinin, and it became almost the default motorway police car. In fact the UK police loved them so much that when GM pulled the plug on the Seni in 1993 the UK government requested that GM keep the line going and build an extra 200 or more right hookers for police and diplomatic use before they brought the Omega on steam. That says a lot, doesn't it? The old CIH engine was redesigned for the 1989 Senator 24 valve which brought more power, 204bhp@6000rpm, but also more complexity, which did impact on their reputation for bulletproof reliability a little. However, in 24V form this smooth-looking flagship would easily top 140mph, with claims from some officers that 155mph was possible. Its 0–60 time of 9 seconds was okay, but not indicative of just how rapid it could be on point-to-point motorway work.

The Senator filled the gap between the ageing Ford and Rover products and the advent of the Volvo T5 in 1995, so its timing was perfect. Just about every force in the UK used them as their Traffic cars except the Met who stayed loyal to the Mk2 Rover 800, which they had to do because the Senator was considered foreign. This was the first time we had seen such dominance since the days of the Austin Westminster. But it wasn't all good news. The Senator, fast though it was, could be a nightmare to drive in the wet, with a very tail-happy handling that some loved but that caught out many a police driver, even at 20mph. The bonnet tended to ripple in front of your eyes at anything over 100mph, which had you literally sitting on the edge of your seat, whilst clutch judder was a constant irritation that was never cured, although many were autos. To many traffic cops the Mk2 Senator was the best Traffic car they ever drove; it was certainly one of the fastest just so long as it wasn't raining, and it is one of the true legends of motorway policing!

Steve Woodward; or shall we make this impersonal?

Vauxhall eventually replaced the Senator with the Omega in 1994 and this car came with a distinct advantage over its predecessor in that there was also an estate model in the range. After several decades of using saloon cars and hatchbacks on Traffic, Volvo broke the mould with the T5 estate and suddenly estate cars were cool again and, hey, you could get so much more kit in the back! Thus the Omega estate also sold in sizeable numbers and the old idea of carrying lots of extra equipment suddenly became a new idea and Traffic men everywhere joined the green welly brigade and drove estate cars! But to be fair, most were saloons; even the Met, who had never even looked at a big Vauxhall before, now used the Omega MV6 on both Traffic and as a

unit on the Diplomatic Protection Group (DPG), whose cars are traditionally painted in red to allow easy identification by foreign diplomats
and their staff. The Omega could have been even more successful than
the Senator but for one thing; the Volvo T5 was the car that redefined
the Traffic car.

The Met's buying policy

The Met were able to purchase Omegas because their buying British
political edict had by now fallen foul of two issues. Firstly, there was
no competitive big car that was wholly UK-made, although even the
Rover 800 they had used for years had been a joint venture with
Honda, so it sort of qualified as British! Secondly, under the ever-
expanding remit of EU competition law it was actually illegal for the
UK government to insist that the Met bought UK-made cars. So,
freed of their shackles, the Met abandoned Rover – as BL had now
become – and bought the best car for the job. The policy had been
effectively crazy for some years already; for instance, the Astra, which
was and is made at Ellesmere Port in Liverpool, had qualified, whereas
the Senator had not, even though they were both made by General
Motors, an American company. Other foreign-owned but UK-made
cars such as Nissan did not qualify. The motor industry had effectively
been global since the 1920s, with Ford and GM both making cars in
the UK and in other countries as well, but the British government
took a while to catch up with this in an understandable effort to
protect jobs at home.

Volvo's cars, it's probably fair to say, are not everyone's cup of tea!
Without doubt the power of the pen has persuaded more than one
person to avoid this marque because one journalist branded the entire
range as 'tank like'. That label has stuck like a limpet ever since and has
been used by a succession of other motoring journalists too lazy to

think up their own description or, more likely, because their actual automotive knowledge is somewhat lacking! Basically, if the police are using your product you must be doing something right. As we have explored here already, the police service can be extremely fussy about the cars they use, especially on Traffic. The Volvo story, however, began back in the mid-1960s.

The Volvo story: how Volvo Traffic cars blazed a trail for foreign cars in the UK police

In 1965, the Hampshire and Isle of Wight Constabulary had a problem. The force was running a number of Mk2 Austin A110 Westminsters and a few similar Wolseley 6/110s on its Traffic Division fleet. Although fundamentally very strong, simple cars, like many of their time they weren't exactly reliable and the all-important spares back-up was even worse, which meant they spent far too long off the road. Police vehicles need to be reliable, and as the strategic road network grew throughout the two counties, this became even more important. The accident rate on our roads was increasing rapidly and the amount of emergency equipment these patrol cars needed to carry also increased, to the extent that a standard saloon car just wasn't big enough. A decision was made to design a purpose-built Accident and Emergency Tender (a sort of pre-Range Rover-type unit of similar thinking to the Humber unit discussed earlier) capable of carrying all the necessary kit to the scene of a crash, quickly and reliably.

Representations were made to Chief Constable Sir Douglas Osmond, and in early 1965, following a series of meetings, the Deputy Chief Constable Mr Broomfield instructed the Chief Inspector, Traffic Division, Len Pearce, to undertake a project to produce an Accident Emergency Vehicle. It had to be capable of carrying a large amount of equipment to be used at accidents, but in all other respects had to remain a normal traffic patrol car.

Chief Inspector Pearce formed a team of people to assist him and these included Inspector Jack Hamblin DFC and Sergeant Les Puckett of the Traffic Division, the Transport Officer Tommy Atkins, the Workshops Admin Officer Cliff Thorn and the Workshops Foreman Jim Fraser. Between them, these people held a wealth of knowledge and experience in car design, engineering and performance. They looked at the various options open to them in the estate car market in the UK at that time and found that most were either too small or vastly under-powered. They ended up with just three contenders; the Series 3 Humber Super Snipe estate, with its six-cylinder, 3-litre engine; the Citroën DS19 Safari estate, 16 feet in length and self-levelling but powered by only a four-cylinder, 2.0-litre engine (why did Citroën never put a V6 in the DS?!) and the Volvo 121 estate, with a four-cylinder, 1780cc engine.

Arrangements were made for demonstrator vehicles to be supplied for road trials and a blue Volvo 120 Amazon (as they were commonly called) was borrowed from Rex Neate Volvo Distributors at Botley, near Southampton. Volvos in 1960s Britain were about as common-place as South Korean cars were in the 1990s, and there weren't that many people who knew much about them. Cliff Thorn also managed to acquire a sample gearbox from Ken Rudd, another Volvo supplier in Southampton. It was examined very carefully and was described as a masterpiece in engineering, finished to a very high standard.

After a few weeks' trial period, the big Citroën DS Safari was found to have the capacity, but not the weight-carrying capability required. It was also unstable at high speed when fully loaded, so this car failed to make the grade. Everybody liked the Volvo, but its 1800 engine lacked power. A second 120 estate was obtained, in white this time because Volvo didn't have a black one. It had a Ruddspeed conversion fitted to the engine and this consisted of twin carburettors, high-lift camshaft and a four-branch manifold exhaust, and this helped the acceleration and top end. On a quiet dual carriageway, with three passengers on board, Inspector Jack Hamblin, a former RAF Lancaster bomber pilot, took the Volvo past the 100mph mark and on to a claimed 116mph. According to legend, only Jack Hamblin was allowed to drive the car at this speed, because he was the only person within the Hampshire Constabulary who had ever travelled at over 100mph on a regular basis, whilst thundering down the runway aboard his Lancaster!

In all departments, the Volvo won hands down. It was a good-quality, solidly built motor car that had a good turn of speed and handled well, even when fully loaded. The mechanics liked working on it and it was good value for money. But it was foreign. No police force had ever bought foreign cars in great numbers before and Hampshire knew that they would probably be criticised for even thinking about it. Nervously, the team approached the Chief Constable Sir Douglas Osmond with the result of the trials and requested permission to buy the Volvo. His reply was, 'You're the experts, if it's the best car for the job, then go out and buy it.'

And so, in June 1965, CHO 621C became the first foreign marked police vehicle to patrol the streets of the UK. But just in case the force received too much flak from the media, it also bought the Humber, registration number COT 778C, and it was decided to place both cars on a further six-month trial, only this time with the Traffic Division. Basingstoke got the Volvo, whilst Eastleigh trialled the Humber. After three months, they swapped over and reports were submitted to Len Pearce.

Consideration was given to respraying the cars black, as had been the tradition with all police vehicles, of course, but a decision was taken to leave them in white, as they would be easier to see at the scene of an accident in poor visibility. Another new concept on these cars was the introduction of an illuminated, roof-mounted fibreglass box incorporating the Police/Stop sign at the rear and the Lucas Acorn blue light. A blue panel across the front of the box had the word 'Police' stencilled in white on it; these were manufactured by Wadham Bros. of Waterlooville. The Winkworth bell was positioned out of sight, behind the radiator grille, on the nearside. The public were left in no doubt that this state-of-the-art police vehicle meant business.

After the six-month Traffic Division trial the Humber was deemed to be unstable at high speed when loaded and was noticeably heavy on corners. Its three-speed column gear change also proved difficult to use at speed. The Humber was kept in service for three years, but all concerned preferred the Volvo, and in May 1966 a second 121 Amazon, FOR 298D, was purchased.

This purchase coincided with two important factors. First, the heavy steel accident signs were replaced by roll-up cloth and vinyl items that buttoned onto fold-up metal frames. This weight-saving idea allowed for the provision of a petrol-driven generator to be carried instead. This powered two Mitralux floodlights, carried on tripod frames, to help illuminate accident scenes on dark country roads. The second and more important matter concerned the audible warning device that was fitted. The cars had been fitted with a Winkworth bell, which wasn't loud enough when used on fast A roads, so the bell was supplemented by the all-new two-tone horns. These electric air horns, made by Fiamm, were much louder than the old bell and proved popular with the crews that drove the Volvos as they appeared to help clear the traffic that much quicker.

During the 1960s and 1970s just about everyone seemed to smoke, but even back then smoking in police cars was a serious disciplinary offence. The standard-fit ashtray on a Volvo 120 is a slide-out unit from the left side of the dash. To facilitate the fitting of the Police VHF radio set, the ashtray was removed. What were our 1960s smoking officers to do? One of them relates the following story:

We were dispatched to Winchester railway station one afternoon to pick up the Chief Constable Sir Douglas Osmond to convey him back to Police HQ at West Hill. Mr Osmond was a very heavy smoker and the first thing he did after getting into the rear of the car was to light up.

'Where's your ashtray?' he enquired.

'Oh, we don't smoke in the patrol car, Sir,' came the reply.

'Yes, I know that,' said Sir Douglas sharply, 'but where is YOUR ashtray?' he demanded.

A rather sheepish officer then presented the Chief with an old Golden Virginia tobacco tin that was bolted to a 12-inch length of plywood, shaped like a boat oar, that sat on top of the transmission tunnel in between the two front seats!

Over the next couple of years three more Volvo 121 estates were added to the fleet: LOR 187F and NCG 236F had larger 1986cc engines, uprated brakes and other improvements and were stationed at Basingstoke and Aldershot Traffic Sections respectively. Both these vehicles covered more than 140,000 miles in their lifetime. There are no details of the fifth car, but these early Volvos laid the foundation stones of a relationship that has now lasted, unbroken, for 50 years.

But the introduction of the Volvo wasn't welcomed by all. There were mutterings in the local press about it, and some members of the public wrote in protest to the Chief Constable. In the 1 October 1965 issue of *Autocar* magazine a Mr A.M.F. Gillan of London SW1 wrote to complain that he had seen a Volvo patrol car in Basingstoke and that it was a crime that the police were 'living it up'. In a future edition on 29 October a Mr Michael Cram from Glamorgan continued the discussion by sending in a photo he had taken of a British-built Mk3 Ford Zephyr 6 being used by the Polis in Stockholm and wondered whether a strongly worded letter had been sent to the Swedish equivalent of *Autocar*!

All these complainants were somewhat pacified when informed that this was still only a long-term evaluation of the product, and, as if to reinforce the point, two Mk1 Triumph 2000 estate cars were purchased. The public unrest about buying foreign cars was only a foretaste of what was about to come. As the force moved into

1971 it was about to be criticised at the highest levels about its vehicle policy. Although there were mutterings in the press about Hampshire buying the Volvo 120s, this was always deflected by replies that it was just an experiment and that the vast majority of the fleet was British built. But Hampshire then bought the 120's replacement, the all-new 144 DL saloon, in the latter part of 1969, and in 1970 and 1971 these were added to with another 30. This was headline news, not just locally but nationally, and questions were even asked in the Houses of Parliament. Pressure was put on Hampshire by the Home Office, who in turn were being bombarded by pressure from the media. The fact was, the Volvo was the best car available at the time and, unlike the more traditional British-built cars, it was delivered on time, was reliable and had an efficient spares back-up. Things came to a head when the Home Office discovered that Hampshire had ordered another batch of Volvos and insisted that the order be cancelled. Volvo complained that the decision was in direct breach of the then European Free Trading Association (EFTA) trading agreement, of which Britain and Sweden were both members. The Home Office backed down and Hampshire kept its Volvos. But the Chief Constable Sir Douglas Osmond continued to receive huge amounts of criticism and eventually issued the following statement:

'The Volvo is regarded by patrol car drivers as the most outstanding high-performance vehicle that has ever been used by this force and there is no British vehicle in the same price range available which stands up so well to the excessive demands of road patrol work. Quite apart from its superior performance, there is no doubt that it is an economic proposition and results in financial saving to the ratepayers. My Police Authority has agreed that we must continue to purchase a number of these cars, but at the same time we also purchase and assess every new British car that is of the same power, speed, performance and price range. This policy has been rigidly adhered to, but we have so far come across no British model which in this respect compares favourably with the

Volvo. The suggestion that it is in some way unpatriotic to invest in the Volvo and not a British product should be seen in light of the facts. There are a great number of factories in this country producing components for export to AB Volvo and the value of the components exported for use in Volvo cars all over the world is far in excess of the value of the complete cars imported. It is in the interests of the British industry therefore that the good will of the Volvo Company towards the use of British components should be maintained and the embargo by public authorities on the use of these cars might well have adverse effects on the balance of payments.'

Brave words indeed. Hampshire stuck by its policy of buying the best vehicle for the job and changed the face of police vehicles in the UK forever. Almost overnight the names Volvo and Hampshire Constabulary became world famous. Slowly but surely, other forces followed suit and bought the Volvo 144 as a patrol car, although in nowhere near as big a number as Hampshire. Norfolk, Cambridgeshire, Derbyshire, Lincolnshire and Durham all experimented with them and each force purchased a couple on 'long-term evaluation'. But for one reason or another they failed to catch on and it would be several years yet before the rest of the country was politically ready to accept foreign police vehicles on a large scale.

So what was the Volvo 144 like as a patrol car? What made it worth all the political flak? Its engine was basically the same 1800cc, 4-cylinder unit used on the 120 series. Improvements to the chassis, suspension and gearbox meant better performance all round, although in their basic form they only had a top speed of 107mph. The cars later received Rudd Speed conversions and this increased top end to around 115mph. They were solidly built and had a quality feel that was not experienced with home-grown products. Above all, they felt safe and gave the driver an air of confidence at speed that few other cars had ever achieved. They were equipped with the newer-style roof box and blue light, two-tone horns, reflective red side stripes and a reflective 'Police' sign on the doors. The white panel on which it was placed would almost glow in the dark, when light was transferred onto it from an approaching vehicle.

Between 1969 and 1974, when the 144 went out of production, Hampshire used no fewer than 90 models, the highest ever number of patrol cars from one manufacturer to date. The average mileage for the cars was 130,000, with one of them notching up 144,015 miles. Better than this, though, was the resale value at auction, where the cars fetched more than £1000 each, in comparison with about £500 for most other makes.

Following the Amazon were more than 90 Volvo 144DL saloons, including the big bumper model purchased by Hampshire Constabulary, and no fewer than 350 of the 200 series from 1975 to 1992. This included the 244DL with its 2.1-litre engine, the 244 GLT with the uprated 2.3i engine, the 240 Police Special (built to Swedish Polis spec) and the 240 GL model. The force also had two Volvo 264DL saloons, with 2.7-litre V6 engines and a manual gearbox (very unusual) to try to combat the threat from BMW. A single batch of eight 360 GLEi saloons was purchased as urban area cars, and in 1994 came the famous T5 range with first the 850T5 saloons, then the 850T5 estates (which were automatics), followed by the 1997 model V70T5 and then the bigger 2000 model V70T5. Currently (2017) the force operates a number of Volvo XC70 D5 estates as ARVs. In between all these the force has also trialled the 740 Turbo, 460, 960, S40, V40 and more recently the V60 D5.

By the 1990s the political landscape had changed somewhat and there were plenty of foreign cars in use within the UK's police fleet from the likes of BMW, Peugeot, Mercedes-Benz, Honda and Daihatsu, to name but a few. But Volvo were about to unleash a car that became an instant legend: the 850 T5.

Just the letters T5 were enough to make the hairs on the back of your neck stand on end, not just with coppers but with the public, too. It all started with the 1994 season of the British Touring Car Championships (BTCC) when Volvo, with the help of Tom Walkinshaw Racing (TWR), entered two Volvo 850 T5 estate cars into that season's race programme. They really caught the public's attention and it was a great advertising idea. A series of political manoeuvres forced Volvo to enter saloons the following year instead, and they gained 12 pole positions and won six of those races! A Volvo winning races at the BTCC was even bigger news in the UK, although Volvos had been raced in other parts of the world for some years.

It wasn't just racing that Volvo was interested in; it wanted a serious slice of the police car market, not just in the UK but Europe-wide. To facilitate that took a stroke of pure genius and they did what no other manufacturer had done previously and gave (yes, gave) an 850 T5 saloon to every police force in the UK. It certainly caught Vauxhall napping, as their Omega was released at about the same time.

The car itself was fundamentally different from anything the police had ever used in the past. It had a 2.3-litre, 5-cylinder turbo-charged engine with front-wheel drive and it pumped out a massive 240bhp, giving it a genuine top speed of 152mph and a 0–60 time of just 6.9 seconds. Those are impressive figures to say the least, and the smiles on Traffic cops' faces were a mile wide, without a moan or a gripe in sight! Actually, that's probably not true; there were bound to be a couple who didn't think it was quick enough. They are traffic cops so they are not allowed to be cheerful . . .

The cars had a very distinctive sound and sort of burbled on tick-over. Put your foot down and it would launch itself like a scalded cat. But there were a couple of issues. Firstly the cars were front-wheel drive, which meant learning a whole new style of driving, especially at high speed. And secondly, the test cars that Volvo sent out didn't have traction control on them, which meant that 240bhp through the front wheels could mean only one thing – plenty of wheel spin, even in the dry. Legend has it that front tyres were being replaced at 3000 miles! But there was a more serious issue in that the cars offered virtually no engine braking.

VOLVO brake issue

The Volvo T5 was a legend for all the right reasons. Except one; the brakes. But before we go into the exact issue it might be worth explaining the difference between the T5 and other mere mortal cars. Up until the mid-1990s, when Volvo unleashed the beast, the main cars used by the police on traffic patrol were generally all of the same genre; an engine of around 2.8 to 3.5 litre, rear-wheel drive with manual gearboxes. Anti-lock brakes were a fairly new concept at this time and the aforementioned cars also came with a good amount of engine braking once your foot came off the accelerator pedal.

The Volvo was somewhat different. It was a 2.4-litre, 5-cylinder, turbo-charged motor with front-wheel drive and most forces opted to use it with an automatic gearbox. Learning to drive a powerful front-wheel-drive car took some getting used to after years of rear-wheel drive, and one of the T5's peculiarities was that there was little or no engine braking; the engine just seemed to keep spinning, like a sewing machine, or for those who ever rode a 1970s Japanese two-stroke motorcycle, it was very similar. So you found yourself relying almost entirely on the brakes. They really weren't up to the job of hauling a car travelling at 150mph down to a stop quickly enough. And it wasn't long before the complaints started rolling in.

To Volvo's credit they were quick to react; after all, they had invested a huge amount of time, money and effort into getting these cars to wear a blue light and stripes, and the last thing they wanted was dissatisfied customers returning their products. Police motorway cars travel at three-figure speeds almost daily, and whilst the public were quite happy with the braking performance of their cars the police certainly weren't. So Volvo fitted larger discs, callipers and brake pads straight out of the box of bits that the TWR team had used during their British Touring Car Championships foray. It worked a treat and the T5 stopped almost as quickly as it accelerated.

However, in September 1997, shortly after the 850 T5 morphed into the S70/V70 T5 model, which was the best version of it that Volvo ever produced, a second problem occurred that grounded every police T5 in the country for a few days. On 4 September 1997 Sussex Police sent out a telex message warning of a problem with the anti-lock braking system whereby if the ABS was activated during emergency braking it worked fine unless it was activated for a second time shortly afterwards (possibly during a harsh pursuit?); then the ABS failed to operate, leaving just standard braking efficiency which could cause the police driver a serious problem when evaluating their braking distances. Sussex Police and Volvo engineers were quick to investigate the issue; there were no nationwide modifications required on this occasion, so the problem was probably an electrical glitch unique to that one particular car, but it shows just how seriously the force and the manufacturers themselves treat such incidents.

In 1997 Volvo made the car even better with a slightly revised model, now called the S70 T5 or V70 T5 (S for saloon, V for Versatile – i.e. an estate). The engines were basically the same but slightly more refined – marginally slower but still a rocket ship on wheels, and, in the opinion of most, it was by far the better car, so much so that in 2017 we know of two still in service with different forces who love them and refuse to part with them, despite both being over the mileage and age limits. It was certainly better than what came in 2000 when the third-generation T5 appeared. A complete body change had seen it put on huge amounts of weight and it was in effect an S80 estate, even though such a car didn't exist. But it had a bigger problem than weight gain, which was its over-intelligent Engine Management Unit. It basically worked by mimicking your driving style; if you drove it flat out everywhere it would allow you to do so, but if you were a sedate driver it would reduce the revs to help maximise fuel consumption and wear on the engine. That was all very well if the car was in private ownership, but a police Volvo T5 could get driven by maybe four or five different drivers in a 24-hour period. If the early crew had a quiet morning and had just pottered about, the car would adjust itself accordingly. However, if the late crew got a shout as soon as they got in the car it would almost refuse to cooperate with the driver and would try to slow things down by pausing at junctions for a second, or if powering out of a bend you could feel it holding back. It was infuriating, and despite nationwide complaints it took Volvo over a year to acknowledge that there was a problem. However, they eventually remapped the ECU and the problem was solved, but it could have seriously affected Volvo's police sales.

In 2007 Volvo produced a Police Special V70 T5 called the Volvo Turnkey project, which was built to Swedish Polis specification with all manner of built-in accessories designed with front-line policing at its core, but the financial crash of 2008 put paid to that one. However, this third-generation car did continue in its basic form; in 2010 it got diesel power and was called the D5, whilst others came with all-wheel drive and were known as V70 AWD. A couple of forces were lucky enough to take on the V70 T6 with a twin-turbo-charged in-line 6-cylinder engine that pumped out a massive 300bhp through its AWD system, but in later form the once King of Traffic lost its crown to BMW, and although the cars aren't quite what they were when first introduced they will forever have their place in history as one of the great Traffic cars.

Talk to any Traffic cop from the T5 era and every single one of them will have stories to tell of its legendary performance. The famous 'Liver Run', which involved Rover SD1s from the Met making a mercy dash from Stanstead in Essex through London's congested streets to a hospital to deliver the organ in time for its urgent transplant operation was really only made famous because the whole thing was videoed at a time when in-car video recording was a brand-new concept, and the footage was given plenty of 'air-time' on programmes like *Police, Camera, Action!*. Those mercy runs are quite commonplace, but they are not made famous because they don't appear on our TV screens. Well, here's one that would have made great TV, but you'll have to be content with a written description.

PC Steve Woodward from Hampshire Police was on early turn in the Portsmouth area. He and his crew mate, PC Mike Batten, were in the middle of their breakfast when the phone rang in the office. It was the Control Room Inspector and he had an urgent job for them: take two cars to the local Queen Alexandra Hospital at Cosham to transport a medical team complete with transplant organs to Bournemouth Airport, some 45 miles away. They took two Volvo V70 T5 estates, both with automatic gearboxes, and drove straight to the hospital. They were there within two minutes of receiving that call. As they arrived at the door they were met by a team of six surgeons who had been up all night removing the heart, lungs and other internal organs from a donor patient and were now heading to Edinburgh to transplant them straight into the waiting patient.

As the six of them hurriedly got into the Volvo patrol cars, Mike made arrangements over the radio to get local units to block off the nearby roundabout to all other traffic whilst another unit closed a huge traffic-light-controlled crossroads that gave access to the M27. With three surgeons in each car, the two Volvo T5s made their way out of the hospital, across the roundabout, through the crossroads and headed west along the M27 where the cars quickly reached their flat-out speed of 145mph and stayed there.

Despite the speeds involved and the concentration required, all the passengers struck up a great conversation. In PC Woodward's car one of the surgeons was heard to say, 'This is fantastic, I've always wanted to do this', which is quite incredible when you consider what he does for a living. PC Woodward then asked what was in the box, now safely strapped into the rear seat.

'Oh, that's the lungs,' said one of the surgeons. 'Your colleague has the heart.'

To which PC Woodward replied, 'That's not possible, I've worked with Mike for years and I know very well that he doesn't have a heart.'

This caused a lot of laughter, of course, but that was tempered somewhat as a motorist moved out from lane two for an overtake straight in front of the leading Volvo. There was no choice but to move left and take to the hard shoulder at three-figure speeds as the dust and debris that gathers on the hard shoulder was thrown up as if hit by a small tornado. As the journey continued, so the two officers were informed over the radio that a Dorset Police Traffic car would meet them on the A31 Dorset border and escort them to the airport.

And sure enough they were met by a rather old Dorset Police Vauxhall Senator, but as the A31 was only two lanes wide instead of three it meant the speeds had slowed to around the 120mph mark, although they increased slightly as the three cars then headed south along the A338 towards Bournemouth. The Dorset car then left the A338 and took the Volvos down a small country lane, which was a real surprise to the two Hampshire officers. At the end of the lane there was a local officer with his panda car holding open a large three-bar wooden gate which looked like it was the entrance to a field. All three cars drove straight through, along a small service road that led onto the runway of Bournemouth International Airport. The three patrol cars then drove flat out down the full length of the runway to the small executive-type jet that was waiting with its steps down and engine running. All six surgeons leapt out, grabbed their precious cargo and ran up the steps for the next leg of their journey.

As the aircraft roared off down the runway the three officers swapped a couple of 'war stories' and Mike revealed that one of his surgeons had slept throughout the entire trip down the motorway and didn't even wake up when he was forced to take to the hard shoulder! The Volvo T5 was fast, stable and comfortable enough for a sleeping surgeon to get forty winks. The total length of the journey was 46 miles and they did it in 21 minutes. You work it out . . .

Let's turn the clock back to 1972. Hampshire Police were still being criticised for buying a few Volvos when neighbouring Thames Valley Police went one step further . . . they bought German! For the same reasons that Hampshire looked overseas, so did Thames Valley, and they bought several BMW 3.0 Si saloons for their Traffic Division. At the time they were running a big fleet of Jaguar XJ6s and Ford Consul GTs to patrol the M1 and M4. The BMWs had a 3.0-litre, straight-six motor with fuel injection and 200bhp on tap, giving it a top speed of 135mph. It was a true driver's car and a near-perfect Traffic car for its day. As if to rub salt into a very raw wound, Thames Valley Police agreed to take part in a BMW advertising campaign and had one of their sleek-looking cars together with a BMW police spec motorcycle watching a speeding BMW 2002 racing past them with the slogan 'It Takes One to Catch One', which drew a sharp intake of breath when the posters and full-page ads started appearing in magazines and papers. But within a matter of months other forces started doing the same, with the likes of Derbyshire and West Mercia Police both using the 3.0 Si.

In 1972 BMW released the very first 5 Series, or E12 as it was known in BMW speak. Launched as a 520, they introduced the 525 2.5-litre, straight-six engines in 1975. Again, it was a driver's car and it had German build quality married to reliability and a decent spares back-up. Strangely, neither Thames Valley Police nor Derbyshire nor West Mercia came in for the cars but, horror of horrors, the City of London Police did. How could they? How dare they? To bring German cars right into the heart of London was surely a two-fingered gesture aimed at the politicians? The only other force to use the E12 was Hampshire, who rather liked the idea of decent reliability.

In 1981 BMW released the E28 5 Series, and the 528i really set Traffic officers' pulses racing. Its 2.8i engine was good for 130mph, and with near 50/50 weight distribution it really was a great driver's car. Except for one thing: the seats. They were incredibly uncomfortable during an eight-hour shift on the road, which gave our Traffic cops something to really moan about! The car was a huge success, though, in City of

London, Fife, Hampshire and Grampian Police. Politically, though, most forces were still very wary about going foreign.

But that seemed to change forever in 1988 when BMW unveiled the 'Ultimate Driving Machine', the all-new E34 5 Series, firstly with a 3.0-litre, straight-six motor and then two years later with the multi-valved 2.5-litre motor that pushed the car close to 140mph. It was the best yet and really put the 5 Series on the world map. As a Traffic car it was almost bulletproof, with mileages of 200,000 easily achieved with little or no reliability issues. For many Traffic cops this was the best they had ever driven, no question. Forces like Hampshire, Fife and now Cheshire were all big fans, with Cheshire also using the new Touring version.

In 1995 the 5 Series morphed into the E39, but this was Volvo T5 time and only the Met Police, who weren't that fond of the Volvo, took the E39 car onto its Traffic fleets. Then, from 2003 to 2010, BMW gave us the most successful car to date; the E60 and E61 530d saloons and estates. All of them were 3.0-litre diesels putting out 230bhp with a top speed in excess of 140mph. It was everything a Traffic car should be: fast, comfortable, reliable and with enough room for the ever-increasing amount of kit. The cars came under the BMW Authority's line, whereby they were taken off the standard production line to be specifically turned into police package vehicles with everything factory installed, so there was no need for the local workshops to do much except bolt a set of number plates onto them. This factory fit included all the necessary extra wiring (and there's a huge amount of it) for the blue lights, which weren't just on the roof now but secreted into the front grille, the rear bumpers and elsewhere. Electronic devices like ANPR, black-box flight recorders, hands-free mobile phone kits, Airwave radio sets, VASCAR and the actual Battenburg livery were all applied in-house by BMW. If the car was for ARV use it was fitted with all the necessary gun safes and racking for other items like ballistic helmets and shields, all of it a requirement to police twenty-first-century Britain. The role of the traffic cop was changing and changing fast. By now the Government had introduced Highways Traffic Officers to patrol our motorways, and their responsibility was to deal with broken-down vehicles, attend any motorway incidents to secure the scene, supervise vehicle recovery and roadworks and do all the non-urgent jobs so that the police were left to fight crime. The Government didn't want the police cruising the motorways any longer but sadly lost sight of the fact that a police presence is also a deterrent to bad drivers. Just about every force in the UK was now using the BMW 530d as its Traffic car, mostly in Touring variants, although a couple of forces like Cheshire opted to use the saloon.

The current F10 and F11 BMW 530d, introduced in 2010, are very similar to the previous models in terms of specification, looks and performance, and 2017 saw the launch of the next generation of Police specification 5 Series cars, which were introduced at the National Association of Police Fleet Managers' Conference and Exhibition.

Perhaps one of BMW's most successful vehicles has been the X5. Its timing couldn't have been better, being introduced just as the Range Rover was being pushed as a super-luxury barge with a price tag to make most of us shudder. The all-new BMW X5, with its renowned 3.0d engine and permanent four-wheel drive, was everything the Range Rover should have been. The X5 was quick but, most importantly, it handled brilliantly with very little body roll – unlike the Range Rover, which often felt like it was about to topple over on some bends! The X5 was half the price of the Land Rover products but certainly wasn't half the car, unless you took it off-road, in which case you might have to call someone with a er . . . Land Rover, maybe, to come and drag you out of the mud – the Chelsea Tractor was a very apt nickname for the car! But as a motorway car on Traffic patrol it was superb and could do almost everything the Range Rover did. The Hampshire Constabulary was the first police force in the world to obtain the services of the X5 in 1993; not even the Germans had them before Hampshire. It wasn't long before other forces followed suit. Now on its third generation, the BMW X5 continues in police service today as a top Traffic car.

CHAPTER EIGHT

POLICE CAR LIVERY AND EQUIPMENT

What was it Henry Ford said when faced with buying one of his or anyone else's cars way back when? 'Any colour you like, just so long as it's black.'

Before the Ford historians write in, I know that's a more complex story than the cliché portrays, but in essence the police were in exactly the same boat as the public when it came to choosing the colour for their first ever patrol cars. If black really was the only option, then black it had to be. I doubt anyone back then gave it a second thought; just being mechanised and mobile was enough for them to think about. In truth, by the time the police started buying cars in any numbers, even pre-war, there was certainly a choice to be had on colour. But as the public bought more and more different-coloured cars, so the police, who have always been sticklers for tradition, stuck with their black cars; it was thought to be more authoritarian.

As post-war Britain got back on its feet and cars became more readily available and affordable to a few, so the police started to buy more and more cars, and, as we have explored elsewhere, they started to form proper Traffic Divisions for the first time. Cars like the Wolseley 18/85, followed by the 6/80 and 6/90, the Riley RM series, MG TCs and TFs, Austin A55, A95, Ford Zephyrs and Vauxhall Wyverns were all fairly popular police vehicles and all of them came in black – you won't see any photos of them in any other colour. The officers that crewed them were hand-picked and they were rather proud of being one of that force's drivers; most of them at this time were probably ex-military veterans who knew a thing or two about spit-n-polish, which is why those same photos will always portray the aforementioned cars with glass-like paint finishes, chrome shiny enough to use as a mirror for shaving and tyres that looked like they'd just been fitted. It has to be said that back then the pressures of policing weren't quite what they are today, which is why they had two hours per shift to ready the car prior to patrol and to wash it down at the end, plus all day Sunday to give the car a proper going-over with a full wash and wax. Not your modern-day polish; mind you, no no, they used the hard bar stuff that required real effort. I was never afforded such luxury and I remember Sundays being the day we hand-washed our cars in the yard with a bucket of soapy water and an old sponge. It was almost as if the public had a wash radar, as typically we would always get called out on a job mid-wash!

By the mid-1950s there were a few rogue forces that started to do things a little differently, with green cars being rather popular. The Cheshire Constabulary, for example, ran a big fleet of Rover P4s, all of them painted green, and possibly even before that the Rover P3s they used might also have been green. The Somerset and Bath Constabulary used dark green Mk1 and Mk2 Jaguars, whilst the Portsmouth City Police had green Morris 1000 dog vans and in 1966 had a couple of green Mk1 Ford Transits. Other colours included grey or blue, mostly on light commercial vehicles, although many coach-built vehicles were still produced in black.

So how did the public know that there was a police vehicle coming up behind them on an emergency call? With no blue lights yet, what was the clue? Quite simply the officers put their headlights on, and that, plus driving at speed, was usually enough for a much more law-abiding public to move out of the way. Of course, there certainly wasn't as much traffic around in those days.

It is difficult, actually it's been impossible, to ascertain a date or

location for who had the first audible warning devices and visual aids
in the form of emergency lighting to aid emergency vehicles attending
the scene of an incident. The first warning devices were Winkworth
bells; these were fitted to the front of cars and had a push-button
electric motor powering the hammer inside a brass-chromed bell that
rang out its very distinctive tone. The bells were produced at the
Winkworth factory in Reading and were fitted to Traffic Division cars,
ambulances and fire engines. Without doubt the bells gave the emergency services a voice with which to warn everyone that they were
approaching – and it really worked. It is likely that the first vehicles to
have them fitted were actually ambulances during World War II and
that wholesale fitting of bells to police Traffic cars started post-war.
What we do know for certain is that some Met Police cars were still
fitted with bells as late as the early 1980s! If you watch episodes of *The
Sweeney* on TV now you'll hear the bell quite clearly on Reagan and
Carter's bronze Ford Consul GT from 1975 when it first aired, right
through to the Mk2 Granadas in the later episodes of around 1979. As
the bells gave way to the two-tone horns, at least one force collected
up all their bells and took them to a local scrapyard where two tea
chests (remember those?) full to the brim with discarded Winkworth
bells were sold for just £2 per box.

About the same time as the bells were introduced a few forces started
to experiment with police signs either placed on the roof of the car or
on the front grille area. Most, if not all, of these signs were made
in-house either by the force mechanics or in some cases by the officers
themselves. Could you imagine a police car with effectively a home-
made police sign? Some of them could be illuminated at night, but all
were the very first police signage that the public had ever seen. Some
of these signs were little more than a metal plate, similar to a registration plate with the wording embossed upon it, whilst others were proper
boxes that could be illuminated from inside. I like to think I would
have knocked up something quite fancy.

One of the other items that adorned the roof or the front grill was
a public address speaker. The Met Police favoured the twin speaker

set-up mounted on the roof whilst most county forces opted for the single cone system, usually placed on the front grille. These PA systems were initially just that, a way for the police to address a crowd of people, say at a football match or public demonstration, or during a search for a missing person to inform the public and give out the person's description. During the late 1960s and 1970s these speakers, which were also fitted to motorcycles, could be seen mounted at the rear of some patrol cars and were actually for the officers' benefit, not the public. Back then, VHF radio sets were big heavy units installed in the car itself and not the personal handheld items of later years. When the officer was out of the car attending an RTA, often he still needed to keep in contact with the control room, so at the flick of a switch the audio transmissions from the radio were heard outside the car via the PA speaker. It wasn't unusual at this time to see a small crowd gather around a police vehicle to listen to the radio traffic and to muse at the pip tones that would accompany it.

Other signs included 'police stop' signs usually found at the rear of the car, often placed inside the rear window, and most of these early units really were handmade, 'Heath Robinson' in their structure and operation, and they hadn't improved much by the late 1970s in some forces. Some of the more sophisticated units were made from metal, often brass, with the wording cut out like a stencil with a thin sheet of glass placed behind the lettering and with a bulb placed inside the box to illuminate it at the press of a button on the dash. But these were few and far between; most were little more than a printed board laid flat on the parcel shelf that was attached to a piece of string that travelled through a set of pulleys screwed into the roof, connected to a handle at the front of the car. As the string was pulled, the 'police stop' board popped up at the rear. At least one force was still using this system on its unmarked Mk2 Ford Granadas as late as 1980.

As you can probably tell by now, equipping police vehicles was all a bit ad hoc, and it's the same when it comes to that most obvious of emergency service kit, the blue light. This is not an exact science by any means, and trying to pinpoint which force was the first to fit them and when has stumped police car historians for many years. As a starter, we do know that the Riley RMs used by the Portsmouth City Police from about 1950 to 1953, and which were detailed in the previous chapter, were fitted with twin Butler driving lamps on the front bumper as standard. The city force removed the lens from the offside lamp and painted the inside of it blue but stencilled the letter P on it so that when the lamps were illuminated the lens shone blue with a white letter P, indicating it to be a police vehicle. The cars were also fitted with a chrome Winkworth bell and a PA speaker, and on the nearside of the dash a small hinged police sign could be flipped up by the observer seated in the passenger side. All these items helped to make the cars some of the most advanced of their time.

The only real link to the fact that police emergency lights are blue goes back to the days of Sir Robert Peel when, in 1861, the Metropolitan Police instructed all superintendents to ensure that the lamps placed outside their police stations were blue, and it has remained the police colour ever since.

We have already taken a good look at the Lancashire County Constabulary's use of the MGA, but it's worth reminding ourselves here just how groundbreaking those cars were in terms of the equipment they carried and, more importantly, the change in colour from black to white. Make no mistake about it, these cars were way ahead of their time. In 1958 the force was tasked with policing a brand-new concept in road travel that today we just take for granted as the norm, but back then this really was headline news. With the opening of the Preston Bypass, Britain's first motorway, cars were now able to travel at hitherto unheard-of speeds without having to slow down for junctions or traffic lights or parked cars. The police were clearly concerned that some people's perception of their driving skills far outweighed their actual ability and that they needed to be policed in a very different way.

The MGAs were equipped with two spot lamps: one on the front wing with a blue lens and one on the rear wing with a red lens. These lamps didn't flash but just emitted a steady blue or red light. There were PA speakers fitted front and rear, a proper illuminated 'police stop' box at the rear, a VHF radio set and police labels on the bonnet and boot lid. Above the grille was another small illuminated police box that had a small orange or red light fitted above it. To make them sound as good as they looked they were fitted with the all-new two-tone air horns that were easily heard above the roar of high-speed traffic.

These were state-of-the-art patrol cars for their time. They were teamed up with another initiative with the introduction of Mk2 Ford Zephyr Farnham estate cars, also painted white, the idea being that the cars could be easily identified during inclement weather on the road. These cars carried all the necessary equipment needed at the scene of an RTA on the Preston Bypass, and to facilitate a quick response the cars were fitted with a revolving blue light placed on the front corner of the roof, together with an illuminated police box with an orange lamp on top of that. To the rear was an illuminated 'police stop' box, two PA speakers, a set of two-tone horns and a VHF radio set. These cars meant business and gave the officers a real sense of purpose. The experiment, for that is what it was, was a huge success, and many of the pioneering ideas that Lancashire introduced are still in use to this day throughout the UK.

Those new illuminated boxes were produced by a local firm called Ferrie Plastics, in Blackpool, and over the next few years they produced a whole raft of glass-fibre boxes in various shapes and sizes for the emergency services. Another firm, Wadham Brothers, of Waterlooville, cornered much of the market in the south of England by producing their own range, and between them they produced most of the boxes needed by the police up to and including the late 1970s.

The Preston Bypass heralded the dawn of the motorway era, and in 1959 the first sections of the M1 were opened in Bedfordshire and Hertfordshire before it was extended still further into Northamptonshire. The Home Office was impressed with the success of the Lancashire experiment and therefore decreed that the Ford Zephyr estates, similarly equipped, should also police the M1. Other forces watched closely, and slowly but surely the transition from all-black cars to white cars for use on Traffic patrol, regardless of whether that force had any motorway or not, started to take place.

The mid-1960s saw enormous changes within the UK Police service with the introduction of Unit Beat Policing and panda cars, officers being issued with the Pye pocket phone, two-way UHF radio sets, foreign cars, the breathalyser, the extension of the area car scheme and a rapid expansion of the motorway network, with the M2, M3, M4 and sections of the M6 being opened.

More and more police vehicles were being purchased to help with increased workloads and responsibilities. Area cars were the local emergency response units and they too were now being delivered in white, had illuminated roof boxes with a blue light adorning the top, two-tone horns added and police labels placed along the sides, all designed to make the cars more visible to the public and to aid progress when on a shout.

In the late 1960s the Home Office conducted an experiment by instructing a number of forces to paint some of their patrol cars black and white. The origins of this idea are difficult to pin down, but perhaps it stemmed from those black-and-white press photos taken of the early blue-and-white panda cars. Or perhaps the idea came from the USA where the use of black-and-white cars had been around for several years. Whatever it was, forces like Durham and West Riding of Yorkshire were tasked with painting a number of their cars black with white doors, bonnet and boot lid. They looked fantastic, it has to be said, with cars like the Mk3 and Mk4 Ford Zephyr, Mk2 Jaguar, Morris 1800S and the Hillman Hunter all getting the treatment. It certainly made the cars look very distinctive, but the trial was fairly short-lived and wasn't tried elsewhere, probably because it was cost prohibitive.

In February 1967 a tragic accident became a game changer. Our motorways at that stage were beset with fog, probably none more so than the M6. Staffordshire Police were responsible for policing a large section of the M6 with a fleet of Mk2 Jaguars that were seemingly quite advanced cars for the duties they were performing. The cars were white with a small roof box and blue light fitted and large police labels in black placed along the sides of the car. It was early morning and the fog was pretty thick as PC Clive Blackburn and his crew mate attended an RTA in lane three of the motorway, stopping behind a damaged car to offer it some protection. As the two officers got out of their patrol car it was struck from behind by another vehicle, and sadly Clive Blackburn lost his life.

After this incident Staffordshire Police set about making their motorway cars a lot more visible and pioneered a number of important initiatives. The small roof box was out and in came a full-width box with a much more powerful blue light that had twice the output of the old Lucas Acorn unit. They painted the rear of the box bright red and added two red reflectors on either side of the large, illuminated, flashing 'police accident' signs. The boot lid was then covered in a red vinyl-type sticky plastic and on the rear bumper a large metal sign, painted white with the word 'police' in red, was very prominent. The side of the car was still adorned with large 'police' letters in black. When the rear doors of the car were opened to access some of the emergency kit that had been moved forward from the boot to help spread the weight, the force used some of that red vinyl adhesive on the inside of the doors to aid visibility of the car still further. This new livery and extra kit made these cars quite distinctive and was much heralded at the time. For once it wasn't done for the benefit of the public but to aid officer safety in an environment that is fraught with danger.

Needless to say, those forces that had motorways running through their county were quick to start looking at the livery on their cars. However, it was the East Sussex Constabulary, a force that didn't have any motorway, that can lay claim to inventing the orange side stripe which gave rise to the expression 'jam sandwich' cars. The Austin A60 Cambridge wasn't exactly a common sight as a police car, but East Sussex ran quite a few of them alongside their Farina model Morris Oxfords. As with the origins of the blue light and the Winkworth bell, the exact details are almost impossible to ascertain, but did the side stripe come about not by design but by some kind of oversight?

Take a look at the photo (above) of the East Sussex Constabulary Austin A60 Cambridge delivered to the force sometime after August 1967 (about six months after the fatal crash on the M6) in its Snowberry white paint but with a standard factory-painted red stripe inside the stainless-steel trim lines. Was this car ordered in white but the force didn't realise it came with the striping as standard? It's the sort of mistake that is easily made. Having taken delivery of the car(s) in this colour combination, maybe the officers and the public alike made comment about how well they stood out, and thus the idea was born to stripe up other cars in the force using adhesive red tape. Certainly the MGB GTs they used from 1966 were initially issued in plain white, but by early 1968 they were all striped. Coincidence? Unlikely.

The idea caught on very quickly and within months Traffic patrol cars everywhere were dressing up in the latest fashion accessory. Hertfordshire Police, whose Mk2 Zephyr estates had made way for Jaguar 240s, had placed a single stripe across the width of the boot lid before going the whole hog and adorning the side of that marque's replacement, the Rover 3500 P6, with proper reflective striping down the sides. The Hampshire Constabulary sent a contingent of Traffic officers up to Hertfordshire for two days of motorway training prior to the opening of the M3 in that county, and they came back with the idea to place striping along the side of their Mk2 Triumph 2500 motorway cars and on the big Mk1 Ford Transit they would use to carry additional emergency equipment. Just about the only force that didn't stripe up its cars at this time was the Met, who left it until 1978, more than 10 years after everyone else had taken up the idea. A few of their late model Rover 3500S cars got stripes, but only as an experiment at a time when the Met was taking delivery of the first Rover SD1 cars, some of which were delivered in Zircon blue, a colour used by the outgoing P6 area cars. They did eventually opt to use a gold stripe bordered top and bottom by orange stripes, which they stuck with for more than two decades.

There was one other force that didn't stripe its cars at all for many years but made their cars so distinctive that there was absolutely no mistaking what they were. That force was Lancashire, pioneers in so many aspects of police vehicle design and technology that it should come as no surprise that they came up with the idea to repaint their Traffic patrol cars in bright orange – all over! The first experimental cars to get this treatment were a couple of old Mk3 Ford Zephyr Farnham estates. The colouring was deemed a huge success, and over the next few years the scheme was used on the likes of their MGB GTs, Mk4 Ford Zephyrs, Mk2 and Mk3 Ford Cortinas, Ford Consul GTs, Mk1 Ford Capri 3000 GTs, Mk1 Ford Transit Accident Unit vans, Triumph 2000s and 2500s and the Range Rover. It has to be said they looked fabulous and are arguably the best-looking patrol cars of all time. But it wasn't cheap to do, obviously, and in the end the scheme

proved cost prohibitive and so it was axed. What a shame the accountants took hold, hey?

The Lancashire Mk1 Transit Motorway Accident Unit vans featured
another adaptation, being adorned with flashing red beacons instead
of blue. But why? Well, there were two reasons. Firstly, it was thought
that red or orange beacons would show up better than blue, particularly in adverse weather conditions, and secondly the bulb used in
beacons at that time was nothing more than a 21-watt brake light
bulb. But using flashing red lights on a moving vehicle was illegal at
that time, so the lights were only ever used when stationary at the
scene. To help facilitate a quick response, the Lancashire vehicles were
fitted with a flashing amber fog lamp positioned above the illuminated
police sign on the front of the car or motorcycle. By about 1972 much
brighter tungsten bulbs became available, so the red lenses were
replaced with blue ones, which also meant that the blues could be
used on the move.

The orange stripe, usually bordered in blue, remained in vogue for
several years until about 1978, when one of the manufacturers of the
material used came up with Saturn yellow, which they marketed as
having better reflective qualities than the red or orange. Ever mindful
of officer safety and the constant desire to make their cars more distinctive, a number of forces opted to use the yellow stripe instead of the
red. Some did it as an experiment but ditched the idea after a bit, whilst
others converted all of their cars over to yellow stripes bordered in blue
(except West Yorkshire, who bordered theirs in green, and Cheshire,
who bordered theirs in red for a while).

From then on it basically became a free-for-all in the striping department, with every force designing its own distinctive livery. It became
an identity issue really, almost tribal in its make-up, with every single
force using a different design. There always has been a sort of rivalry
between forces, and I remember it well. Even the basic stripe made way
for diagonal multicoloured stripes in some forces, with others using a
variety of hitherto unheard-of colour combinations to help make their
vehicles that much different from their neighbours. As a guide, here's

a brief rundown on the difference between neighbouring forces from the early 1990s, starting in Devon and Cornwall with a yellow stripe bordered in blue with blue and white chequers, Dorset with a similar scheme, Hampshire with red and white diagonal stripes, Sussex with a yellow stripe bordered in red, and Kent with a yellow stripe bordered top and bottom with a wide blue border. Meanwhile, Surrey stuck with a plain orange stripe with no border and Merseyside used a green and blue stripe. Gloucestershire Police had red and blue diagonal stripes whilst Lancashire invented one of the better-looking and practical designs (of course they did!), with blue and yellow arrowed chevrons along the side to act as direction signs at the scene of an incident (genius). Warwickshire, on the other hand, used two shades of red whilst North Wales had green, red and yellow stripes. For those of us who are interested in police vehicle history this was a rich seam of photographic opportunity, but the Government had other ideas and things were about to change.

In the early 1990s the Government instructed the Police Scientific Development Branch (PSDB) to commence research into a national police vehicle livery, in part to stop the aforementioned free-for-all that was going on and also because they were being pressurised by Brussels to make our police vehicles recognisable to visitors to the UK no matter where they might be in the country. One wonders whether that same pressure was brought to bear upon the French and the Germans?

In 1992 the PSDB liaised with the Association of Chief Police Officers (ACPO) Traffic National Motorway Policing Sub-Committee to devise a new, highly conspicuous livery to be used on police motorway patrol cars. The parameters were set to include cars operating on a motorway, so that they should be visible throughout the day and night and be clearly identifiable as police vehicles at a minimum distance of 500 yards away from an oncoming road user. At night the minimum illumination is usually provided by the oncoming vehicle with head-lights set to dip beam. This minimum distance condition should apply during daylight hours in rain or mist (but not fog) and without the cars roof lights in operation.

The PSDB in collaboration with ICE Ergonomics then developed what has generally been referred to ever since as the Battenburg livery: a series of blue and yellow blocks placed along the sides of patrol cars. Between 1994 and 1995, a number of experimental versions of this new livery were trialled by some forces, including the Met, Surrey, Thames Valley and Suffolk Police. Some of these experimental schemes showed the blocks placed in the centre section of the car whilst the front and rear sections were covered in reflective silver panels with a thin red stripe running through it or outlining the profile of the car. But in the end they settled on the whole side of the car being covered in the Battenburg livery, and in September 1997 the ACPO Traffic Committee approved the scheme. What? A date by which we can claim a new police scheme actually commenced? Well, that's a first!

Although the scheme was 'recommended' by ACPO, it wasn't made compulsory; in fact, it took some forces almost 20 years to comply, but eventually the scheme was adopted nationwide and not just on motorway patrol cars but on virtually all police vehicles, including section cars, vans, motorcycles and other units – even marine vessels. But the scheme didn't get the approval of the world-renowned Transport Research Laboratory (TRL) in Berkshire. In 2001, following extensive testing, the TRL concluded that the Battenburg livery masked the true shape of a car and that motorists would take over two seconds longer to react to it at motorway speeds, which could compromise police officer safety. As if to prove the point, they decorated one of their response cars, a Mercedes E-Class estate, in luminous orange along the car's flanks, rear tailgate, wheels and profile. Now where have we seen that before? Oh, and instead of blue lights they used orange! But the TRL's recommendations fell on very deaf ears and Battenburg lives on, not just on police vehicles but also on ambulances, Fire Service vehicles, Highways Agency units (which is a Government con to trick the public into thinking that a Highways Agency vehicle is a police car) and just about every security company and even some breakdown recovery trucks. Which, when you look at it logically, detracts entirely from the original concept.

In the early to mid-1980s our car manufacturers were trying to make their vehicles look ever sportier and gave them rear spoilers, usually made of a soft black rubber compound; think Ford Escort XR3i or Ford Capri. The marketing strategy suggested that having such spoilers greatly aided downforce and thus improved your car's performance. It wasn't long before the after-market goody-makers starting bringing out their own versions, and soon every young male had to have one on his car to stay in fashion. And so did the Old Bill. Really? Yes, really. Out went the little police stop box bolted to the rear bumper or incorporated into the roof box and in came the big boot spoilers. They were made from fibreglass and finished in black (mostly, although some were in white) and they incorporated the flashing police stop signs. They were actually quite practical and looked good on the likes of the Ford Capri 2.8i, the Mk2 and Mk3 Ford Granada, the BMW 528i, the Rover SD1 and the Mk1 Vauxhall Senator, with some even being adapted for use on Range Rovers. But they were expensive to make, as each spoiler had to be custom made to fit each individual model of car and, as with everything else, the bean counters started asking questions about the cost, so the scheme died.

Another expensive and fairly short-lived accessory was the stem light. The original idea was to elevate the blue light by a few feet so that approaching motorists could see it from further away when the patrol car was at the scene of an accident. The light was placed on top of a pneumatic pole that was bolted to the floor of the car and went up through the roof. In normal mode the blue light could be used like any other, but when required the pole was elevated by pressurised gas and up went the light to a height of about 10 feet. This could only be done when the car was stationary, of course. When the scene was cleared, a valve was turned anticlockwise and the light descended to its normal position. Later models saw a large roof-mounted box, almost circular in shape, fitted with a huge blue light that was mounted on a sprung steel coil, which meant the workings for the whole system were kept within that box and didn't intrude inside the car itself. An additional feature included the fitting of floodlights under the blue light to help illuminate a darkened scene from above, which proved very useful. However, few things in life are said to be policeman-proof, and so it proved with the stem light. Police car history is littered with tales of officers forgetting to lower the stem light before driving off and bending the pole at 90 degrees and writing it off. Or even forgetting to prime the thing before pressing the 'up' button, then watching in horror as the pole reached its full height at 30mph and the expensive lighting unit continued its journey skywards before falling back to earth and smashing into a thousand pieces! Range Rover and Transit Accident Units were the vehicles usually fitted with stem lights, although they could be seen on the likes of Ford Granadas, Land Rovers and Vauxhall Senators in some forces. Surrey Police had a rather nice Mk2 Ford Granada 2.8 estate for use on the M25 in the early 1980s; it was fitted with one of the aforementioned rear spoilers on the back of the roof that had two blue lights mounted on top of that, twin spotlights at the front of the roof and a huge stem-light system in the middle of the roof. It was incredibly top-heavy and tended to roll rather too much on bends, thus it quickly gained the nickname of the General Belgrano.

Around 1979 and into the 1980s, two things changed the sight and sound of our police vehicles: full-width light bars and yelp and wail sirens. New York had arrived in Britain. As with previous changes, it's nigh on impossible to say who was first with these systems, but they do seem to have coincided with each other. It sounds a bit strange nowadays, as we are so used to hearing the sound of modern-day siren systems, but back then it was literally headline news. Letters were sent to the papers from disgruntled members of the public deploring the idea that the British police were using American-sounding sirens, and I suspect these people were direct descendants of those who wrote similar letters complaining about the use of foreign cars in the mid-1960s!

The sirens were often referred to as tri-sound because they not only had the yelp and wail siren but could also blast out the old two-tone sound. The siren was particularly useful on motorways and other fast roads where the projected sound could be heard from a lot further away than the two-tones. The officers loved them because the path ahead could be cleared so much quicker, and within a fairly short period of time all emergency service vehicles, not just police cars, were being fitted with them. Believe it or not, there is a slight difference in tone between the sirens used by police, fire and ambulance services. To facilitate the change from, say, yelp to wail the system is wired into the standard-fit horn button on the steering wheel, which is why you can often hear the horn being used briefly as the siren transfers from one sound to the next. It's where the term 'BLAT' came from. The use of 'blues and twos' – blue lights and two-tone sirens.

The American-style light bars were a revelation in style and efficiency. To have police cars fitted with a single blue flashing light one day to multiple lights in both blue and red, programmed through a control box inside the car to give different combinations of light, was nothing short of sensational. The light bar system spawned a new generation of suppliers, with the likes of Federal Signal, Premier Hazard and Code 3, to name but a few, all producing ever more dynamic and efficient units. But they were incredibly expensive, often ten times the price of a single blue light, and forces were under pressure to comply with new European legislation that called for all cars used as emergency response (i.e. all traffic and area cars) to be fitted with at least two blue lights. So a number of forces opted to make their own units by placing two of their old blue lights at either end of a standard Thule roof rack. It was simple, cheap and actually worked rather well. But as the costs for such units came down so all emergency response units were fitted with them. The lights themselves were initially halogen-lamp powered, which eventually gave way to LED lights that use virtually no power and so don't drain the battery if the car is switched off at a scene. Eventually much smaller lights, often referred to as repeater lights, were fitted into the grill or the rear windows, as inside rear lights or as a covert fitment elsewhere on the car, with some units being lit up like Christmas trees. These were all fitted as additional lighting to the light bars.

In the early 1990s some more adhesive material found its way onto our cars with two additional items; one for officer safety, the second to aid blossoming air support units. The former consisted of yellow and red reflective diagonal stripes being applied to the rear of all police vehicles to prevent them being rear-ended during poor visibility or at night. The latter consisted of roof numbers being applied together with a large orange dot or other symbol to help Air Support Units identify cars as police vehicles and to ascertain which force they were from, in the case of cross-border operations. Each force had been assigned a force code number (PNC number) many years earlier for administrative purposes and these numbers were now applied to the roof, usually at the rear and sometimes with the cars' call sign labelled at the front of the roof. For example, a Hampshire car's PNC number was 44 whilst Dorset was 55 and Cleveland was 17. The symbols that were also on the roof consisted of an orange dot to signify that the car was a Traffic car (later changed to a black dot), whilst a black triangle showed that the vehicle was a Dog Section unit, a black square signified a public order van, and a black asterisk signalled that the car was an armed response vehicle (ARV).

In more recent years yet more adhesive symbols have been applied to those same ARV units, with bright yellow asterisks being stuck in the corners of windscreens front and rear to let other officers know that the car now arriving at their scene is a gun ship and that everything will be okay. I drove those exact ARV cars to the scenes of many incidents, and fellow officers would know exactly who we were.

Modern-day police cars now carry in or on them a huge amount of electronic equipment and devices to aid the fight against crime, including hands-free mobile phone systems, black-box flight recorders, data trackers, hands-free digital radio units with voice activation, dot matrix warning signs, drug-driving detection devices, digital breath test recorders and several other devices. However, two systems stand head and shoulders above all of these. The first is Tracker, a system to detect the location of stolen vehicles. If you had a Tracker unit fitted to your personal car and it got stolen then the good people at Tracker UK triggered the system and the car started to send out a silent radio transmission. Meanwhile, those police vehicles within about a 10-mile radius of that signal had the Tracker receiver device alerted that they were within touching distance of a hot stolen vehicle. The code on the unit was transposed into a registration number and the four Tracker aerials on the roof of the car then acted like a compass to help guide the police towards the stolen car. If that car was still on the move, it could still be difficult to locate but certainly not impossible. If the car was static, the process was somewhat easier and the system was responsible for some quite remarkable success stories where a Tracker car had been stored in a warehouse or barn with dozens of other stolen vehicles. For a police officer a Tracker activation is an exciting thing! I've had many in my time. The reason it's so great is because you're pretty much guaranteed to find the vehicle and likely nab the bad guy, too. I remember a particular Tracker activation that went off in Cheshunt; we were guided towards the stolen property, which turned out to be a whacker plate on the back of a flat-bed transit van. The firm were so fed up with the equipment getting stolen they installed trackers into even their smaller hand tools. Needless to say, the two chaps in the Transit never saw us coming and were banged to rights having just stolen the tool from the roadside.

The second and much more significant item of equipment is Automatic Number Plate Recognition (ANPR), which is about as important in fighting crime as fingerprints and DNA have become. The results that ANPR has brought to the police service have been quite staggering. Up until its invention and distribution to the police from around 2005 onwards, all police officers had to rely on a number of things to detect crime linked to vehicles. It might be a tip-off about a certain car or person driving that car, it might be pure luck in stopping that car at that time, or, more often than not, it came down to a copper's nose! That gut feeling that something wasn't right, didn't look kosher or didn't fit the area, and most of those gut feelings came down to the experience of the officer concerned. The Yorkshire Ripper was arrested because the two officers knew something just wasn't right about why he was where he was. The rest of that story is history.

ANPR removed much of that 'luck' and turned that copper's feeling into digital fact. Basically it works like this; if there is a marker placed on the Police National Computer (PNC) about a vehicle and that vehicle then passes a police vehicle equipped with ANPR, that marker is flashed up onto the screen on the dash immediately, often before the officer has actually seen the car itself. Those markers can include anything from out-of-date tax to no insurance or MOT to driver known to be disqualified, a persistent drink or drugged driver, to cars used to commit burglary, drug supply and a whole host of other criminal activity, including acts of terrorism. It's not unusual to leave a patrol car at the side of a busy road to go and speak to the driver of a car you have just stopped and return to the patrol car, say five minutes later, to find that your ANPR system has logged half a dozen suspect vehicles that have passed you in that short space of time. It's that good – it was an amazing tool to me during my short career in the police.

The police service never stands still and is often at the forefront of modern technology devised to help its officers in the fight against crime. Without doubt one of the best tools at their disposal on a daily basis is the car, so new ideas will continue to be developed to get the performance possible out of them and to keep police one step ahead of the criminal.

CHAPTER NINE

COMMERCIAL BREAK

I have to say I struggled with how to deal with commercial vehicles in police service in this book, as there are so many and they are so varied that they warrant a volume in their own right. However, it seemed wrong to ignore them completely, especially as there is some crossover between cars and commercials. Is the rural Mini Panda van a commercial vehicle or a personal car? It is quite clearly a commercial for tax purposes, but the design is based on a car and carries out a police role which in most forces was performed by a car. So in the end I decided to include an overview chapter to give a flavour of how important these vehicles have been.

The police have a long history of innovation and practical mechanics when it comes to finding solutions to a problem. The Service needs vehicles that very few others would find a use for, thus mainstream manufacturers were often not interested in financing the research and investment required to convert standard vehicles into specialist police units. So it fell to people within each of Britain's 42 police forces (more than that prior to the 1967 amalgamations of several city and borough forces into larger county forces) to find a solution, and they either converted the vehicles themselves or commissioned a local firm to carry out the work to their design specifications. In twenty-first-century Britain, however, the roles are almost completely reversed, with the manufacturers or specialist conversion firms such as MacNeillie, Kinetic, Coachman, Fame and several others undertaking the work to a very high standard on behalf of the police and other emergency services. This in part is due to legislation, warranty issues, health and safety and, of course, cost.

The number and type of conversions required was and still is as varied as the vehicles upon which those modifications are carried out and include dog section vans, personnel carriers, accident units, underwater search units, mounted section units to transport horses, armoured vehicles, mobile police stations and exhibition units, catering units, armed response vehicles, cell vans and covert observation vehicles, to name but a few. So let's concentrate on just a few of the more interesting examples to give you a flavour of the sheer numbers involved.

A wonderful Nottingham City Police Bedford PC Utilicon, which was a van converted into a 7-seater station wagon by well-known coachbuilders Martin Walter Ltd of Folkestone, stops a Vauxhall HIX Twelve-Four in 1947.

The earliest form of police transport, of course, was the horse-drawn cell, often referred to as the Black Maria, a term that stuck for over a century, years after the mechanisation of the human race and the relegation of the horse to much more leisurely activities. But where did that term emanate from, and why? It may surprise you, but it actually hails from the USA.

On 31 January 1902, the *Amador Ledger-Dispatch*, the local newspaper of the Amador and Calaveras area of Jackson, California, carried the origin of the term 'Black Maria'. It read thus:

> When New England was filled with emigrants from the mother country, an African-American named Maria Lee kept a sailors' boarding house in Boston. She was a woman of great strength and helped the authorities to keep the peace. Frequently the police invoked her aid, and the saying, 'Send for Black Maria' came to mean, 'Take him to jail'. British seamen were often taken to the lockup by this woman and the stories they spread of her achievements led to the name of Black Maria being given to the English prison van.

An article in the 22 October 1912 edition of the *Lodi Sentinel* attributed the term 'Black Maria' to being derived from a nickname for a specific African-American woman living in Boston, Massachusetts, during Colonial times: Maria Lee. Allegedly, it all began when Maria brought three drunken sailors to the lockup at the same time because they were too much trouble to keep in her boarding house. She continued doing this and became of regular help to the police, especially when sailors in the area got so out of hand that even the police couldn't subdue them!

The City of St Louis Police Department purchased its first Black Maria in 1850 because it was too difficult for patrolmen to walk their suspects back and forth to jail. It was a horse-drawn carriage, also painted black, with the carriage acting as a secure prison cell complete with iron bars on the windows and doors. Years later, on 9 April 1866, another Black Maria was purchased, and the phrase is used in the minutes of the board meeting of the St Louis Board of Police Commissioners.

With the passage of Robert Peel's Metropolitan Police Act in 1829, the concept of policing was firmly entrenched in England. When the second Metropolitan Police Act was passed in 1839, constables were no longer employed by the magistrates and became a police organisation instead. Somewhere between 1839 and 1841, police wagons came into use to aid officers on foot and horse patrol.

It's interesting to note that the term Black Maria survived, much like that of panda cars, years after the original had ceased to be, and is still used to this day.

When the police started to use motor vehicles, some of the very first ones were of commercial vehicle status, like Albion, Thorneycroft and Crossley, and they made ideal prisoner transport once the officers and maybe a couple of local tradesmen got stuck in to convert the rear box into a secure cell with barred windows, bench seats, shackles on the floor and secure locks on the doors.

One of the most famous conversions from those early days came when the Liverpool City Police were trying to quell rioting during the transport workers' strike of 1911, in which thousands of people went on the rampage for several days after weeks of tension and unrest. By using large, heavy-duty planks of wood, the officers built themselves a sort of armoured Trojan horse to encase the rear of the lorry upon which this unit was built, and also the cab to protect the driver. Towards the top of the box were windows placed at head height to enable the officers to observe the rioters below.

This earliest of personnel carriers was a huge success, allowing the police to gain access to the centre of the trouble without officers getting injured and by using less manpower. It was an ingenious design, and as we look at the vehicles that followed the general design concept, you'll see that it hasn't really changed that much in over 100 years.

During the Blitz of World War II, one of the cities that got hit the hardest, outside of London, was Portsmouth, the home of the Royal Navy. As officers from the Portsmouth City Police went about the business of rescuing people from bombed-out buildings and securing premises, they and colleagues from the newly formed National Fire Brigade needed refreshments during the long hours they worked. With no natural unit available they managed to acquire (although just how they did so is not recorded) an old Dennis single-decker bus and then set about converting it themselves into a rather smart-looking mobile canteen unit. War reservists and ladies from the WRVS poured out the tea from huge metal teapots into those enamelled tin mugs and no doubt served up a few hearty meals of stew and dumplings, or bowls of soup and plates of sandwiches.

These two examples show the ingenuity of some of the forces, but what's clear is that there was certainly no actual policy involved and the craftsmen required to undertake any such conversion work would have been found amongst their own ranks or maybe sourced very locally. So from an historical point of view it has proved difficult to find accurate reference material because most if not all of the commercial vehicles used up until the end of World War II were in-house conversions, and it wasn't until about the mid-1960s that things started to take on a slightly more professional look. One thing we do know is that, pre-war, coffin makers often also worked on wooden coachbuilt conversions; a similar skill set was needed and ironically they had similar supply and demand criteria. A good riot could be good business for coffin makers and commercial vehicle body builders alike . . .

Prisoner transportation, as previously discussed, was and always will be the sole responsibility of the police. By the early 1950s the Metropolitan Police were using a modified Morris Commercial PV van, in black of course. The standard vehicle, in production from 1946 to 1953, was powered by an OHV 4-cylinder engine of 2050cc, through a 4-speed manual gearbox. The body had a wooden frame, made of ash, with aluminium-clad sides, and featured sliding side doors and a canvas roof with the two rear doors opening full width. However, the police version differed considerably and featured a one-piece rear door and a metal bulkhead, and the canvas roof was replaced by a steel one, obviously to make the 'cell' completely secure. The rear section had bench seats bolted down the sides. On the front of the cab was a bell and above the windscreen an illuminated police sign. Much of this conversion work was undertaken by the Met's own craftsmen based at their Main Repair Depot (MRD) at Northolt. These Morris Commercial PV vans seem to have only ever been used by the Met and were possibly amongst the very first bulk orders of commercial vehicles ever purchased by the police, with some of them still in service in the mid-1960s! Each station within the capital was assigned one of these vehicles as their 'station van'.

One of the few professional converters that a number of forces used in the 1960s and 1970s was Wadhams, of Waterlooville, Hampshire. They already had a lot of experience in producing ambulance conversions, particularly on the Morris Commercial chassis, and so they set about producing a secure prisoner transport vehicle. A lot of work was needed to make it secure, with mesh screens on the windows, metal panelling on the inside of the doors, a raised metal floor, bench seats, two crew seats, a dividing partition and a sliding door to give the crew access to the prisoners if required. This sort of vehicle was most probably used to transport convicted prisoners from court to prison in the days when the police undertook such duties, before the Prison Service and then private companies like G4S took over.

When Ford introduced the Mk1 Transit in 1965 it was tailormade for police use in so many ways, but it really found its niche as a Station van and has remained a firm favourite ever since, through all of its incarnations. The short wheel base, low-roof model was ideal. It was a go-anywhere vehicle that could be driven by just about anyone with reasonably competent driving skills. With very little modification the rear load area could be converted into a secure cell to collect prisoners from the street and convey them back to the station, just like Maria Lee did all those years ago. As the Transit's fan base grew and it morphed into the Mk2 it really did hit the big time. Just about every force in the UK used them for the same purpose and just about all of them were modified in different ways. The Met, for example, only used the van with sliding cab doors, whilst other forces opted to use an adapted crew bus type with side windows in the rear, which were barred to allow officers to see the prisoners inside the cell before they opened the doors. The glass in these windows was often replaced by Perspex, which didn't shatter. It wasn't unusual to cram in, say, 20 prisoners in one go at football matches during the violent decades of the 1970s and 1980s!

By the time the Mk3 Transit was introduced in 1987 there were a couple of radical changes on the horizon. The police were experimenting with diesel vehicles for the first time, and when it came to ordering the next batches of commercial vehicles, diesel became the number one choice rather than petrol. The second change came with the functionality of the vehicle itself. Until now, the police had used two types of Transit; the station van type as already described, and a personnel carrier or mini-bus type to transport officers in bulk to incidents or operations, and such like. This necessitated buying and maintaining two types of vehicle, with the latter perhaps not being used often enough to fully justify its purchase. There was also a major issue around health and safety when it came to the transportation of prisoners and officers in the rear of a van, which were fitted with wooden slatted seats and no seatbelts.

The answer came with the introduction of what became known as the 'multi-role' vehicle. The new Mk3 Transit was a bit more versatile than its predecessor and could be purchased in long wheel base form but with a high-top roof, and this unit formed the basis for the multi-role. In the mid-section of the unit were placed five seats; three rearward facing and two forward facing with a table in between. All the seats were fitted with seatbelts and there was enough storage space for riot shields, water carriers, first aid kit and the all-important Max-pax coffee and hot chocolate drinks! The rear section had a proper built-in secure cage capable of holding two prisoners seated opposite each other. The five officers (plus three up front) could be transported in relative comfort and safety whilst engaged on an operation, or perhaps on a Friday or Saturday night patrolling their local trouble spots with the added extra of an on-board cell, should it be required, removing the need to call up another van. This concept worked so well that it is still the standard used today, although the big Mercedes-Benz Sprinter has stolen a large slice of the cake away from Ford in this department, with the likes of Vauxhall's Movano, VW's Crafter, Peugeot's Boxer and the Iveco Daily series all vying for the same business. Much of the conversion work to fit the cell and specialist storage and seating is done in-house by the manufacturers themselves these days, but there are a good number of specialist firms like MacNeillie, Kinetic, Fame and several others who have all come up with innovative ideas of their own to produce the ultimate multi-role carriers for police use.

In rural areas large multi-role vehicles just aren't required, but transportation of the occasional prisoner is. So what do you do if your prisoner is non-compliant or violent? You can't insist that he behave himself when seated behind the driver of the Ford Focus patrol car and hope that he complies! So a number of manufacturers, including Ford, Vauxhall and Peugeot, now produce a car or small van like the Transit Connect or Peugeot's Partner with a single secure cell fitted in the rear, or, if it's a car, half the rear seat is removed and the cell, complete with its own seat, is built in. This means that the officer is safe and the prisoner is secure and also safe. I have transported many violent people in the back of vans and seen some crazy things that violent offenders have done – often when they're under the influence. Without the transport facility of the cage it would have been impossible to safely move these people. This adaptation has, simply, saved the lives of offenders and officers alike.

So much for prisoner transportation, but what about the dogs? Well, much like humans, in years gone by they were merely placed into the rear of any old commercial vehicle with no creature comforts (no pun intended), and it was down to the individual officer to perhaps do something to his mode of transport to reduce the risk of serious damage to what is basically a 'living' work tool, a police resource and a valuable one at that; it takes many man-hours to train a police dog, and this all has to be paid for. A former colleague who was a police dog handler of some repute took delivery of his new police vehicle in the early 1960s: an almond-green Morris Minor van which came nicely equipped with a VHF radio set, Lucas Acorn blue light, two-tone horns and two illuminated police signs on the roof. But for his German Shepherd dog, aptly named Greife (German for 'to bite'), there was nothing: just a bare metal floor. So Terry got some wood and panelled the floor to make it much less slippery, added a wire mesh grill between the load area and the cabin to prevent the dog entering the cabin during heavy braking, and he added some hooks for leads and storage for Wellington boots and other essential items.

I suspect Terry's story is not unique during this time or even before that. Vans commonly used as Dog Section vehicles included by the Morris Minor, Austin A35, Mini and all of the Ford Escort range – from Mk1 through to the late models of the 1990s. Along the way we've seen other slightly larger vehicles in use from time to time, including by the Hull City Police who used Bedford CA vans, whilst the Glasgow City Police carried up to six dogs in the rear of a Mk1 Transit van whilst towing a twin-axle box trailer behind it containing all their necessary equipment.

But by the turn of the twenty-first century things were about to change. Police forces started to replace vans with estate cars, properly equipped with built-in cages at the rear, mostly supplied by the manufacturers themselves in a similar move to the way in which they'd approached the multi-role vehicles. The cars were faster, a lot more comfortable for both officer and dog and, most importantly, came equipped as standard with air conditioning, whereas many vans at this time simply didn't. Why was this so important? There were two very high-profile deaths of police dogs left in the back of their vans on hot days, and, quite rightly, a huge public outcry followed, so the police service was left with no choice but to update the facilities for the dogs themselves. To ensure that the public are aware of these changes you will now see signs placed on the sides and rear of Police Dog Section vehicles that clearly state that the vehicle is air conditioned. The air conditioning can be left running with the dog inside even when the car is unattended by the handler, by using a method first put on Traffic cars a few years earlier. The 'run-lok' system enables the officer to leave the vehicle running whilst it is stationary at an incident for long periods of time with the blue lights running to protect the scene and the personnel working on it and keep the car fully locked. It's a simple but very effective system.

Another initiative borrowed from that City of Glasgow Police idea involves the use of multi-dog units whereby a large van, usually a Mercedes Sprinter or long wheel base Ford Transit, is fitted with around six cages plus storage for all the kit within the van, to transport the officers and dogs to a pre-planned operation, usually a football match or other large public gathering. This system facilitates the use of just one vehicle instead of six, which means less fuel used, only one parking space required and all officers and dogs ready to deploy at the same time when required.

At one point during the late 1990s and into the early 2000s a couple of forces experimented with fast-deployment dogs from the back of a Volvo T5. The idea was that the dog handler was given a Volvo T5 fitted with a quick-opening electric tailgate operated at the push of a button from the dash. The dog unit would latch onto the lead patrol car in a pursuit following a ram raid or similar crime, and when the occupants decamped from the car and started to run, the dog handler would hit the button, out jumped the eager German Shepherd that had been trained to run down the side of his patrol car, past the next one, then chase and detain whoever he'd seen running from the scene. The theory sounds great but the idea didn't catch on, which basically means it didn't work, and I suspect there are probably one or two stories to be found as to why!

Dogs aren't the only animals that the police need to transport around; horses are still in use today with a good number of the UK's forces, and their role has hardly changed over the last 100 years or so.

The horses of the Mounted Section are used mainly at public order events like football matches, demonstrations and during the late 1990s to help police the two opposing sides in the fox-hunting incidents that took place in parts of the countryside. The horse is a huge asset (literally) to policing and usually a welcome sight for the officers on foot, especially if confronted with an angry and hostile crowd; there are few people brave enough to stand in the path of a charging horse.

But getting the horse to the scene requires specialist vehicles and over the years we have seen these units develop from humble beginnings to today's rather more sophisticated vehicles. It goes without saying that farmers, horse traders, show jumpers, race-horse owners, private individuals and many others have been well catered for over the years, particularly when it comes to specialist manufacturers being able to supply 'off-the-peg' vehicles specifically to transport horses, albeit at huge financial cost.

The police have used a fairly wide variety of such vehicles over the years, from the likes of Bedford, Thames, Leyland, Ford, Mercedes and Iveco, to name but a few. Most of today's units have been converted by companies such as Coleman Milne and command six-figure price tags, so needless to say it's a huge investment and a long-term commitment. The vehicles have to meet all manner of legislative requirements to protect both the animal and the human, and are basically a mobile stable with all the necessary fitments on board to ensure that the horses are properly cared for during transit. Forces such as the Metropolitan Police, City of London, Thames Valley, South Yorkshire, Greater Manchester, Nottinghamshire and South Wales Police all have good-sized Mounted Sections and they are often deployed to assist at events policed by neighbouring forces that don't have such units. In 1993 the Lancashire Constabulary took delivery of what must rate as one of the largest police vehicles of all time, and they got it for free. The Leyland two plus two articulated truck was originally the unit that carried the land speed record car Thrust 2 and was donated to the force by Sir Richard Noble. It was subsequently converted into a rather nice luxury pad for Lancashire's police horses.

One of the stranger-looking commercial vehicles that the Police have converted for their own specialised use is the observation unit, which is just about unique to the Metropolitan Police, although their neighbours in the City of London Police did copy it for a while. As we all know, if you want to demonstrate about something you generally take your complaint to London, having amassed enough support to make it worth your while. The capital also has a large number of football clubs that have to host supporters from all over the UK, which in most cases will necessitate a police escort from, say, the nearest railway station to the football ground and back again. The Met are very well versed in policing large crowds, whether peaceful or otherwise, and are called upon several times a week to undertake such operations, but to gain maximum control the officer in charge of such mass gatherings needs to be able to see as much of it as possible. So in the early 1970s they took an old Morris LD van that had been a station van for several years and converted it into a control unit. The technicians at MRD in Northolt cut a large hole in the roof then fabricated a wooden frame, glazed at the front, back and sides, and mounted it on top of the roof. A small platform was then bolted to the floor on which the officer in charge of the operation could stand and peer out of the observation tower rather like a tank commander in his vehicle's turret. Inside the van a couple of officers would be seated at a built-in desk manning the radios and issuing instructions to those officers on the ground. The van would generally lead the demonstration to help control the pace and direction of the crowd.

The black-and-white dome mounted on top of the observation tower is always a point of interest. A joint policy decision was made by the Met Police and the London Fire Service and London Ambulance Service to make their command vehicles clearly visible to their own staff when deployed at a major incident. The 1970s saw the IRA bombing London on several occasions, necessitating all three emergency services to attend in large numbers. So the agreement was that the domes on Police Command Units would be in black and white, the Fire Service in red and white and the Ambulance Service in green and white.

These early control units were replaced in 1976 by no fewer than six Mk1 Ford Transit vans that were totally unique to the Met. They came with a 2.0-litre petrol engine married to an automatic gearbox which wasn't a standard Ford fit. They were converted by Maxeta in Cambridge, and with the van set in first gear it could crawl along at a steady 3–5mph, the same pace at which humans walk, which meant that the driver didn't have to keep dipping the clutch or braking to reduce speed, as it did when using the old Morris LDs. On a long march this obviously helped reduce driver fatigue and kept mechanical wear down, too. The vans were fitted with a unique power socket underneath the front of the vehicle that allowed it to be plugged into the mains when not in use. This was to power a small electric pump that kept the engine oil warm and circulated so that in the event that the vehicles were required urgently they were literally ready to go straight out. They also carried a small Honda generator on board to help power all the electrics including the three VHF radio sets and a whole lot more. Apart from the driver it carried a crew of three – two radio operators and the Commander for the event. Maxeta completed all the work, including the observation tower, dome and interior fitments. All six vehicles were based at Lambeth and remained in service until the mid-1990s. On retirement two ended up in the Sponden Museum, which eventually sold them on, with one now in a private collection in Germany and the other purchased at auction by a film company for use in the movie about the Iranian Embassy siege of May 1980. The Transits were replaced by two hybrid vehicles custom-built by a company called W and E. The Fiat Punto-based petrol engine was hooked up to an electric motor, but they proved so unreliable that within 18 months they were decommissioned and the whole system was scrapped.

At the beginning of this chapter we mentioned the personnel carrier built by Liverpool City Police in 1911. Since that time things have moved on a bit in terms of the vehicles put into service, and especially the materials used; no more planks of wood, that's for sure! Over the years the police have utilised a number of vehicles to transport its personnel around en masse. The Met Police, for example, had a large fleet of Bedford OB coaches painted dark green and these were later replaced by buses supplied by General Motors. Other forces purchased former coaches with varying degrees of success. But moving 30 or 50 officers around at the same time wasn't always necessary and most of the time only a handful, say 6 to 12 in one vehicle, was sufficient. One of the earliest purpose-built vehicles was the Fordson E83W Utilicon, built between 1938 and 1957, a small (by today's standards) estate variant of the E83W van. It had an 1172cc side-valve engine producing 10hp, giving it a top speed of just 40mph – and that was before you put another four or five burly coppers in the back!

During the 1950s and 1960s the Austin and Morris J Series minibuses found favour with a lot of forces, followed by the Morris LD. Then in 1965 Ford launched the first Transit van, which was quickly followed by its minibus variant. Capable of carrying three up front and nine in the rear on proper seats, it proved to be an instant success with the police. It was quick, easy to drive, reliable, had plenty of space, and with a blue light on the roof made a bit of a statement about its intended role. It achieved some notoriety in 1979 when during an anti-Nazi League demonstration in Southall members of the Met's Special Patrol Group (SPG) arrived in their dark blue Mk1 Transit carriers to quell much of the public disorder. During the ensuing fighting an activist called Blair Peach was knocked unconscious and died the following day. Although it was never proved, the finger of suspicion pointed firmly at officers from the SPG, and shortly afterwards the unit was disbanded. But the term SPG became ingrained in the public psyche, and at public order situations for many years afterwards, even outside of London, there were often shouts of 'Look out, here come the SPG' as extra officers arrived in their Transit carriers. The phrase became so ingrained in British anti-establishment culture that it was used as the name of Vivian's pet hamster-type animal in the BBC sitcom *The Young Ones!*

In 1978 Ford updated the Transit range and launched the Mk2 model, which was basically the same but with a new front end and revised engines ranging from the acclaimed 2-litre Pinto to the much-loved 3-litre V6 Essex engine. It was faster, handled better and was a lot more refined than its predecessor. As we moved into the turbulent and violent 1980s, the police Transit minibus really earned its place in history. This was the decade of mass organised football violence, both in the stadiums and out on the streets, the year-long miners' dispute, the equally long Wapping print workers' strike, CND marches, the Greenham Common 'Peace Camp' (which was anything but peaceful) and serious rioting in Brixton, Birmingham, Toxteth and Tottenham. At each and every one of these events the police arrived in their Mk2 Transit carriers, equipped with mesh grills, riot shields, petrol bomb skirts and under-chassis fire extinguishers.

It became a symbol of that decade as, week in, week out, dozens of the things could be seen snaking their way across the country to the next trouble spot. Mesh grills placed across the front windscreen were a new idea to help protect the cab from missiles hurled at them, and when the grills weren't in use they were wound back up into their rest position via rails placed on the front section of the roof. The side window glass was replaced with Perspex that didn't shatter on impact, and a system of fire extinguishers placed under the chassis meant that if a petrol bomb exploded beneath the carrier the fire could be extinguished at the press of a button from inside the cab. During the incredibly violent Wapping dispute, more than 300 police officers were seriously injured, including a number who sustained life-threatening injuries when sharpened metal fence posts were thrown like spears straight through the sides of the Transits, hitting the occupants inside. With immediate effect the interiors of all police carriers nationwide were lined with polycarbonate panels that were virtually impenetrable. Just as those Liverpool officers had done in 1911 with their planks of wood, so the Met and others had to do with its modern equivalent some 75 years later.

Owing to the necessity for rapid deployment of Support Group officers in these vehicles, a method of getting the 12 officers in and out of the vehicle in double-quick time was devised. It was called 'embussing' and 'de-bussing' and it was practised for hours. Everyone had their place within the unit; you had to get in and out in a certain order and the record was something like 2.2 seconds! When it was done in public it did look very impressive. It's before my time, but young officers in that era spent hours, days, weeks and months inside our Mk2 Transit carriers during the 1980s. They slept in them, drove them at some scarily high speeds, ate their takeaways and doggy bags and sipped lukewarm Maxpax hot chocolate in them and banged their heads on the exit doors. But they loved them; it was a home from home for many, and they are remembered today with great fondness by those who served during that period.

The Mk2 Transit was superseded by the Mk3, which we have already discussed, but the Metropolitan Police moved away from them for a while and opted to use the Leyland Sherpa 400 series carrier with its 3.5-litre V8 Rover engine. It was a bit of a beast with ample power but wasn't the most reliable of things. The Met later went back to the Transit before settling on the Series 1 Mercedes Sprinter for a decade or more. Today they are using something quite unique, based on a London Ambulance Service body shell made from carbon fibre and dropped onto the larger 7-ton chassis from the Mercedes Sprinter. Because of the extra weight, the brakes have been upgraded to help haul the carriers to a halt. They are powered by the standard 3.0-litre V6 diesels pumping out 248bhp through an automatic gearbox. The Met commissioned dozens of these carriers in 2007 at a cost of £78,000 each, which sounds rather expensive until you take into consideration that they have an estimated lifespan of at least 12 years. The spacious interior can seat 10 officers, and there is racking for shields, with lockers at the rear to store other much-needed equipment. They were initially issued to the Tactical Support Group (TSG) and later to the boroughs for policing their respective areas. All the carriers are finished in metallic silver with the standard Met striping along the sides. The carriers are said to be lovely to drive and very comfortable for the crews who use them. The City of London Police has near-identical carriers but finished in dark blue.

In twenty-first-century Britain most county forces use the standard Mercedes Sprinter carriers complete with windscreen grills, blacked-out windows and with a two-man cell in the rear; they are a very purposeful unit.

In the first decade of the new millennium a good number of forces had to close rural police stations following a round of government cutbacks, but those areas still required policing in some way, and so the mobile Rural Police Office was born. They were generally based upon a Peugeot, Citroën or Fiat six-wheeled chassis with bodywork built by the likes of Fame, Coachman or MacNeillie. The interiors varied according to force specification but usually had a desk and some seating to enable written statements, crime reports and other documentation to be completed when members of the public popped in to see a police officer when the unit stopped in their village, maybe once a week or monthly. They carried all the necessary paperwork, public information and advice leaflets, and most had a small kitchen area and possibly a mobile cassette toilet on board. But within 10 years these units also faced the axe during the next round of cutbacks, and many rural communities lost virtually all contact with their local police forever.

In 2000 the Road Deaths Investigation Manual (RDIM) was placed onto the Statute books to ensure that all road deaths were treated as homicides until proven otherwise. The process followed the principles laid down in the Murder Manual that CID had used for decades. One of the recommendations included the equipping of a functional mobile office to be taken to the scene of any road fatality or potential fatality so that it could be used as a base, a control centre, a point of reference, an office for all those involved in the investigation process. Because the RDIM sets out the manner in which the police have to investigate such incidents, it often necessitates the closure of the road for maybe six hours or more and not, as some journalists and members of the public think, because the police are being bloody-minded about it all. Most forces have opted to use near-identical vehicles as their Roads Policing Incident Command Units. They are mainly Mercedes-Benz Sprinter vans equipped with a table and seating for four, a couple of wipe boards, a TV and DVD player (for watching CCTV footage), full radio comms, a cubicle cassette toilet, a microwave, kettle and fridge, with the rear of the unit jam-packed with all the necessary extra equipment for road closures, scene security, forensic suits, evidence bags, extra lighting powered by an on-board generator and a whole host of other items. The vans are basically a self-contained unit able to stay at the scene of any incident for 24 hours or more if necessary.

Finally, coppers like their food – and we aren't talking doughnuts here! Keep the police fed and watered and they'll happily walk over hot coals for you. If they are denied such sustenance, watch out, because boy will they moan about it. It's understandable, given the length of duty they are expected to perform at an incident or operation, and this will often see them being on their feet for 12 hours or more. But just how do you feed, say, 50 of them without them leaving the area to go back to a station? Simple, you bring it all to them. You may recall that at the beginning of this chapter we saw how the Portsmouth City Police converted an old bus into a mobile canteen to feed their officers at various bomb sites throughout the city. Although that conversion itself was somewhat unusual, the concept of providing much-needed refreshment was not. I remember attending the Potters Bar rail crash. It was a scene of utter carnage. Soon the area was filled with police and we worked long hours that day. The arrival of the catering van was essential for a short break, if only for a swig of tea and a bite of a sandwich, and then willingly each officer returned to restart where they left off. It made me realise an army truly does march on its stomach.

A number of forces tried a variety of vehicles with varying degrees of success. The Hull City Police had a large Morris Commercial complete with serving hatch dishing out hot drinks, whilst the Hampshire Constabulary in the late 1970s had a Ford A series van complete with on-board ovens that were supposed to cook pre-packaged frozen meals in foil trays. The trouble is the ovens weren't great and tended to burn the pies at one end and left the other end still frozen! It gained the nickname the 'Forced Feeder'. In the 1950s the Metropolitan Police introduced their first catering unit based on a Morris LD van. It gained the nickname 'Teapot 1', and the name stuck. In fact, the unit was never allocated an official call sign, and everyone just knew it as Teapot 1. The LD gave way to a black Mk1 Ford Transit articulated truck. Those of you who know your Transits will already have raised an eyebrow because Ford didn't make an artic chassis for the Transit, so this unit was clearly converted by someone, although it's not known who. In the 1990s the Met purchased two Dodge cabs with articulated trailers and converted those into mobile catering units, so the Met now had Teapot 1 and Teapot 2 and both were a very welcome sight for those officers deployed on the ground. They even had the wording 'Teapot 1' emblazoned across the cab roof so that officers knew exactly where it was. Modern-day policing is somewhat different, of course, and following the huge cutbacks in public spending since 2008 the catering department was disbanded and put out to private tender. An independent company is now responsible for supplying Met officers with refreshments using two converted Transit vans with serving hatches. The company refused point blank to have the moniker Teapot 1 or 2 placed on the vehicles, but they couldn't argue about the vehicles' unique three-digit fleet letters being applied to the side of the vehicle, because all Met vehicles have to have them. It just so happens that these vehicles got the letters KFC and BBQ as their fleet numbers and apparently the bosses have never noticed. We won't be telling them, right?

CHAPTER TEN

PERFORMANCE CARS

Police car history is littered with stories about the Old Bill using high-performance sports cars. But why? Is it because they rather fancied themselves driving the latest convertible down your local high street whilst wearing a pair of mirrored shades and showing a bronzed arm on the window ledge? Or did a couple of Chief Constables, much the worse for wear at the bar one night, start playing that very male game of 'Mine's bigger than yours, Jack' and it sort of got out of hand? Probably not. Maybe there's a more rational explanation. And to put it simply, there is. The answer is speed. That's it in a nutshell: speed. No matter what era we are talking about, whether it's a 1930s Invicta or a twenty-first-century Mitsubishi EVO X, the basis upon which the police feel it necessary to purchase certain performance cars comes down to its level of performance in comparison with the competition at the time. And by competition we mean the bad guys. Whatever it is they are driving, the cops need to drive something as quick or preferably much quicker!

That quip about the two Chief Constables seated at the bar and sipping their Martinis might not be that far from the truth, though. Not the bit about their levels of testosterone, but more about them comparing notes about a particular local issue they might be having. This is the basis upon which the decision to buy certain cars is based;

it will be a local problem, not a national issue, that will have senior officers searching for a solution. The media, local politicians and the public will all have contacted the police to complain about a certain automotive issue that concerns them. All of them will be asking the same question; what are you doing about it? By the time it's reached that stage the guys on the front line will already know there's a problem and they will be struggling to cope with it, given the tools at their disposal at that time. No doubt they too will be banging on the guvnor's door demanding a solution. So the pressure is on.

The cars we'll be looking at here will all be driven by Class 1 Advanced Drivers, all of them Traffic Officers capable of squeezing every last drop of performance out of anything with four wheels. I was fortunate enough to have been placed on my advanced driver course while I was in the force, so I know that it's no use giving a Subaru Impreza Turbo to a policeman who only passed his driving test last week and expecting him to pursue a couple of bank robbers down the motorway at 120mph. That's only going to end in tears. Instead, the task at hand will be given to the Traffic Division or the Roads Policing Unit, as it's called these days, or a special unit will be formed to combat the problem, made up of advanced drivers only. These local issues may only occur during certain times of the day, usually at night if it's crime-orientated, so the cars are often utilised as mainstream Traffic cars at other times to get good value out of them – and you can bet those same politicians, public and the media will be just as quick to complain that the police are now living it up in their posh sports cars and why aren't they out there catching rapists and murderers!

Performance sports cars were traditionally open-top two-seaters that basically carried the same engine as their saloon counterparts, but of course they had much lighter body work, and thus the performance was increased considerably. Obviously some go-faster bits were added, like twin carbs, higher-lift camshafts and other mechanical items, which all helped give the cars a certain edge. This state of affairs remained until the 1960s, when suddenly speed was democratised both in terms of car body type and the money needed to own a really fast car. It's not some-

thing to discuss in detail here, but it was basically the advent of
homologation specials for international rallying and racing that
prompted manufacturers to produce range-topping vehicles in just
enough volume to satisfy the rules of the day. BMC's senior hierarchy
even objected to the Mini Cooper, the real trailblazer in volume terms,
as they thought the cars would be difficult to sell. Of course, the Mini
Cooper sold like hot cakes because it was fantastic fun to drive and
carried a name that had taken Jack Brabham to two F1 World
Championships. Ford followed suit with the similarly F1-linked Lotus
Cortina and suddenly the whole world realised these cars were not just
'Win on Sunday Sell on Monday' loss leaders but could actually make
money in their own right. The same thing happened in America, where
Henry Ford had used racing and record breaking to sell his cars from
his very earliest days, and cars like the Ford V8 Coupe and later the
Hudson Hornet (there is a reason Dean Moriarty drives a 1949 Hudson
in Jack Kerouac's *On the Road* – the car had become an American icon,
even then) winning in NASCAR events, and the subsequent 1960s muscle
car power race saw huge sales on the back of performance. It's amusing
to note that in 1986 Ford executives were even nervous about producing
another car with an F1-related name, the Sierra Cosworth (25 years after
the Mini Cooper), which of course also sold out almost immediately,
so much so that, just like the Mini Cooper, it became a mainstream
model. This class of performance-biased sports saloon was cemented
by the arrival of the Golf GTI – between 1976 and 1983 more than
450,000 GTIs, about 8 per cent of Golf production, were produced and
no manufacturer ever questioned the financial or marketing wisdom of
making a sporting version of their basic car again. The police were not
slow to take note of this trend, and by the late 1960s the combination
of the increased standard equipment the police were carrying (cones,
signs, etc.) and the availability of high-performance sports saloons meant
the pure two-seater sports car's days were numbered in police hands.

The first UK police to identify the need for truly fast cars, and most
importantly get the funding for them, were London's Flying Squad,
'The Sweeney' as they have become known. They had pioneered the

use of vehicles for actual police work as well, as we've seen, but by the mid-1920s the Crossleys were becoming out-of-date and slower than the cars they were apprehending.

To solve this, police engineer Major T. H. Vitty, the Met's Chief Engineer, did some research and eventually ordered six (research by the Lea-Francis Owners Club has uncovered seven chassis numbers supplied which does make you wonder if one was written off in an accident) Lea-Francis 12/50 L Type 4-seat Tourers with coach-built Cross and Ellis bodies direct from the Coventry firm. These cars were all supplied in Mole Grey, although it's believed some were later painted in other colours to prevent criminals recognising them too quickly. They only weighed around 750 kilos and had well over 50bhp from their twin-port 'Brooklands specification' 1.5-litre Meadows engines so were quite fast for their day, and could achieve a maximum speed of around 75mph. However, their design philosophy was that of a lightweight car, indeed their adverts referred to them as such and that was ultimately one of their weaknesses (literally), for police work as the Flying Squad were regularly using them to push other cars off the road and, being lighter than the cars they were trying to apprehend, they would often come off worse. They also had reliability issues principally around a weakness in the 'Star Gear' design of their differentials which led to back axles failing when being used in extremis. Lastly, because they followed, shall we say, a Lotus philosophy, officers had to drive them hard to achieve and retain speed, which made them liable to go out of tune quickly.

The six Lea-Francis 12/50s were delivered in May 1927, and a small group of officers led by George 'Jack' Frost initially spent a week at Brooklands being specially trained to drive them. This included measuring tyre adhesion on different surfaces in different weathers plus doing handling and high-speed exercises. They then entered service with no police markings and were initially greeted with enthusiasm by their drivers. They were used with their canvas roofs in place partly to conceal the officer's uniforms and then very bulky radio telegraphy gear, but mainly because the radio aerial was concealed in the canvas itself.

The 'Leafs', as they became known, covered huge mileages served with great distinction and were much loved, but after eighteen months of experience the squad realised that lightweight sports saloons were not what they needed. Sadly, none of the seven have survived.

What they needed was a car with bigger lungs and a heavier chassis, so when it was being driven fast for long periods it was not under extreme stress, and when attempting to ram bad guys off the road it was heavy enough to hold its own; a philosophy that has continued to this day, although ramming has thankfully been consigned to the history books.

The first really high-performance British police car which set the template for all others to follow in this kind of role was the 1928 Invicta 4.5-litre High-Chassis Tourer that I was lucky enough to see at Gaydon, that actual car. It was an honour. In today's often hyperbolic terms it would be called a 'high-speed interceptor', and as such it is a very important car in UK police history.

Invicta had been founded by racing driver Captain Noel Campbell Macklin (and yes, he was the well-known post-war racing driver Lance Macklin's father) in 1925 after two other short-lived attempts at car making had failed for various reasons. Macklin was a maverick inventor in more ways than one who later founded the Railton marque, examples of which were also used by the Flying Squad. He was knighted just after World War II for his services to the military in the field of boat building after he had founded Fairmile Marine, a company which revolutionised how small- and medium-size motor launches were designed and built using proprietary parts that allowed them to be produced quickly and flexibly by many suppliers; suddenly this became vital in 1939, and the company produced over 1,800 boats during the war, including some of the famous Motor Torpedo Boats. Sadly, he passed away in 1946 at the age of 60.

In the 1920s Macklin was very keen to make what he called a 'top gear car', a machine with the flexibility and torque to go from low speed to around 100mph in one gear. It must be remembered that in the days before synchromesh-gears the concept of most normal modern race engines, high power from a narrow power-band very 'cammy' engine enabled in use by a 6- or even 8-speed gearbox was an anathema. Gear changing was slow, and a faff best avoided if possible, thus some designers sought to produce cars which needed the minimum of gear changing, because these long, slow gear changes were uncomfortable on the road and cost lap time in racing cars. Macklin was also a great steam car enthusiast and of course they, like the new electric cars, have instant torque from rest so he was keen to reproduce, as closely as possible, this characteristic in a petrol car. In fact, he was still experimenting with designs he could produce to re-popularise the steam car as late as 1922, although he abandoned those ideas before moving to the commercial stage.

Although Invictas were only produced from 1925 to 1933, the firm's impact and legacy was greater than that of many companies which lasted three times as long. They produced not even a thousand cars in that time but claimed overall victory in the Monte Carlo Rally in 1931, when a car driven by Donald Healey won numerous other motorsport events and claimed many records, often with sisters Violette and Evelyn Cordery at the helm; these indomitable supporters of the marque claimed two Dewar Trophies in Invictas, one for driving 10,000 miles around the world in 1927 and another in 1928 for covering over 30,000 miles in 30,000 minutes around Brooklands over a 2-month period.

By 1928 the Metropolitan Police's Flying Squad had learned more about the criteria they needed. On numerous occasions they had rammed bad guys and come off worse in their much loved but fairly small Leafs because they were simply not substantial enough to do the damage needed to stop a felon. This is before seatbelts were invented, when steering columns were tubes that could stab you in the chest. These men were fearless heroes dealing as best they could with a new phenomenon, vehicle crime. They investigated various possibilities but started to work with Macklin as an advisor on vehicle purchase and driving techniques. It was natural he would recommend the car he'd developed, although it should be noted that much of the design work was done by the well-known engineer Willy Watson, who later worked with W. O. Bentley at Lagonda and designed the engine which ended up in the Aston Martin DB2. The Invicta used the then new Meadows

4.5-litre straight-six OHV unit which developed over 100bhp and had a high-torque, low-revving character which suited urban work and came close to Macklin's steam car aspirations.

In late summer 1929 the Flying Squad's legendary expert driver, Jack Frost, tested Invicta's 4.5-litre racing car, which had recently raced at Le Mans, at Brooklands and, wonderfully, on the Westminster section of the Thames embankment, which 'B' Division officers closed off early one morning for some high-speed runs. How times change...

He was very impressed with the car's sheer pace, its handling and road manners, praising especially 'its tremendous acceleration factor of 10-to-60mph in less than 10 seconds'. An order was placed for one car, others would follow. This first car was actually built as a chassis in very late 1928 but not delivered until late summer 1929. It was one of the first 4.5-litre Invictas built, being in effect a hybrid of the 3-litre chassis and Meadows' new, larger engine. Only months later Watson designed a new, heavier duty chassis to match the larger engine.

It proved its worth very shortly after entering service though, and indeed became quite famous as the supercar that could outrun any criminals' mount, something which acted as a deterrent itself. The car cost the police £632, at a time when the list price was £985, so even then the marketing value of police sales was recognised. Captain Macklin, a man who served his country with distinction, may well have felt duty-bound to help combat the scourge of the smash-and-grab raids then becoming popular.

The incident which made the car's reputation happened in the early hours of 24 July 1929, shortly after it entered service. Patrolling with five detectives aboard, but driven by Frost, the Invicta's crew stopped at a late-night coffee stall in Kennington, and one of their number, Detective Constable Bob White, spotted three known felons in a large Vauxhall 30/98 – a car similar in performance to the Invicta. The police followed at a distance, with two more officers in one of the older Leafs following. The Vauxhall's crew drove across Westminster Bridge, not realising they were being followed, and eventually stopped outside Millington's Tailor's shop on Victoria Street. The gang backed the big

Vauxhall up to the shop's front and started to attach chains to drag the steel grille off. Frost had hung back but now drove straight towards the gang who realised what was happening just in time and sped off. Frost was eager to try his expensive bit of kit in a genuine pursuit for the first time and in an interview some years later said, 'I knew it was time for the fun to begin.' I'd like to have met that man.

The two cars thundered through Tothill Street, along Petty France and into Buckingham Gate where the Invicta drew level. At over 50mph Detective Inspector Ted Ockey dropped the illuminated MP sign and sounded the gong, but the Vauxhall refused to stop.

Jack Frost's report is very illuminating:

'As we were both speeding level at around 50mph through the night streets. Ockey spring from our car to the Vauxhall. Holding on with his left hand he grasped one of the crew around the neck but the Vauxhall increased speed again and two of the crew hit him on the head with a heavy jemmy. As I wrestled with the wheel to keep my car level, we prayed that Ockey would be able to jump back to safety, but another member of the crew battered his knuckles with an iron bar so that he fell off. By God's grace I managed to avoid him as he collapsed into the road. That split second I have relived too often in my memory.' [sic]

Amazingly Frost managed to brake and swerve around the stricken Ockey but lost ground to the gang. Their savage treatment of his friend and colleague spurred him on, though, and he drew closer again. As they slowed down to negotiate the tight right-hand turn into Ebury Street, Frost, who was surely enraged, accelerated and rammed the big Vauxhall amidships. It turned over and landed against a wall with a sickening thud. The driver, Charles Lilley, was arrested almost immediately; the other two, Alfred John Head and Alfred Hayes, made a fight of it using a gemmy and iron bar, and tried to escape through a block of tenements but were subdued and dragged into Gerald Row Police Station by the Flying Squad team.

Ockey was taken to hospital and was off work for two months. Remarkably the doctor who treated him, Dr A. M. Barlow, testified that his extraordinarily course and thick head of hair had saved his life! His

This centre section showcases some of my favourite police cars and pictures.

We start with an MGA and Ford Zephyr Mk2 Estate, the two cars that started motorway policing in the UK. I was lucky enough to examine them courtesy of a PCUK club meeting at Gaydon Museum when researching this book and loved them. The MGA is the one you see as a course car at the Goodwood Revival, a genuine ex-Lancashire service car. The Ford originally policed the Bedfordshire area and has been restored by Ford UK Heritage.

MGs had a glorious history with the police force and the MGA was, arguably, the peak of that. However, Ford's Zephyr Mk2 and, especially, Mk3 became legends, ranking as one of the all-time great police vehicles in a variety of roles. Ford printed a brochure for the police specification version, which is beautiful. One of the joys of doing this book has been discovering the lovely period brochures I never even knew existed.

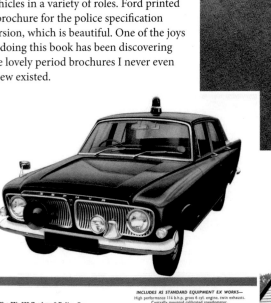

The Mk III Zephyr 6 Police Car

PRODUCED SPECIFICALLY FOR POLICE DUTIES

INCLUDES AS STANDARD EQUIPMENT EX WORKS—
High performance 114 b.h.p. gross 6 cyl. engine, twin exhausts.
Centrally mounted calibrated speedometer.
Alternator. Nylon high speed tyres.
Heavy duty suspension and wheels.
Reinforced front seats.
Fitted wiring looms for police signs and equipment
(loud-hailer, gong and roof beacon not supplied).
Plus many other features
Turn to back cover for detailed specification.

Continuing the Ford theme, the company used this wonderfully bucolic image to promote its first car ordered for the new Panda Car concept, the Anglia 105E.

Sad news.

POLICE

POLICE

We must say we think our policemen
are wonderful. (Having just sold them over
200 Escorts.)
However, what's good news for us is
bad news for you.
An Escort Panda works up 53 bhp. And
looms out of the blue at around 80 mph.
Enough we think, to score the daylights
out of the over 70's.
Of course, we don't pretend it's as hot
on a chase as our Z cars.
But there are times when an Escort can
be an awfully difficult beast to shake off.

What little it loses on the straight,
the positive rack-and-pinion steering
and road holding gains on the bends.
Or, in traffic, the nifty gear change
may even give it an edge on two or three
litre jobs.
(Crossflow head engines with 5 bearing
crankshafts are very handy for that kind
of rough play.)

And at around 35 miles to every gallon
it goes a lot further as well.
So it'll still be there to run you in when
you run out.
After a short test drive we asked
P.C. driver Loveless to comment.

"I was impressed with the size and
space for the occasional passengers."
We also noted that there was room
enough for a quick heel and toe in his
size twelves.
When all's said, it does rather ruin
any theories about the Panda being the
Mr. Plod of the police force.
Which is fair warning.
Unless of course, you getaway people
fancy a police escort.

ESCORT *Ford*

The Escort replaced the Anglia and continued to be popular as a Panda in Mk1 and
subsequent Mk2 form. It is tempting to remember the RS Escorts that dominated rallying
and racing and saw police use as high-speed tools, but Ford UK, as their legendary
Chairman Terence Beckett famously said, did not exist to make cars but 'made cars as
a means of making money'. That meant having a ruthlessly targeted motorsport and
advertising programme aimed at selling cars, and this advert beautifully shows just how
a discounted police order can be used to sell to the public.

The archetypal panda car is,
however, the Morris Minor.
Despite being launched in
1948 (and thus a very old
design by the time panda cars
came along), it was loved by
the force and public alike, and
the police continued buying
it in large numbers until
production ceased in 1971.
Most were in Bermuda Blue
and White, but there were
variations, including those
driven by The Lothians and
Peebles Constabulary, whose
Minors were in the darker
Trafalgar Blue and White
specified by their then Chief
Constable, Willie Merilees,
who wanted them to stand out
from the lighter blue of the
otherwise similar Edinburgh
City cars.

For a car so small, the Mini casts a big shadow over both the motor industry and police cars. It was used for anything from picking up lost pigs in rural areas as a small station van or panda, to high-speed pursuit work in Cooper S form. The van version is often overlooked, which is why I've chosen to feature it. In some ways it's the best Mini; still small yet big enough to be truly versatile. This example was bought by the North Wales Constabulary in 1978 and was one of a batch of 30. Its call sign was WA320 and it has been beautifully restored by an enthusiast.

Surprisingly the Austin A40, which is much more practical than a Morris Minor and shared its engine and gearbox, didn't catch on as a panda car outside of the West Midlands. It earns its place here because it is arguably the first hatchback police car, and I really want to know why this picture was taken in Birmingham by Austin's publicity department. A horse, some tower blocks and an A40? It's like the beginning of a bad joke.

One of the surprises of researching this was how important the Crossley Tender was. They were the first police vehicles used for policing duties rather than moving high-ranking officers around, but in my mind they will always be the Lesney Model of yesteryear I had as a kid! If I'd not had that model I may not even have known what one was, so I include it here, as if the RFC/RAF had not ordered so many there wouldn't have been any to issue to the Metropolitan Police after the War was over. This is a picture of my man Lakey's model; he's kept his in a display case under his anorak, of course …

The first fleet of 'foreign' police cars – like this Volvo Amazon Estate – caused quite a political rumpus at the time, but they proved their worth and started a relationship between the UK police and Volvo which survives to this day.

Early radio experiments being conducted in 1949 or early 1950 on a rare Vauxhall Velox LIP – the 2.3-litre OHV straight-six-engined version of the Wyvern LIX.

Petrol head coppers – and believe me, there are many – remember Ford's performance cars in police spec, and the king is still the Escort Cosworth, in my book. Amazingly they came with an optional helicopter as this image from the Northumbria Police shows!

I love a Lotus Cortina, it's worth restoring one for the intake rasp alone, so couldn't resist using this lovely Ford PR shot from 1969 showing two officers map-reading next to their Ford Cortina Lotus (as it was called by then).

While both high-speed Fords have police car hero status, the Granada Mk1 and 2 were much loved by traffic divisions all over the UK. The Mk1 seen here in 1973 is wearing Consul GT badges, as many police cars did, but everyone still affectionately called them 'Grannies'.

As the Granada got larger and Ford's concept car proved too expensive, the Sierra Sapphire Cosworth provided a new favourite Ford for traffic officers. Fast, capable and just the right size, it was universally loved and, considering its motorsport origins and high power from a 2-litre engine, surprisingly reliable.

The Austin A60 Cambridge and its badge-engineered brethren served in a huge variety of roles over a 15-year service life. This picture, taken in Deptford in the mid-sixties, proves accidents happen to police drivers. You can see the double-line Cross-ply tyre skid marks. For the truck fans that's an Albion Claymore with a Thames Trader in the background.

The A60 may also have had an influence greater than you could possibly imagine. When BMC stylist Syd Goble designed the different badge-engineered 'marque' version of the same basic bodyshell he added a stripe for the Austin A60 version. The Snowberry White car came as standard with a stripe in Embassy Maroon. As far as anyone can work out, this is the first (or one of the first batch) of UK 'Jam Sandwich' liveried police cars. Did it come about initially because they came from Longbridge like that? We may never know, but it's the nearest this book will get to a conspiracy theory.

A Daimler SP250 'Dart' patrols in London. Their mere presence changed the nature of speeding and particularly motor cycle 'ton-up' thrill-seekers around West London in the early 1960s.

Curiously the police didn't use Imprezas until 1998, when two Prodrive modified Vehicle Crime Unit cars entered service with the Humberside Force for reasons similar to the deployment of the Darts. At this time Humberside had the worst vehicle crime figures in the UK, but nothing on the road could outrun an Impreza, as the local criminals soon found out at their cost. An awesome driver's car I've always loved and one of the cars I'd most like to have used in my time on the force.

The Ford Transit has been the backbone of the police service for many years in a huge amount of guises, not least of which is the station van. This picture shows the Suffolk Constabulary Orwell Division Headquarters in 1970 with their V4 petrol Mk1 example being loaded with equipment. Sadly the old rural police headquarters are now as rare as Mk1 Transits.

This is a Ford Transit Mk2 petrol Special Branch anti-terrorist van, which has been restored. When de-fleeted from the police it was bought and used by a film company and appeared in numerous films, including one about the Iranian Embassy siege.

The famous mobile control unit – but the Transit was also used as a public facing vehicle such as this LWB Mk2 on duty during the opening of the Orwell bridge in Ipswich in 1982.

My favourite police car from my time in the force was my Volvo T5, and I was pleased to be reunited with this preserved example. They were great cop cars: fast, capacious, super-reliable and well backed up by a manufacturer who really cared about gaining police fleet sales and was prepared to go the extra mile to keep us coppers happy. Brilliant!

The Rover SD1 is one of my all-time favourite police cars because it was the car I aspired to drive. This is the actual camera car from the famous 'Liver Run', the high-speed medical transplant drive we recreated on *For the Love of Cars*. It still has the camera that famous footage was shot on.

This particular 4.5-litre 'High Chassis' Invicta Tourer was the first Invicta used by the Metropolitan Police Flying Squad and as such really marks the beginning of high performance police cars. It was a super-car of its day, only one step removed from being a sports racing car for the road, and I loved it. The police made a number of small modifications to it and one survived its restoration, a Tapley Brake Gauge, which measures deceleration and allows the driver to make corrections to the wheel accordingly, although whether anyone would actually look at this gauge during an emergency stop is something you have to wonder about.

OK, I know it's not a real police car, it's not even quite a real MkIII GXL, but my mate Phil would never forgive me if I didn't feature it here, and let's be honest, it's cool, isn't it? The police did use Mk3 Cortinas by the bucket load, but very few drove UR Audi quattros, which is why the Cortina got the nod.

In fact, the MkIII, IV and V were used in a huge variety of roles, in both marked and unmarked guises. This picture of a MkIV leading a band during the 1984 Ipswich Town Carnival proves that the job has many aspects and some very pleasant days.

Of all the cars ex-traffic coppers talk about, the Vauxhall Senator B is one of the most fondly remembered, especially in later 24-valve form. Popular at the time when forces were starting to procure jointly, they continued to police our motorways long after they had been replaced because the production line was actually kept open to build an extra 200 UK police Senators, some of which were held back for over a year before entering service. One of the all-time great motorway cars.

The Omega more or less carried on where the Senator left off, and both are legendary in the police world. This Grampian Police example watches over the snow-covered landscape, but the Omega was also used by the Diplomatic Protection Group (DPG) in London, who always use red cars so they can be easily distinguished.

I love MGs and could have nominated any number, but although its police career was limited, this restored BGT V8 is one of my favourite preserved former police cars. MG faded out of police use because saloon cars got faster and the amount of required equipment increased, meaning sports cars were less practical.

However, the name did make a low-key comeback in the early 2000s with the Rover 75-based MG ZTT. This is in use with Warwickshire Police on the M6.

As MG's star waned, a new name was coming into the UK police car market that would eventually grow to become a major player, BMW. The famous 'It takes one to catch one' advert from 1973 is a landmark in car advertising, and it certainly told the world that BMW were after the police market.

It takes one to catch one.

On two wheels or four there's only one thing that compares with a BMW – another BMW. You get a formidable combination of power, roadholding and instant response.

The BMW 3·0Si has a 3 litre, six cylinder fuel injected engine producing 195bhp. Maximum speed is in excess of 130mph. 0-60 is in 7·6 secs.

The BMW 75/7 Motorcycle has an air cooled, 4-stroke flat twin 745cc engine producing 50bhp. Maximum speed is 109mph. 0-60 is in 6·4 secs.

Drive a BMW. After all, if you can't beat them, join them.

For the joy of motoring.

By the time this E28 generation 528i went into service with the Hampshire Constabulary, under call sign CM-07 in 1987, BMW were starting to make serious inroads. This car was based at Cosham Police Station (SETA, south eastern traffic area) covering the M27, M275 and A3M. Police livery was also moving on, and this boot-box is shaped as a rear spoiler to try and minimise its disruption.

Today BMW products are used a lot, especially the 3 series as an area car and the X5 as a motorway car, although the 5 Series is also popular and is seen here being used to pull over a suspect. BMW have a huge share of the market and supply their vehicles with the complex electronics used in modern policing.

I wanted to include this picture in colour, because although the Chrysler 2-litre never really made its mark as a police car, this demonstration car marks the end of a great tradition of Rootes group Humbers, which formed the backbone of police and government service. This Humber Super Snipe is an example that was easily the biggest and most capacious estate car of its era. It formed the basis for many equipment-carrying vehicles before the Range Rover took on this kind of role. It also offered possibly the most luxuriant walnut dash ever seen in police service.

CAPRI RANGE

1, 2, 3 & 4. Capri 2.8i

I've avoided including too many fantasy PR exercise cars in this section – it's a book about real police cars after all – but we all agree that the recent Ford Mustang is a fantastically good-looking car, don't we? So seeing that Ford have made up a police demonstrator and loaned it to various forces, it counts. I doubt any will be ordered, although it is a bargain in bang-per-buck terms, but it does look cool as …

The Capri was developed as a European Mustang, using a similar principle: developed as a stylish coupe economically using the mechanical package of a top-selling saloon. Perhaps surprisingly, it had a long career with the police in the UK, not just with Bodie and Doyle. From its launch in 1969 it was used in surprisingly large numbers and continued to serve until a couple of years after its demise in 1986. Ford even made up brochures for police specification Capris. This 1985 2.8i was with Greater Manchester Police and has been preserved by an enthusiast.

The Riley RME is the oldest known car equipped with a blue light and the oldest surviving radio car, so we had to see it in colour. Such beautifully elegant cars, I've always loved them.

I could have nominated any number of Jags, the Mk2 especially being a car that democratised speed for both the police and criminals, but really when it boils down to it those 3.8-litre Met Police S-types were just so cool, weren't they? And this is the coolest picture: Westminster Bridge, a Met Police Traffic Car (the black ones were Area cars) circa 1969 complete with Mickey Mouse roof lamps. It was posed by the Met's photographic department but is still wonderfully evocative.

The Wolseley 6/110: THE iconic police car. Tough, reliable, fast (it shared an engine with the Austin Healey 3000) and capacious: the sight of a black 6/110 gladdens or freezes the hearts of a certain generation, depending on which side of the law you were on!

Not many police cars make it into movie posters, but if there were to be any it would be the 6/110. You can read about the film in Chapter 15, but the poster is just wonderfully bonkers, isn't it? A sultry (later Dame) Judi Dench lying on a pile of money while a 6/110 speeds towards her giving her a haircut... love it.

I was lucky enough to drive this Range Rover and loved every minute. The Range Rover was the right car for the job at the right time, and no car has ever transformed police operations the way it did.

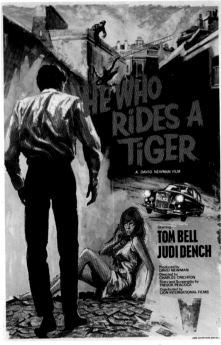

A Range Rover closing the M6 at Hilton Services in the late 1970s.

A Cambridge Constabulary example from 1975 lit evocatively by its own stem light.

Modern police cars are so complex that special factories now build homogenised vehicles that are identical for every force in the UK, excepting their local insignia. This is the production line of (and I quote from the press release) 'arresting Vauxhalls' in part of the vast Luton plant which converts over 2,500 vehicles a year for Emergency Service use and is the largest police car factory in Europe.

One of the reasons why shows such as the Brooklands Emergency Services day are so popular is because of the variety of cars and liveries that individual force procurement produced, as this group of historic police cars owned by members of the PCUK Club, owners of former police cars, shows.

amazing act of bravery, in which he had certainly risked his life, led to him being presented with a cheque for £15 from the Reward Fund by the Metropolitan Magistrate. He was quite rightly given the King's Police Medal for Bravery in the New Year's Honours list. All the officers involved received Commissioner's Commendations and smaller cash rewards.

The felons were all found guilty and sentenced to between three and five years respectively. Most importantly, the British Police's first truly high-speed pursuit vehicle, this lovely Invicta, had very forcibly announced to the world that the police would chase and capture criminals. The arms race of performance between the criminals and the police began that night.

Major Vitty, the Met's chief engineer, was delighted with the Invicta, and more were ordered. This set a trend, and the Flying Squad were soon using other high-performance cars such as Lagondas (which used the same 4.5-litre Meadows engine) and Bentleys. This led to another nickname among the pre-war criminal fraternity (prior to one made famous on television), the 'Heavy Mob'.

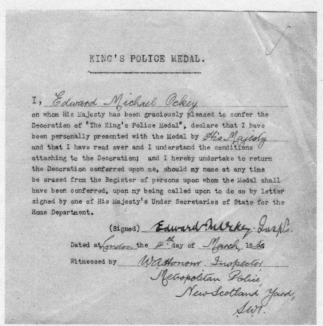

The letter, signed by Ockey, accepting his King's Police Medal.

Some of the very earliest performance cars used in large numbers by the police were MGs, or as they were in the days of correct grammatical contraction (it stands for Morris Garages, as I'm sure you know), M.G.s. The complete history on police use of these cars is detailed in the brilliant book by Andrea Green, *M.G.s On Patrol*, which not only looks at each and every model ever used by the police but also lists the registration number of each car and to which force it was allocated. Pre-war cars included the various Midget models and even the VA Tourers. The Lancashire County Constabulary in particular loved their MG cars, with the likes of the two forces in Sussex, Kent and Warwickshire following closely behind.

Post-war, as most forces started to form their Traffic Divisions, the obvious choice for many were MG products because, bangs per buck (budget was an issue even then), nothing came close to an MG in performance terms for the money that they cost. The TC was one of the first cars announced after World War II (and it's important to note that MG played a noble part in the war, being involved in, among other things, the development of the tanks known as 'Hobart's Funnies'), costing only £480 in 1945 when few cars were ready to go into production anyway and those that did were generally pricey. Police officers were sent by train, armed with a set of number plates, to the factory in Abingdon to collect their new cars. The most favoured at this time were the MG TC and later the TF. For their day these cars were pretty quick, with top speeds of around 80mph with stopping provided by efficient, well-designed drum brakes. MG were keen to help supply ready-built cars to the police and could supply and fit radio equipment, aerials, illuminated police signs, PA systems, fire extinguishers and other kit as required, so that the cars could be utilised almost immediately. Although these cars weren't being used to combat a local issue, as described previously, they were required to combat the menace of the motor-car-driving public whose skills behind the wheel in the pre-1930s were merely learned as they went along, with often fatal consequences. MG did not give the cars away but they were competitively priced because they understood the publicity value of having the police seen in their cars, and the Abingdon-based team, backed as they were by the Nuffield group, were perhaps more public spirited than some manufacturers in feeling it was their duty to supply the police. They were proud to provide forces with their cars and had a reputation for hands-on modifications when requested and also for sorting warranty issues generously for police customers. This went a long way and definitely helped to keep police forces loyal to their products.

By the 1950s our roads were becoming a lot more congested, and MG, after some Austin Healey-induced BMC corporate delays, introduced the beautiful MGA 1500, a two-seater sports car that took the marque to a whole new level. With a top speed close to 100mph, their performance was as good as their appearance. Once again, the Lancashire County Constabulary bought these and the later Mk2 MGA, with its bigger 1600 motor, by the bucket load. In all they had more than 50 of them, some in black whilst others came in white – which was a first for any force. Lancashire pioneered many policing initiatives, including the use of female officers driving cars. This was unheard of anywhere else at this time and the force received a lot of publicity about it, not all of it positive. Contrary to popular belief, the black MGAs were not used solely by WPCs and the white MGAs by male officers. In truth, the black cars were used as beat cars and driven by those officers who had qualified at standard B level of driving, whilst the white cars were used by the Traffic Division and driven by officers who had passed the advanced A level driving qualification. Thus it was even rarer to see a female officer driving a white MGA, but only because very few of them volunteered for such duties.

The white MGAs performed a role in 1958 that was another first; they policed a new type of road. The Preston Bypass, which later formed part of the M6, saw the MGA being used alongside the Mk2 Ford Zephyr Farnham estate car to police this four-lane carriageway. The cars needed to be highly visible to the public to help deter those whose driving standards exceeded their ability, and they needed to be further equipped with a range of kit over and above those black beat cars. So the traffic MGAs had a blue spot lamp fitted to the front offside wing, a red spot lamp fitted to the rear offside wing, twin PA speakers, an illuminated police sign to the front, an illuminated police stop box to the rear, two-tone air horns, a VHF radio set crammed into the boot and an assortment of first aid kits, blankets and a couple of police accident signs squeezed in there as well. For their day these cars were without doubt some of the very best-equipped patrol cars in the country.

Restored MGA 1600 Lancashire Police Traffic car

592 KTB is a 1959 MGA 1600 from the Lancashire County Constabulary. As one of the white MGAs, it could only be driven by a Class 1 Advanced Driver, whilst black cars were only driven by officers with a Class 3 certificate. This car was restored to full Police specification in the 1990s, using the correct original parts from the Lancashire Constabulary stores. It's regularly used as the Clerk of the Course car at the Goodwood Revival historic race weekend each year, as well as at many other events around the UK. It was a real pleasure to have some time to examine the car.

The MGA was later replaced by the MGB and Lancashire's love affair with the marque continued unabated. They bought dozens of them, again some in black but most of them in white, and they basically performed the same duties as before. When MG introduced the MGB GT in 1966, other forces started coming on board, in particular the Sussex Police, who had more than 30 of them. With a slightly larger boot than the MGB, a modicum of extra kit could be stowed to make the cars a little more practical, especially with their hatchback-like tailgate. The cars were ideally suited to female officers, although male officers also got to drive them, because they were by the standards of the time quite easy and lightweight to drive. Other performance cars of the period prior to powered steering and brakes needed serious driver-brawn to get the best out of them.

When the 3-litre MGC was introduced it was the Metropolitan Police who suddenly became interested and bought more than 30 of them for traffic duties. Again, all the publicity photos taken at the time show the cars being driven by female officers. In part this was a deliberate ploy by the Met to encourage women to join the force, with the notion that if you did you'd get to drive one of these fancy new sports cars. I suspect the reality was somewhat different! Meanwhile, Lancashire opted to use the MGC GT model instead, as it came with a hard top and practical boot storage. The white paint had now given way to fluorescent orange, making the cars highly visible and, it has to be said, rather attractive. However, the car's performance must have been hugely reduced with the fitment of oversized roof boxes that had all the aerodynamic qualities of a semi-detached house. In the early 1970s MG popped a V8 into the MGB GT shell and sent the car out on trial with a number of forces, with only Thames Valley Police buying a couple as unmarked Traffic cars. MG was hugely successful at a time when the police needed a good-quality, fast sports car to undertake a variety of policing needs that standard saloon cars just weren't capable of, and more exotic sports cars were too expensive.

In 1959 the Metropolitan Police were being hammered by a new craze that was sweeping the capital in particular: Café Racing. Young men on their high-powered Triumph, Norton and home-made Triton (a Triumph engine in a Norton frame – thought to be THE business at the time) motorcycles would gather at the local café to show off their bikes and their new girls before putting a record on the jukebox and then racing along a predetermined route, usually down to a nearby roundabout and back again, before the record finished. Needless to say, the road-accident rate went through the roof and the police just couldn't catch the bikes whilst driving their Wolseley 6/110s and similar Austin Westminsters. The answer came with the introduction of the Daimler SP250, or Dart as it was originally called when launched at the New York Motor Show in April 1959. Sadly, at that very show the American Chrysler Corporation complained that they owned the Dart name and asked the BSA group (who, ironically enough, owned both Daimler and the Triumph and BSA motorcycle brands that the SP250s were being purchased to chase!) to rechristen it. Daimler, needing to do this in double-quick time, plumped for the project code, SP250 for Sports, and the 2.5-litre V8 engine. I was fortunate enough to film a scene at the Ace Café in my Channel 4 show *For the Love of Cars* where I met both a retired officer and former boy racer for an interview. Remarkably, they recognised each other from back in the day! We ended the scene with an amazing drive in an original black Daimler Dart, still with all its police gubbins. What a fantastic day that was.

The Dart was styled, uniquely one has to say, by Jack Wickes, and could never be described as conventionally attractive, but it was cool, both then and now. Its fairly lightweight, fibreglass body covered a chassis which was copied almost directly from the Triumph TR3A (something which Daimler, amazingly, got away with, despite Triumph making some complaining noises – BSA did not own Triumph cars), but the jewel in the car was its 2.5-litre V8, which gave the car real performance. The engine had been designed by Edward Turner, an engineer who was well known to the bike world, having designed the Ariel Square Four (and if you are not familiar with that but fancy some engineering-based head scrambling, have a read about it, it's fascinating) and Triumph's famous parallel-twin line which began in 1937 with the 5T Speed Twin and lasted, in various forms, until Triumph motorcycles finally folded in 1983. The Daimler V8 engines were gems, and were actually made in two very different forms: a smaller 2.5 litre and larger 4.5 litre, which was used in the larger Majestic Limousines – both were similar in design and appearance. The 90 deg V8s were pushrod designs with cast chrome-iron blocks and aluminium head/crankcase using part-hemispherical combustion chambers, and they certainly owed much to Turner's study of both Cadillac and Chrysler V8s. Unlike the

American units, though, they loved to rev, just like Turner's bike engines, and remarkably the smaller unit was red-lined at 6000rpm but had a good spread of even torque and was fabulously smooth as well.

Of course, the SP250 line was truncated by Jaguar's purchase of Daimler from the BSA Group in 1960; it's often said that William Lyons was nervous of the restyled SP250 that Daimler was known to be working on as it might steal E-Type sales. Lyons respected the V8s, so he made a successful Daimler-badged version of the Mk2 using the 2.5-litre version and at least one MkX prototype with a big-block 4.5-litre version, which apparently 'flew' according to period Jaguar test drivers but sadly never made production. However, the main driver of Jaguar's acquisition of Daimler was Daimler's underuse of its large factory sites and Lyon's desperate need for expansion into those larger factories.

The Daimler badge and those V8 engines were a bonus. The 140bhp@5800rpm gave the SP250 a top official speed of 121mph and a 0–60 of 10.2 seconds, although legend has it those figures were conservative; it was certainly more than capable of keeping up and catching the errant 'Ton Up Boys'.

One of the most popular venues for this café racing was the Ace Café on London's North Circular road, and although the place fell into disrepair in the 1980s and 1990s it has since been completely restored (ironically by a retired police motorcycle Traffic officer!) and is now a thriving venue for bikers and car enthusiasts alike, with the added bonus that it serves the best breakfasts in London.

In all, the Met employed some 26 Daimler SP250s from 1960 and they were loved by the crews who used them, so much so that, even though the car went out of production in 1964 after only 2,650 had been made, the police continued to use them until 1968. Amazingly, around a dozen have survived, which is a remarkable testament to their quality and the enthusiasm that exists for the car. Surrey Police and the Cambridgeshire Constabulary also used the Dart for similar issues that they were facing in their force areas.

In 1966 the Met replaced the Daimler with the Mk1A Sunbeam Tiger V8, another high-performance car of its day with a 4.3-litre Ford V8 fitted, giving it a top speed of 120mph and a 0–60 of around 8.6 seconds. The basic car was a slightly remodelled Sunbeam Alpine that had also seen service with the Leeds City Police. The thumping big V8 motor gave the car plenty of torque, but the lack of power steering, as with the earlier Daimler, made driving around London's urban streets somewhat heavy on the arms. By the time the Tiger was introduced, the Café Racer culture was dying out, but a more affluent public were now buying cars like the E-Type Jag and even the occasional Ferrari, and they weren't afraid to use them. The Rootes group car was updated in 1968 when a new 4.7-litre V8 replaced the 4.3-litre unit and the car was re-branded as the Mk2. However, it was never sold in the UK, with all 623 cars made being shipped to the USA, even though they actually built 633 cars, with only 10 being right-hand drive and six of those built especially for the Metropolitan Police. Whilst three of the original Mk1A cars have survived, only one of the six Mk2s appears to have done so.

The Austin Healey 3000 isn't a car that springs to mind when you are thinking about police cars, is it? For reasons best known only to the old Devon Constabulary, they decided that the big sports car was the ideal motor to deal with high-speed motorists and yet more of those 'Ton Up Boys' on the A38 and A30 roads in the county.

As Devon Constabulary officer Aubrey Slade now recollects, it was an interesting car with which to police the roads.

Aubrey Slade

I joined the Devon Constabulary in 1949 and after six years on the beat I joined the Traffic Division at Torquay, where my first patrol car was a Ford V8 Pilot with its side-valve engine and three-speed gearbox. Fuel consumption was just 14 mpg. The Fords were subsequently changed for Austin A90 Westminsters, which proved to be more economical and faster, although far more skittish on the rear end than the Pilots. In fact, one of the crews put two concrete blocks in the boot of their A90 to improve rear-end traction. Over the next few years I got to drive the Austin A70 Hereford and the Austin A95 Westminster on patrol.

In 1961 we began to hear rumours that two Austin Healey 3000 sports cars were being purchased for the Traffic Division. After much speculation Superintendent Rundle of HQ Traffic Department announced that John Gear would be getting one of the cars to work the A30 Honiton area and I would get the other one to police the A38 Cullompton Road. So John and I, armed with a set of number plates each, set off by train to the factory at Abingdon to collect them. The plates were 117 HTT and 118 HTT and we proudly drove the cars all the way back to Exeter where they were given the once-over by the Chief Mechanic Albert Smith.

I found the big Healey to be strong and well built, it was a joy to drive and the performance was excellent. The handling was predict-able, although being without power steering made it quite a handful compared to today's standards. The exhaust system, however, was extremely vulnerable as there was only four inches of ground clearance and over rough roads the twin rear tailpipes and silencer had a tendency to scrape the road surface and were frequently damaged, necessitating their replacement.

The cars weren't marked up in any way whatsoever, and one night we were tasked with looking in on a local Motor Club Night Rally. The start point was the car park at a local pub, and, as I drove in, the Rally Marshall slapped a couple of adhesive rally numbers onto the doors! When I told him who we were he looked rather sheepish and apologised, saying he thought we were competing in the rally. Most of the rally people were really friendly and very interested in the big Healey, with one of them saying, 'Fancy being paid to drive a car like that.' He had a point.

As stated, the cars weren't marked in any way and the only kit they carried was a Winkworth bell behind the grille, which was pretty useless. I've seen even motorcyclists looking down at the engine on their bikes thinking it was rattling when we came up behind them and put the bell on.

I suppose the Healey 3000 was not really a very suitable car for police work, but having said that, I found it to be the finest and most enjoyable car that I have ever driven, even though many of today's more mundane cars will out-perform it in speed and handling.

Both 117 HTT and 118 HTT have survived to this day, although no one currently knows who the owners are or where they live.

The Devon Constabulary later had the services of a couple more sports cars in the shape of the Triumph Vitesse complete with illuminated police signage, two-tone horns and a blue light on the roof. The cars were finished in turquoise blue.

Triumph sports cars were almost as popular as MGs; in fact, depending on whom you talk to, some would say even more so. I've owned both marques, but as an MG Midget was my first ever car I have a soft spot for MGs. Although they weren't sold in quite as large a number, several forces opted to use Triumph's sporting range, especially if they already had previous experience using the firm's bigger saloons and estates. One of the more unusual choices was the Bournemouth and Dorset Constabulary's decision to use the Triumph GT6, which quickly gained the nickname of the 'County Camel'. Dorset in 1967 was very rural. It had no motorway (in fact, it still hasn't!), and only one small section of dual carriageway on its border with Hampshire. All its other roads were A and B roads cutting through vast swathes of countryside. But that's not to say it didn't share some of the same traffic problems experienced by the rest of the country; it still had its fair share of accidents, speeding vehicles and drink drivers. The new drink-drive laws had been introduced the previous year together with the first generation of breath test devices, consisting of the old green box and glass tubes method. At that time, Dorset's Traffic fleet consisted mainly of Mk1 Triumph 2000 saloons followed by the later Mk2 model, so the force already had a strong link with the brand. The forces' Traffic Division consisted of 80 officers, only four of whom were female.

Now there does appear to be some conflicting evidence as to how and why Bournemouth and Dorset ended up buying a small number of Triumph GT6 patrol cars. One theory is that Triumph persuaded the force as some kind of promotional gimmick, with the proviso that they were driven exclusively by the four WPCs. However, it seems unlikely that a motor manufacturer would be in a position to dictate how a force might use its cars, so I suspect that this theory could be slightly inaccurate, although it's possible the idea came from the force and was embraced by Triumph more enthusiastically than originally planned! Theory number two centres around the fact that the force needed a car that was both fast and agile enough to cope with the county's road system; a sort of first response unit, until the slower, fully equipped Triumph saloons arrived. Theory number three has a ring of truth about it and coincides with other national trends, in that the force was having difficulty in recruiting female officers into the job, and what better way to advertise the glamour side of things than using the cars as an advertising tool, i.e. 'join the Police, drive a Triumph GT6', which is possibly where the first theory may have become a little blurred.

The car itself is obviously quite small inside, and with room at a premium it also made physical sense to have the cars driven by WPCs only. It would be difficult to imagine that two six-foot-plus men, sitting together in such close proximity, would be anything other than very uncomfortable, and they would have complained bitterly, as Traffic men are inclined to do at the best of times!

So where did the expression 'County Camel' originate, and why? More mystery surrounds this expression, but perhaps the most obvious conclusion is that it came from the Met (what a surprise!) because they have always referred to officers working outside of London as County Mounties, so calling a hump-shaped police car from the sticks a camel does make some kind of perverse sense! Or did it have something to do with the WPCs who drove them? You decide.

At least three GT6s were employed by the force with an Mk, which a reputation for tricky on the limit handling, model being stationed at Dorchester and a later Mk2 car based in Bournemouth. There was at least one Mk4 car from 1973, which looks like it was actually crewed by two male officers! Maybe some reverse equality was at play here, or maybe they just loved the sound of that lovely creamy-smooth 2-litre straight-six motor. With its 0–60 time of around 10 seconds and top speed of 110mph, it's difficult to understand quite why no other forces appear to have used the GT6; it was a very competent sports car, for a camel . . .

The GT6 wasn't the only small Triumph to take a seat at the police dining table. The Stoke-on-Trent City Police had a couple of Triumph TR3s in black with red leather interiors. The cars were fitted with a hard top during the winter months, and because the car was so low down it meant that the blue light had to be placed on top of a 10-inch pole in order for it to conform to the lighting regulations of the period! It did make them look rather odd. A few years later the Cheshire Constabulary had a couple of Triumph Spitfires, and again it would appear that these were used mainly by the WPCs of the force. Legend has it that the Avon and Somerset Constabulary had a couple of Triumph TR7s finished in green, although one source is adamant that in fact there was only one car and that it was one of the few TR8s made on test for a couple of weeks. This does make sense in a way. Triumph got very close to launching the TR8 as a volume car, and only pulled the plug at the last minute, so their attempt to sell one into the police has a ring of truth. What history it might have made! We all know how good the Rover V8 was in police hands, and if the V8 Lynx (a sort of Scimitar GTE-style hatchback coupe based on the TR7, which is on show at Gaydon Museum) had been produced (and it was close), it would surely have been a popular Traffic car.

The Salford City Police and their neighbours in the Manchester City Police had the services of the Triumph TR4 and TR4A for general

Traffic patrol. It would appear that these cars weren't purchased to combat any particular issue but rather that they were quick enough to deal with just about anything that moved. Introduced in 1961, the TR4 had a 2138cc in-line 4-cylinder motor (shared with stablemate the Standard Vanguard and, in a very different form, the Ferguson TE20 tractor) that pumped out 100bhp, giving the car a top speed of 109mph and 0–60 in just over 10 seconds. The slightly revised TR4A model that arrived in 1965 had 4bhp more but similar performance. However, it featured a wholly new chassis with independent rear suspension. The cars were very well liked by the officers who crewed them, and during bad weather they were even allowed to put the optional hard top on the car. The Salford City cars looked particularly good with a roof box and blue light on top and with the force crest emblazoned on the doors. The only other force to use the TR4A was the Southend-on-Sea Constabulary, who had a bit of a problem with bikers racing up and down Southend seafront and, just like the Met's use of the Daimler Dart, the Triumph helped elevate the problem. Without doubt Triumph sports cars contributed a huge amount to policing the UK's roads and may also be responsible for recruiting more than one or two women into the job at a time when this was less common.

By the mid-1970s brand-new, open-top two-seater sports cars were fast becoming a thing of the past, as saloon cars were capable of notching up similar or better performance figures. With Triumph that meant just one car; the Dolomite Sprint, which came out in 1973. Here was a four-door, four-seat saloon car with a 1998cc, 16-valve engine (the world's first mass-produced 4-valve-per-cylinder engine) capable of 119mph with a 0–60 time of just 8.4 seconds. It wasn't called the Sprint for nothing, and the Nottinghamshire Constabulary ran a fleet of eight 'Dollys' in their Traffic Department, whilst Sussex Police had an unmarked red Dolomite Sprint in its fleet.

Triumph actually prepared a Sprint demonstrator car that went to a number of forces as an alternative to their larger 6-cylinder cars, which were eventually going to be phased out. West Midlands, and it's believed at least one or two other forces, placed orders for these, but the cars were never delivered because of strike action and the orders were eventually cancelled as the forces needed cars delivered. Perhaps this was another missed opportunity, and although Nottingham loved its Sprints they did suffer from the headgasket woes they were famous for, so perhaps the strikers actually did the police a favour. Great car, though, and one I've always had a soft spot for.

One of the most successful cars ever built, of course, was the Mini – we have covered its more obvious use as a panda car elsewhere, but it was also a rather popular sports saloon when graced with a Cooper badge and mechanicals. Two forces in particular loved the Mini Cooper: the Liverpool and Bootle Constabulary and the Manchester City Police. Other forces like Durham, Birmingham City and the Met Police had a few on their books, but the two north-west forces had the biggest fleets, and a good number of the Liverpool cars have survived to this day. One of the Mini's primary roles early on was as a motorway car designed to catch those who were breaking the recently introduced 70mph speed limit or, during the 1970s oil crisis, the temporary 50mph speed limit that was imposed on the motorway. Top speed was between 85 and 91mph, which doesn't sound fast by today's standards but for its day it was seen as a quick motor. Mind you, travelling at 90 in a Mini must have felt like double that, given the noise intrusion you get into the cabin. As an urban Traffic car, though, it really came into its own. Theft of cars in the early 1970s was starting to become a very common crime (you won't find it referred to as 'joy riding' here. It was theft, pure and simple; it gave no joy to anyone, but rather heartbreak, distress and grief to many). Car thieves, disqualified drivers and burglars will do their utmost to get away from police when they are behind the wheel. They have no morals and never obey the rules. They won't think twice about going through red lights, the wrong way down a one-way street, travelling in the opposite direction around a roundabout or driving on the pavement if it gives them half a chance of escape. Police, of course, do have to obey the rules, and in an effort to tip the scales cars like the Mini Cooper came into their own. Fast and incredibly agile, the Mini could drive down alleyways where other cars wouldn't fit, was quite capable of driving across wet grass like a rally car, and in the hands of a Class 1 Advanced police driver was just about impossible to lose. For many a car thief it was game over. Okay, so they weren't exactly comfortable to work in for eight hours a day, and putting a non-compliant prisoner in the back was nigh on impossible, but the Mini Cooper definitely deserves its place in police car history as an iconic motor that many of those tasked with driving them in anger now look back on with great fondness. Mostly, especially if they weren't too tall!

As the Liverpool and Bootle Police morphed into the Merseyside Police so they replaced their little rally-bred Mini Coopers with a second generation of sports saloons, also with roots in rally, that became instant legends; the Mk1 Ford Escort Mexico. The Mexico was launched in 1970 and was so-named following the car's success in that year's London to Mexico World Cup Rally (Ford were brilliant at cashing in on such success). Ford's motorsport boss Stuart Turner had recced the route thoroughly and took the advice of one of his star drivers, Roger Clark, that the way to do this event was with two-man crews in lightweight nimble Escorts, not three-man crews in larger cars as other manufacturers were doing. Clark also stated that reliability was far more important than performance and that the Escorts they entered should use Ford's simple but ultra-reliable 'Kent' OHV engine from the Cortina 1600E rather than the complex twin-cam Lotus engine or the new, and equally complex, Cosworth BDA. Clark was also mindful that he and co-driver Ove Anderson and had lost victory in the 1968 London to Sydney Marathon after their Mk2 Cortina Lotus had dropped a valve, probably because of low-octane petrol, after they had led for 8,500 of the event's 10,000 miles. It's an aside, but I've always loved the fact that, immediately after that event, Ford manager Walter Hayes sent Clark and Anderson a telegram that read: 'Please don't commit suicide, because we need you. You could not have driven any better. Many thanks for a magnificent effort.'

Sadly, Clark was no luckier on the World Cup Rally, being crashed into by a local driver and retiring, but teammates Hannu Mikkola and

Gunnar Palm won the event using what is now probably the world's most famous Ford Escort, FEV 1H. The Mexico road car was based on the rally car's concept; the heavy duty rally-bred Type 49 Escort body-shell with RS1600 suspension and brakes but with the simple crossflow engine rather than the RS1600s BDA. It was built in the same 'Advanced Vehicle Operations' (AVO) factory as the more expensive RS Escorts. It only had 86bhp, but the nimble lightweight (1965 pounds) Mexico had a small frontal area that gave it a top speed of over 100mph and a 0–60 time of 10.7 seconds, and the handling was sublime. As Ford had a plant at Halewood it made perfect sense both politically and in policing terms to use the car as a weapon against the car thieves and other mobile criminals. Mind you, the full-width illuminated roof box the cars were forced to carry must have knocked at least 5mph off the top end, if not more. But the cars were hugely successful and revered by those lucky enough to use them on patrol.

The Mexico gave way to the Mk2 Ford Escort RS2000, and of course Merseyside Police snapped them up to undertake the same role as their predecessor. The rear-wheel-drive 2.0-litre Pinto engine had better performance figures, and with a top speed of 112mph and a 0–60 time of 8.4 seconds, this really was a quick car in its day. Between the Mexico and the RS2000, these cars built a reputation for Ford that has not been bettered. The early cars were fitted with that same roof box from the Mexico, but common sense prevailed in the end and they were replaced with just a single blue light. A number of unmarked Q Cars were also employed, including a bright orange one. As a police officer, driving an RS2000 must have been a real joy. To be able to use the car's full potential whilst drifting the rear end (they were very tail-happy) through the streets of Liverpool with two-tone horns blaring away under the bonnet must have meant cops with smiles as wide as the Mersey itself.

Whilst the Merseyside Police were happily bombing about in their RS2000s, so their colleagues some 50 miles up the M62 in Manchester were quite content with their Ford Capris, but they too were being blighted by the car thieves and so they formed the TASS unit (Traffic Area Support Services), part of whose role was to pursue and arrest the aforementioned villains.

Ford's biggest rivals to the RS2000 during the 1970s were the more expensive Cosworth-fettled Vauxhall HS (and later HSR) Chevette, and the Talbot Sunbeam, which initially came from Chrysler Europe after they had purchased the Rootes Group. The police service in some areas had used the standard Talbot Sunbeam as a panda car, again because it was cheap, reliable and robust; of course the old Rootes Group still had many enthusiasts among the police and this probably also helped. Forces like Sussex, Avon and Somerset, Cumbria, the Met and Greater Manchester Police used lots of them for that role.

The Talbot Sunbeam Lotus project was conceived and directed by Des O'Dell, who was head of motor sport for Chrysler Europe at this time. He saw potential in the small RWD hatch, which used a cut-down version of the Avenger's platform at a time when most cars of this genre were becoming front-wheel drive. It was his aim to replace the Hillman Avenger Tiger with a car that could beat the RS Fords and the Vauxhall HS Chevettes in international rallying. O'Dell was confident that the Sunbeam had the chassis to do so, but he didn't have the engine to compete. This is where an old connection was to prove very useful. Wynne Mitchell, O'Dell's assistant, had been at college in Coventry with Mike Kimberley, now Managing Director at Lotus Cars in Norfolk. Lotus had been supplying their own 2-litre DOHC engine in quantity to the Jensen Healey sports car, but as Jensen Motors had ceased trading in 1976, Lotus were actively attempting to sell their engine elsewhere. Lotus readily agreed to supply a basic 2-litre Type 907 engine (as used in the Elite and Eclat) plus a rally-tuned engine for competition use. The road-going version subsequently turned into a 2.2-litre unit designated the Type 911 of the following specification – a 4-cylinder DOHC unit of 2172cc with twin 45mm Dellorto carburettors, a 16-valve, all-alloy cylinder head, producing 155bhp and giving it a top speed of 125mph, with a 0–60 time of just 6.4 seconds! These were impressive figures and would stack up even today, making the car the fastest production small saloon of its time. The suspension was tuned by Lotus, of course, with stiffer rear springs and dampers. The exhaust system was also designed by Lotus. Production cars then got fancy sports interiors, special alloy wheels and 'in ya face' paint jobs, most notably the all-black model with wide silver band along the sides (although there were other colours), together with a splash of Lotus badging. Hey presto! An instant hit and a genuine rival to the RS2000. The team won the World Rally Championship for Makes in 1981, and Henri Toivonen and Paul White won the 1980 RAC Rally, so it worked in rallying as well, although the Audi quattro made all cars of this type obsolete soon after.

Rolling chassis were built alongside all other Sunbeams at Linwood, near Glasgow, before being transported south to Lotus in Norfolk. Here they were fitted with the Lotus engine and then mated to a 5-speed ZF gearbox. The work also included modifications to the body shell, fitting a larger radiator and alloy wheels. Space was at a premium so a satellite operation was mounted at Ludham Airfield, some 20 miles away. A dedicated team of 16 employees was drafted in by Lotus and work began in 1978 to build the first production cars. Once assembly was complete, the cars went on another journey to Chrysler, in Coventry, for final checking before being delivered to the dealerships. This method of production continued at a considerable rate until the summer of 1981 when production ceased.

In 1980 the TASS unit at Greater Manchester Police needed a number of fast, good-handling Q cars to help quell the upsurge in car crime where ordinary Traffic cars were having difficulty keeping up with stolen vehicle pursuits. When the unit was first formed they struggled to find cars for them to use and actually had to make do with a number of old Ford Escort estates once used by the Radar Team! So they ordered six Talbot Sunbeam Lotus cars and had them modified still further – not in the performance stakes but in an attempt to make them as nondescript as possible. The cars were therefore stripped of all their go-faster stripes so beloved of the standard models and delivered in six standard Sunbeam colours: red, white, green, brown, honeysuckle yellow and plain black. The standard alloy wheels were retained to keep the modified brakes band, and the important bit under the bonnet wasn't touched either. The cars were then fitted with a Britax magnetic blue light, two-tone horns, a police/stop sign on the rear parcel shelf and a VHF radio.

The official registration numbers were MNF 941W to MNF 946W and each car was given three different registration numbers in total, together with corresponding tax discs. All the cars were moved around the five different Traffic areas every three months to prevent the local toerags from becoming too familiar with them. The five Traffic areas were: No.1, Rochdale, Oldham and Ashton-under-Lyne; No.2, Wigan, Bolton and Bury; No.3, Manchester city centre, North Manchester and Salford; No.4, Motorway Unit; No.5, South Manchester, Stretford and Stockport. The black car was retained by the police driving school for driver training. Towards the end of their careers the Sunbeams were swapped over with Merseyside's Escort RS2000s just to cause even more confusion amongst the criminal underclass!

Overall the Lotus experiment worked very well, although there were quite a lot of reliability issues to contend with – well, it was a Lotus after all! But as a work tool designed to undertake a specific task, they were excellent. They were eventually replaced by the Ford Escort XR3i range and in the late 1980s by the Cosworth Sierra (remember the TV programme *X-Cars*? This model was GMP based and a direct descendant of the TASS unit). Two of the six original TASS cars have survived; the green car and the red one are owned and have been restored by the same enthusiast.

There is an amusing anecdote to the forming of the TASS unit and their acquisition of the Talbot Sunbeam Lotus, in that having formed this covert unit they then invited the local press along, who splashed a nice photo of the cars across their front page! Whoops.

In the early 1980s Ford revealed the Escort XR3 and after about 18 months gave it fuel injection, thus creating the XR3i, another gem from the Ford stable that a number of forces used as a fast sports-orientated Traffic car. Although no specific task seems to have been recorded, it would be right to assume that these cars were again utilised to combat the car thieves and other criminals. The CVH 1.6-litre fuel-injected motor drove the front wheels and the car was capable of 118mph with a 0–60 time of 8.6 seconds. It's somewhat ironic that the police should choose this car to combat the car thieves, given that one of their favourite targets was also the XR3i. Forces like Merseyside, Durham, Dorset and Cambridgeshire Police all used the car as a marked Traffic car.

Another one of Ford's faster cars of the same period was the Sierra XR4i, but only one force appears to have used them as marked cars and that was Devon and Cornwall Police, and they only get a mention here because you may recall that they were used in the opening credits to a brand-new TV series being aired at the time: *Crimewatch UK*. Such is its importance to police vehicle enthusiasts that Corgi even made a model of the 'Crimewatch Sierra'.

In the early 1990s the Metropolitan Police had identified another issue that was causing concern. They now policed a large number of motorways entering or circumnavigating the capital – the M25, M1, M3, M4, M11 and the M40 – and with this came motorists who simply refused to abide by the 70mph speed limit whilst using high-powered cars of their own. We aren't talking about drivers doing 10 or 20mph over the limit but those hell-bent on three-figure speeds whilst racing other like-minded motorists, or those whose reckless driving was putting lives at serious risk.

Before we look at the detail it has to be said that rumour and legend within the police service are pretty close stablemates; coppers are brilliant at starting rumours and even better at clinging onto the legends, whether they are true or not! And so it is with the subject matter here. In 1994 the Met decided that it needed a number of semi-marked performance cars to use as speed enforcement units on the aforementioned roads. A brave selection of vehicles not normally associated with police work was chosen, including the Vauxhall Calibra four-wheel-drive turbo (yet another rally-bred vehicle, although its rallying programme never materialised as intended), the Rover 220 turbo coupe (or Tomcat, as it's known), a Vauxhall Astra GTE, the BMW M3 coupe (finished in yellow because BMW refused to paint them white, or is that yet another legend?) and a Porsche 968 Club Sport (CS).

The rumour mill at the time went into overdrive and it was made clear that there were two BMWs and no fewer than six Porsches and that the specially selected officers were sent to Germany on special driving courses, courtesy of Porsche, to ensure that they were trained to get the very best out of the cars. There was very little factual information available, which is unusual given the pedigree of the vehicles we are referring to, although there was a brief article in the *Daily Express* on 23 May 1994 announcing the Porsche's arrival. The Met are also quite good at taking official photos of important vehicles, but none appears to exist of the Porsche, although there is one of the BMW beneath Marble Arch, which was always a favourite spot for the Met's photographic branch to take such photos. As research material is sparse to say the least, do the rumours and legends stand up? Before we reveal the answer, let's talk about the cars themselves.

The 968 was introduced in 1992 and the lightweight 968 CS was its fastest and most hardcore incarnation, which came out a year later. It had a 2990cc, 4-cylinder, 16-valve DOHC engine that pumped out 240bhp at 5500rpm. It was the last of the family of front-engine water-cooled 4-cylinder transaxle Porsches that had commenced with the 924 (which had initially been designed to be a VW) in 1975. A top speed of 153mph and a 0–60 sprint time of just 6.1 seconds ensured it could keep up with most traffic, although in police hands it apparently returned a mere 15 miles to the gallon. There were three types of 968, including the base model, the 968 Turbo S and the 968 Club Sport. In total just 2776 were built worldwide, so they are quite a rare beast. The CS model differed from its brethren because it had its interior ripped out to make it much lighter and therefore a much sportier car. There was no leather trim, no air-con, no stereo, electric sunroof, windows, seats or mirrors, and no central locking either. Even the sound-deadening material was left out to help save weight, so road noise was greatly increased. It was a true sports car. It is said that once you have driven a 968 CS you won't want to drive anything else, you'll love it that much. The downside for the officers who crewed the cars was that the ride was harsh and skittish, which, given the amount of time they spend on patrol, must have proved rather tiring.

Back to the rumour and legend then. The first of these that we can quell is that there weren't six Porsches but one, which is rather disappointing really. No doubt this rumour came about because the car was passed from one Traffic garage to the next, which may have given the impression that there was more than one. The Met didn't purchase the car either; in fact, all five performance cars were loaned by their respective manufacturers. The second legend to fall by the wayside is rather more amusing, in that the selected drivers didn't get an all-expenses joy ride out to Germany to thrash the new toys around some race track for a week. In reality the 'special training' consisted of a supervisory officer from the police driving school at Hendon testing a number of Advanced Drivers to ensure that they had kept their skills up to date. If they had, they got to keep the keys to the Porsche. It proved to be somewhat divisive as only a handful of officers passed the selection process, although the total number of applicants is not known.

The Porsche and the BMW M3 were the favoured cars to drive (well they would be, wouldn't they?) but they did have their drawbacks. When officers were dealing with the public, the drivers spent more time asking questions and looking at the police cars than they did worrying about the penalty points that were coming their way! There were a lot of envious glances from members of the public and police officers from local stations and neighbouring forces. There were also a lot of adverse comments aimed at the crews, such as, 'It's not right, it's the tax payer footing the bill' and 'why don't you lot go out and catch proper criminals', blah, blah, blah – and that was just from other coppers! In the *Daily Express* article it was stated that 'motoring organisations had criticised the idea and said it could encourage more high-speed chases in which young car thieves have died'. And an AA spokesman is quoted as saying 'Such high-performance cars can be a dangerous temptation for joy riders wishing to match their skills against the police. The officers will have to assess each situation to see whether there's a danger to the public. We are obviously concerned.'

These oh so typically limp-wristed responses were rebuffed by Inspector David Jones from New Scotland Yard's Traffic Division, who stated that 'The Porsche is not a toy, although that is what the cynics will say. It is designed to give us a sporting chance against those who ignore our requests to stop after they've done something wrong, whether it is reckless driving or ram-raiding.'

The Porsche was finished in white, complete with colour-coded white alloy wheels, and it was decorated with standard Met Police striping along the sides, as were all five of these semi-marked cars. There were no police signs front or back and no roof-mounted blue lights, so that the unsuspecting bandit drivers didn't see too much in the way of a police car in their rear-view mirrors. The only blue lights fitted were repeater lights set into the fog lamp recess under the front bumper. The *Express* article showed the car sporting a small Maxim light bar on the roof, but this was there purely for the photographer.

So were the cars a success? Yes, apparently, but the experiment was never repeated. After 18 months or so the cars were all decommissioned and returned to their respective manufacturers. It is known that the Porsche, registration number L323 JMO, has survived and is currently in the hands of a Porsche collector.

The scourge of the mid-1990s was ram-raiding. Well-organised criminal gangs were stealing high-performance cars, sometimes two or three at a time, and then reversing one of the cars at high speed straight through the window of a shop premises where they would leap out of the car, grab whatever goods they had targeted (usually high-value clothing or cigarettes), chuck it into the back of the same car or one of the others waiting outside and then drive off at speed. The whole thing was usually over in less than a minute. By the time the police arrived in response to the alarm going off, the thieves were long gone. If by chance a patrol car picked up the trail, the thieves would simply outrun them. It was the age-old problem that standard patrol cars just weren't capable of keeping up with some of the thieves' chosen set of stolen wheels. So a number of forces who were being hit particularly hard by the ram-raiders took the Ford Escort RS Cosworth onto specially formed units to combat this new fashion in criminality.

The Escort Cosworth looked like it was based on the terrible 1990 MkV Escort, a car that was a disaster for Ford. At launch it was an awful car, one of Ford's worst ever, and the Blue Oval were heavily criticised by launch road testers for the car's lumpy ride, terrible steering, rough engines, cramped and poorly packed interior space, bad seats and it just being, well, a bit er, not very good. Even the styling was at best okay rather than good, but this was important as at the time the Escort MkIV was the UK's and Ford Europe's best-selling car so it *really* mattered in monetary terms. The MkV was facelifted twice; first rather hurriedly in 1992, then more comprehensively in 1995, and it was generally accepted that the 1995 changes made it an okay car, although in 1998 the Focus came out and the Escort name died.

However, this has very little to do with the Escort Cosworth – at least in mechanical terms. Ford knew they had a problem with the new

MkV Escort and needed an image builder even *before* they launched it. That helped get the money for the Escort Cosworth project, but the fact is that this vehicle is unique in motoring history as a car that was designed *from scratch* to be a winning rally car that was then turned into a civilised road car and made in volume; it's usually done the other way around. The project started because Ford Motorsport boss Stuart Turner (him again; I bet he never thought he'd figure so much in a book about police cars) knew the Sierra Sapphire four-wheel drive was basically a good international rally car with a great engine but was just too big. At a planning meeting in 1998 he said, 'Why don't we see if we can take the platform and running gear from a Sierra four-wheel drive, shorten it, then see if an Escort body will fit on it.'

That single line from Turner was mocked by engineers who said it could not be done, but then went away and did it anyway! The seed was sown and 'Project Ace', as it was called, started at that moment. It was developed by a small team headed by John Wheeler as chief engineer at Boreham, Ford's works rally HQ, and was launched in 1992 to rave reviews. It had an enormously long front-line rallying career (winning 10 WRC events between 1993 and 1997) and sold in much bigger numbers than a normal homologation special, with 7,145 being built, before production at coachbuilder's Karrman ceased in early 1996, really because escalating insurance costs had made this fantastic car almost impossible to sell.

As it was basically a Sierra four-wheel drive with an Escort body, the car used the same superb Cosworth-designed turbo-charged YB 2.0-litre engine, mated to a five-speed manual gearbox and permanent four-wheel drive. The cars were given a huge whaletail rear spoiler (which designers originally wanted to be even bigger) and a deep front spoiler that between them produced huge amounts of downforce for a road car, which, combined with good suspension geometry, ensured that the cars stuck to the road like glue and cornered like a house fly. With 227bhp on tap, the car's top speed was 137mph, so not quite the fastest thing to hit the road at this time, but its 0–60 time was a very impressive 5.8 seconds. In the hands of a Class 1 Advanced Police driver the odds were stacked in favour of the Old Bill once more – and the bad guys knew it.

Cosworth

The whole Cosworth-badged fast-Ford era started because Keith Duckworth (his partner was called Mike Costin) saw potential in developing the Ford 2-litre 'Pinto' OHC engine into a twin-cam race engine. Duckworth was no fool! He produced the prototype Pinto-based twin-cam and left it casually in a prominent place on the floor when Ford people came to visit, knowing they would spot it! If he'd proposed it, the company politics and bean counters would have slowed it down, but by leaving it around as a curio they thought productionising it was their idea, meaning that it flew past all that! The Ford Cosworth YB, as it was known, proved very strong and tunable in testing, and as Ford was losing in Touring car racing with the now very old V6 Capri, they pressed the green light button to try to get back on top.

Duckworth had successfully produced a race engine from Ford blocks before, as the Cosworth-designed and built BDA used in Mk1 and Mk2 Escorts (and in very vague terms the DFV F1 engine) had their origins in the Kent crossflow engine used in the Cortina GT and Escort Mexico.

Thus the original Sierra Cosworth was launched in 1985 as a homologation special mainly aimed at saloon racing, which it dominated globally, producing 700plus bhp on occasion . . . However, the car also did well on the rally stage and became legendary as a police Traffic car in all its forms.

Adding Cosworth to the car's name was logical because Cosworth and Ford had grown up together and the Ford funded DFV, which had linked the names in people's minds and had won 155 F1 races – still a record for one type of engine, and it will always remain so, I suspect. It's also important to realise that Cosworth built high-performance engines for all sorts of firms, though GM, Mercedes, some others and the police used them in many arenas; for instance, the Scorpio-Cosworth 24V V6, which is featured in the Traffic Car section of this book.

I'd highly recommend Graham Robson's book on Cosworth; it's an amazing story well told.

One of the getaway tricks the criminals used was to deliberately drive down streets paved with 'sleeping policemen' (no pun intended!) at very high speed, knowing that the pursuing patrol car would invariably have to slow down to prevent permanent damage to the underside of their vehicle, something the criminal couldn't care less about, given that the car he was trashing didn't belong to him anyway. So police spec Cossies were fitted with metal skid plates underneath to protect the car's vital organs and once again give the coppers a slight edge. Forces like Humberside, Cumbria, West Yorkshire and Northumbria all took on the ram-raiders with Cosworth Escorts and achieved a very high success rate.

Humberside Police had previously pioneered the use of high-powered sports saloons when they took the police vehicle market by storm with the introduction of the very first Subaru Imprezas ever seen in police trim. The two cars had been specially prepared by rally car specialists; Pro-Drive had uprated the suspension, added the aforementioned skid plates, stiffened the chassis and included head sets with built-in voice-activated microphones attached to the head rests which left the driver free to keep both hands on the wheel when driving at high speed. It was a purpose-built tool and a very successful one at that. Over the next few years a good number of other forces had the Subaru Impreza in all its variants, including West Yorkshire, North Yorkshire, South Yorkshire, Essex, Hampshire, Cleveland, Nottinghamshire and the Greater Manchester Police, who all used the cars for the same reason; to combat the criminals using high-powered stolen vehicles.

Subaru's biggest rival in both the World Rally Championship and the marketplace for super-fast sports saloons has always been the Mitsubishi Lancer Evolution series of cars (commonly referred to as the EVO), but in police spec it wasn't until the EVO IX and latterly the EVO X became available that a few forces like South Yorkshire, Humberside and in particular Essex Police really took the Japanese supercar onto its fleets. The 2-litre turbo-charged engine officially produced 280bhp, giving it a top speed of 142mph. However, we all know it was considerably more than that because the Japanese manufacturers had at this time imposed a voluntary limit of 280bhp, which they adhered to in literature only . . . The cars were made famous when they featured in the TV series *Car Wars* that was centred on Essex Police's fight against criminals using fast cars to commit crimes.

Meanwhile Humberside Police formed its Road Crime Unit and armed its Mitsubishi EVO Xs with ANPR, and also gave it a bigger brother in the shape of the awesome Lexus IS-F, a 5.0-litre V8 that produced an incredible 417bhp, giving it a top speed of 170mph and a 0–60 time of just 4.6 seconds, which should be enough to scare the pants off just about every criminal in the north-east. Lexus, of course, is the luxury arm of Toyota and allegedly received its moniker when top-of-the-range Toyotas were crated up for shipment to the USA and marked as Luxury Edition Xport United States. Most of their cars these days use hybrid drivetrains, but they have made the occasional fire-breathing supercar, and none more so than the IS-F. The big saloon looks rather ordinary really, although the quad exhaust pipes give the game away somewhat, and when that big V8 roars into life you just know it must be capable of extraordinary performance. Humberside Police have two IS-F cars in its Road Crime Unit, which, in combination with the EVO X, have been hugely successful in combating the criminal fraternity.

In twenty-first-century Britain most modern cars are faster and better performing than just about every car we have discussed here, so apart from the likes of the Impreza, Evos and IS-Fs there are very few sports cars left that can really outperform most other half-decent cars. Ford, though, has a bit of a tradition in coming up with decent performance cars, and in the Focus range they have produced a number of Focus STs that are the direct descendants of the Mexico and RS range. The 2007 model Focus ST saw three forces use them: Dorset, Surrey and the West Midlands Police, all of which had them as ANPR-equipped urban pursuit cars. The ST was fitted with a 2.5-litre, 5-cylinder motor borrowed from the Volvo T5 during Ford's brief tenure of Volvo. The car had 225bhp on tap and a top speed of 150mph. With a decent handling package and chassis tweaks, it was a huge success and it's a real surprise that only three forces used it. With only a few styling modifications on the exterior, like the front and rear spoilers and fancy alloy wheels, the Focus ST was a real wolf in sheep's clothing.

CHAPTER ELEVEN

IT'S GOT COP SHOCKS

The opening few minutes of *The Blues Brothers*, surely up there as a favourite film of everybody reading this book, confirms just how ingrained the idea that police cars are heavily modified is in most people's psyche, right across the world. Elwood's speech from the driver's seat of his 1974 ex-Mount Prospect, Illinois, Police Patrol Dodge Monaco, delivered languidly after they have casually jumped the rising Chicago 95th Street river bridge, can be quoted by petrol heads the world over:

> *'It's got a cop motor, a 440 cubic inch plant, it's got cop tires, cop suspension, cop shocks. It's a model made before catalytic converters so it'll run good on regular gas. What'll you say, is it the new blues mobile or what?'*

That car goes on to be the film's central heroic figure, sacrificing itself just after it has delivered the protagonists, against all odds, intact and on time. In reality, police car history is littered with myths and legends about the police tinkering with the engines of their patrol cars to make them go faster. How many police officers have been at an incident when a little kid tugs at the sleeve of his fluorescent jacket and says, 'Ere, mister, is your police car all souped up?' Or worse, the adult, wearing an old corduroy jacket complete with elbow patches and a dodgy-looking side parting, will raise a knowing eyebrow and then proclaim that his best mate's uncle's brother's dog has got a friend called Jack, 'Do you know him?', who works down at the police work-shops and 'he was telling us all about what you lot do to your cars to make 'em go faster'. You know the type.

It would be unfair to say the public is obsessed by how fast a police car can travel, but people are mightily curious about it and the stories that come attached to certain cars bear almost legendary status. It's fair to say the police sometimes added fuel to this fire, because having the crooks believing they had nothing that could keep up with a police car sometimes made the job a little easier! TV and film car chases of E-Types being (quite incongruously) caught by large Wolseleys containing four burly actors also added a little to this myth.

Some people have conjured up an image of overall-wearing policemen, sleeves rolled up, covered in axle grease, bent over the engine bay of a patrol car, tweaking all the go-faster bits, adding a new dump-valve to the turbo-charger and then taking the car out and giving it a damned good thrashing! Where have all these fables come from? Do the police really need to add performance-enhancing drugs to the engines of their cars, or is there another reason why the public think that there is something fundamentally different between their car and a supposedly identical car used by the police?

Cranhill Arts / glasgowfamilyalbum

The myth has a basis in fact, but only just. From the very beginning, police cars were being modified by the manufacturer to police specification before being delivered, and this was often more about reliability and ease of use than outright power. Manufacturers were keen to help, because seeing the police in their car was good advertising and also because in the early twentieth century both smaller family-owned concerns, and the population in general, felt a greater degree of civic duty than is perhaps evident today; the country had pulled together through one world war and was about to do so again.

```
CONTINUATION OF CAR RECORD FOR CHASSIS No:Class -      NO.20130 - ENG:No.15140
Additional warning light fitted to N/S of instrument panel - unwired.
2" longer steering column fitted.
Bracket for siren fitted to front spring bolts.
Tecalemit hydraulic type nipples fitted throughout chassis in place of the
standard type grease nipples.
Certified speedometers fitted to all five cars.
Sheet of copper fitted between the aluminium bulkhead and asbestos.
Engine exhaust system and petrol tank bonded to earth with ¾" metallic
braiding.
Rubber grommets fitted to all wires connecting generator,receiver,and
transmitter units.
Phosphor bronze spring shackle pins fitted.
Latest type clutch fitted.
Special section radio rail fitted - for fitting of receiver and transmitter
units behind the dash.
Transmitter key supplied and fitted to under-side of instrument panel.
Wooden switch block 7"x3½"x½" fitted to inner N/S of the body.
Hole made for aerial lead-in at the side of screen pillar - allowing ¼"
clearance from metal framing.
Receiver control head fitted to inner-side of instrument panel.
Fn type button fitted on dash to operate siren.
Lucas S.W.3 type screen-wipers fitted.
Door hinges fitted with grease nipples.
Scuttle vent made fixed - welded.
Pneumatic squabs fitted all round.
Web plates fitted to centre pillars.
Rear lid lock bolts shortened.
rubber sealing strip fitted to boot lid.
Wings and running-boards treated with Bitumastic paint and Anti-Stone
material.
Pocket fitted to N/S of scuttle.
Shock-absorber covers removable.
Grab handle fitted to N/S scuttle for passenger.
Dash light fitted under scuttle N/S,for passenger reading.
Client's No.Plates.C.U.3.7 fitted.
(Cornercroft De-luxe Rear)(Bluemel Front).

         (IGN.43.
KEYS:-(
         (BOOT:L.2557.
```

Remarkably, the Alvis Car Company has survived and is still based in the Midlands in premises it has occupied since 1967, and because the company has had a continuous history the records are intact. The firm is still making cars in very small numbers as well as restoring and modifying original models, and through to 1966 (when car production stopped so Alvis could concentrate on military vehicles) produced high-quality cars which were fast, innovative and reliable for their day. Alvis' cars were used extensively by the police all over the world, including Australia, and the order documentation opposite for five Alvis Speed 25 cars supplied to the Glasgow Police shows just how much alteration was done before the cars left Coventry for Scotland. The changes range from hydraulic grease nipples for easier servicing to rubber grommets around the wiring to prevent chafing, from more comfortable seat cushions for long shifts in the car to 4-blade fans to ensure they would not overheat no matter how long they were left idling when dealing with an emergency situation.

As we've discussed earlier, by the late 1940s and early 1950s the police had started to use the motor car in fairly large numbers and by then most forces in the UK were just starting to establish a Traffic Division. Many cars at this time were still of pre-war design, and consequently of pre-war reliability. Just keeping the thing going was sometimes a battle in itself. The police needed reliable transport and the servicing and repair of their vehicles was undertaken by their own staff. Most forces had some kind of workshop facility with perhaps one full-time mechanic who would supervise the policemen whilst they undertook the actual service. On average two hours per shift were spent cleaning the car and checking the oil, water, tyres, etc. Then once a week the car would undergo a comprehensive check-up and minor repairs were carried out to mechanical parts, which might take all day. Every 6,000 miles the car would undergo a full decoke or even a complete engine change; a new engine would be put in while the old one was rebuilt on the bench ready for the next car that came in. During all this maintenance the crew and the mechanics would get to know each car's weaknesses. Where the manufacturers either couldn't or wouldn't rectify a persistent fault, it was down to the officers and mechanics themselves to come up with something to keep the car on the road. And more often than not they did. Many of these modifications were made to simply make the car either more reliable or more robust and weren't speed-related at all. Twin batteries, bigger alternators (or dynamos on earlier cars), good-quality ignition coils and better earthing systems all helped with the electrics. Stronger dampers, beefier springs or raised torsion-bar settings all assisted with the car's road holding. These were essential modifications to ensure that it remained a practical and reliable tool.

Inevitably as they tinkered with the mechanicals some of the performance items were changed, in part to increase the car's power, but this was always tempered with reliability in mind. It was no good having the fastest police car on the block if it broke down as soon as it turned the first corner. It would be impossible to list every modification ever made to every police car, as each force would have a different idea about what they wanted or could afford, and those modifications that were made were often dependent on the local service conditions – urban or rural, motorway or housing estate. Some cars during this period did start to come with a few factory-fitted police specifications, but in general most were carried out within the force over a period of time based on previous experience. It has to be said that many police mechanics were better qualified than the people who designed and built the things in the first place!

To give you some idea of the scale of modifications made during the early 1950s, let's take a look at the archetypal patrol car of that generation: the Wolseley 6/80. Although used by a good number of the UK's police, it wasn't without its problems and not every force will have undertaken the same modifications, so for the purposes of this chapter we'll just concentrate on the cars used by the Metropolitan Police that were usually supplied by the Woodcote Motor Company of Epsom. They were supplied with a heavy-duty dynamo and associated modified wiring loom, but the 6/80 supplied to most if not all police forces had virtually all of these modifications done afterwards in the police garage to make the cars more reliable, including changing the pistons and rings for superior specification components made by companies such as Wellworthy, Duraflex or Hepolite. Series 1 cars were adapted to fit improved Series 2 cylinder heads. The oil-bath air cleaner wasn't favoured so the oil-wetted mesh-type of a non-standard component was used instead. Although the standard SU carburettor was used, much work was carried out by SU *and* police engineers to enhance its performance and allow the 2215cc OHC straight-six to breathe more efficiently. The SU petrol pump was non-standard and uprated with a higher flow rate to cope with the increased performance and heavy-duty points so it didn't 'stick' as standard ones were prone to do. Exhaust valves burning out were a constant problem on the 6/80 and police engineers spent a lot of time searching for a solution. The SAS (Special Armoured Seat) valves gave about 18,000 miles' service, and facing the valve seats with Stellite did increase valve life, so it wasn't long before the police engineers decided to Stellite brand-new Trainco SAS valves so that they would then give 30,000 miles of service. Special output coils with heavy-duty ignition leads and in-line interference suppressors of very high quality were fitted. A Lucas high lift cam with Lucas points and rotor arm with a special Runbaken condenser were considered a better option than the standard fitments. The only true performance modification was a rear anti-roll bar and larger-section tyres, both of which improved driveability at speed. These were only a few of the remedies applied to just one type of car, and they were essential because this wasn't tinkering that these people were doing; they were keeping an important item of the tax-payers' money in a healthy condition.

The Metropolitan Police, unlike most other forces, had an engin-eering development department whose sole purpose was to evaluate vehicles for procurement and then work out a list of standard police modifications and specifications; often mixing and matching from the manufacturer's parts catalogue or developing new aftermarket components. This department, in the days when roads were clearer, were exempt from speed limits as long as they displayed a notice in the rear window saying 'Metropolitan Police Vehicle Under Evaluation'. As cars became faster the engineering department's role became more important and they started to assess the vehicles in a different way, looking not for ultimate performance or cornering speed, but for predictable and benign handling characteristics. The reasoning was that lap time on a race circuit can be gained from the performance of the car being driven on a near-ideal line set up for the corner in the perfect way. Police cars being used in an urban environment don't have that luxury; they must make rapid progress yet be able to deal with the public doing the unexpected and hazards appearing from every direction, from dogs to children running after footballs. The team thus prioritised a car that actually has a low breakaway skid point but that can still safely be brought back, rather than stiff cars that don't break away until very near their adhesion limit, but when that finally happens can leave the driver as a mere passenger. So they would deliberately explore the limit of the car on a circuit and, once comfortable with it, deliberately put violent steering or braking inputs into the car to see what happens. A car that has violent lift off over-steer would not be purchased. Interestingly, both the drivers who did this testing for many years had considerable race and rally experience: Peter Yates and Richard French.

Reliability is one thing but often the modifications were made for safety reasons. Another archetypal police car of its generation was the 1968 onwards 3500 version of Rover's successful P6, with its bombproof Buick-derived V8 engine loved by just about every Traffic cop ever tasked with driving them flat out on Britain's growing network of motorways. And therein lay the problem; when Joe Public gets behind the wheel of his shiny new Rover he knows it has a top speed close to 125mph and, if he's feeling brave and only if the wife isn't sitting in the car and if he's absolutely sure there are no police about, he might just put his foot down (only once, mind you!) to see if it really will go over the magic ton. And within 30 seconds he'll be back down to near the legal limit, wearing a smug smile on his face. But police motorway cars and police drivers are often required to drive at three-figure speeds every day of their working lives and it was at this time that they discovered that the Rover had a tendency to scare even the most experienced officer to death.

The Rover P6 V8 was an impressive car; with a top speed of 125mph it was by far the quickest patrol car ever used at that time. As a motorway patrol car it was ideal, except for one thing, on the M3 crews would experience something that was both frightening and very dangerous. As soon as the Rover hit 90mph, it would change lanes all by itself! The officers at Basingstoke complained bitterly to the Transport Manager who in turn passed on their comments to the Rover factory. Their Fleet Sales Manager, Stan Faulkner, stated that he had been selling Rovers to police forces all over the world and that no one had ever complained about poor handling in the past, stating that it wasn't the car, it was Hampshire's drivers who were at fault. Sergeant Nick Carter took exception to this verbal lambasting and challenged Stan Faulkner to come down to Basingstoke to witness the problem for himself.

A few days later he duly arrived, together with a Rover engineer, and Nick Carter took them out onto the M3, where at an area known as Black Dam the car suddenly veered from lane one across to lane two at 90mph. Mr Faulkner was invited to try it for himself, but declined and stated that the tyre pressures must be wrong. They were checked and found to be correct. The front suspension was then inspected to ensure that the correct units had been fitted and these too were found to be in order. The rear suspension was then scrutinised to see if it had been set up correctly and that also passed the inspection. The car was then taken to a weigh bridge where it was found that the front axle weight was half a hundredweight (28k) lighter when the car was loaded than it was when unloaded. Further tests were carried out and it was discovered that when all the emergency kit was placed in the boot, the front of the car was raised by three inches! The three concluded that the combination of high-speed aerodynamic lift and excess rear-biased weight probably meant that at 90mph or more, the front wheels would actually leave the ground, hence the car would have little or no steering!

The men from Rover went back to Solihull scratching their heads and the traffic boys took all the kit out of the boot and dumped it on the back seat. A few weeks later, Stan Faulkner returned to Basingstoke with an orange Rover 3500S V8, which had been fitted with the most hideous steel spoiler, right across the front of the car. It had been welded to the front apron of the car and almost touched the road surface. Loaded with kit, the car was taken out and was found to handle very well indeed, with none of the previous issues being apparent. The car was left with Basingstoke Traffic Section for a month and they were asked to play around with the spoiler to obtain a compromise between aerodynamic efficiency and reasonable looks. Over the next few weeks the spoiler was cut back with a hacksaw until it looked reasonable, but still allowed the car to handle well. The result was that Rover then produced the spoiler in glass fibre and fitted them to all production V8 models, including those police issue cars that had previously been sold without them. So the next time you look at a Rover P6 at a car show or maybe on your neighbour's driveway, take a look under the front bumper, and if it's fitted with a front spoiler you can impart your knowledge to the owner that the police were directly responsible for that aspect of the car's design. Pretty cool, heh?

The Rover P6 became infamous for another reason, too. The tyre manufacturer Dunlop produced a run-flat tyre system called Denovo which was fitted to P6 V8s, among other BL cars, and several police forces trialled the system in 1973, prior to it being offered to the public the following year. The Denovo was a low-profile tyre that could sustain a complete loss of air pressure at high speed with almost no effect on the stability of the car. The tyre was then capable of travelling limited distances at speeds of up to 50mph in deflated condition, and after repair would be fully operational again. Its secret was a special lubricant/sealant carried in canisters attached to the wheel rim inside the tyre, which was released automatically when the tyre deflated. This lubricant reduced friction, kept the tyre temperature down, sealed any small punctures and partially re-inflated the tyre, all in the space of a few seconds. The tyre had to be fitted to specially made two-piece wheels covered with plastic trims of a radial spoke design that made them easy to recognise. That was the theory anyway. In practice the actual system worked very well, almost too well, because the police drivers who experienced high-speed deflations on the motorway were completely unaware that anything had happened and continued to drive the car at speed. Dunlop, of course, recommended speeds no greater than 50mph after a deflation/re-inflation/seal event, which is all very well if you know it's happened. The first that most police drivers knew of any problems was when the tyre disintegrated and pieces of it were seen flying past the window! On one occasion the impact was so great that the front wing of a patrol car was ripped right off!

Dunlop supplied the police with the Denovo tyres free of charge, which was just as well because they worked out at twice the price of a standard tyre, and even if they didn't get punctured, they tended to wear out a lot quicker anyway. But the biggest problem as far as the force was concerned was that every time one of the devices was activated the wheel had to go back to Dunlop and a new tyre had to be fitted in its place by a Dunlop fitter. This meant that some considerable delays were experienced during out-of-hour periods, like night shifts. After testing the tyres for eighteen months, the police reverted to standard road tyres.

However, that story has a second chapter because the West Midlands Police, whose area included Dunlop's HQ, had another similar experience with their SD1 motorway cars, which were supplied in 1980/1981 with supposedly improved 'Denovo 2' wheels and tyres featuring a moulded-in gel wall rather than the aerosol canisters used in the original. Earlier West Mids' SD1s had used Michelin TRX alloy wheels and tyres, which had performed very well, but within months of the Denovo 2 equipped SD1s entering service three had been written off because the tyres were breaking up, the final one of which ended up upside down on the embankment next to the Chelmsley Wood junction of the M6 because a tread had stripped while the car was being driven steadily in traffic below the speed limit. The traffic officers driving the cars, who had not been involved in the purchasing decision, began to object, and within a week the Denovo 2s had been replaced by the Michelin TRXs. The problem was similar to the original design, which was that, while the internal gel wall sealed the leak fairly quickly, it gave the driver no indication from the car's handling (and remember these are well-trained police traffic patrol drivers conscious of these things) that the pressure had dropped. If a puncture was picked up early in a shift, which was quite common on glass-strewn hard shoulders, the pressure would drop during sustained high-speed running and the tyre would break up. The Denovo 2 also had another potentially dangerous and frustrating problem even when it wasn't punctured. During the sustained high-speed running that police cars are sometimes forced to do, the gel-wall layer heated up and became very fluid. When the car was parked up this gel fluid all dripped to the bottom, so when the crew got back in the car after an incident – usually because another radio call had demanded them to be somewhere else – their car appeared to have square wheels and went thump thump thump. Two or three miles running at 50mph was enough to clear it back to 'round', but it was as frustrating as it was unnerving. Safe to say, both incarnations of the Denovo, which had promised so much, went un-mourned by police traffic divisions even though the regular punctures that motorway cars were suffering was one of the product's original inspirations.

This 1995 Jaguar X300 generation XJ6 police demonstration car I'm driving shows the kind of thoughtful modification the police would ask for. It was loaned to various forces for evaluation and eventually used for some years by the Hendon driving school. It was supplied, at Warwickshire Motorway Police's request, with a rear wash wipe system so that when rapid reversing up the hard shoulder was called for, the screen could be cleared. Jaguar's Police spec cars were badged PS, which stood for Police Specials, but many in the industry knew them as 'Smart Cars' because the factory-based officer responsible for government fleet sales to the police, royal family, diplomatic protection and the military was for many years a chap called Pat Smart. It was a standing joke in the industry that he'd got the job because he had the right initials. In fact he worked for Jaguar for over 48 years, giving sterling service to the company in many sales roles, having started as an apprentice in 1953.

In 1970 Rover introduced the 3500S version of the P6 V8, which was basically the same car with a manual gearbox. Although they had experimented with the super-strong ZF 5-speed gearbox used by Aston Martin and other supercar makers at the time, they had eventually gone into production with a modified version of their own 4-speed manual box from the 2000 P6. Although the gears were shot-peened, the layshaft fitted with stronger taper-roller bearings, the case finned for cooling and gearbox oil pump was fitted, it still wasn't really up to the torque of the V8. Even in normal civilian use dealers were fitting magnetic sump plugs in order to pick up the swarf the gearbox generated . . . From a police perspective the problem appeared to be that the unit was splash-fed oil when put into reverse, and if this were done in a hurry, as is often the case with police vehicles, it would lock solid. A firm in Maidstone was eventually found to sort this and they did a roaring trade in fixing and modifying police Rover gearboxes from all over the country!

The P6 V8 had another weird fault in police service, which was ludicrously simple but took a few weeks to diagnose. The car had 'Sunrise-pattern' clip-on hubcaps that were fitted over the entire wheel. When it first entered service it suffered greatly from flat tyres, which were found to be caused by cut, or in a few cases completely severed, tyre valves. Investigation proved that under hard acceleration these wheel trims were 'slipping' and their sharp edge was cutting the valve stem! The solution was straightforward: remove the hubcap and run the black wheel with no trim, which gave the liveried P6s a slightly more sinister aura. Some forces did, though, adopt the 'bolt-through' Rostyle-like trim used on the V8S, which obviously could not spin.

However, there is one car that is legendary in police circles because no cure was or ever could be found: the ADO 28, better known as the Morris Marina. The year 1973 is likely to go down in police-force history as the year it made one of its biggest automotive errors. With Japanese cars and other imports starting to take a hold, the police were plagued by political pressure to buy British, and many forces succumbed.

The choice was limited, so many ordered Morris Marina 1.8 TCs as area cars. The Marina was based loosely on the Morris Minor component set and was a last-ditch attempt by BL to keep the Morris name alive as a 'technically conventional' car, while the Austin name was used for the advanced front-wheel-drive hydragas suspension Issigonis'-inspired range of cars. Stylistically it was developed by a team partly recruited from Ford and led by Mk2 Cortina stylist Roy Haynes and was very much up to date, but the looks and glossy launch film (made by Ridley Scott's company, apparently!) disguised a chassis that was never capable of safe high-speed use. BL built at least two development Rover 3.5-litre V8-engined Marinas for evaluation and motorsport use, which makes me shudder even now, but the police never expressed any desire even to trial one!

However, having it seen to be used as a police car was obviously an important market for British Leyland. Unlike most of the Twin Carb versions sold to the public, which were two-door coupes, the police variants used the four-door saloon body shell, but with the more powerful MGB-derived engine, giving them a top speed of about 100mph. The cars' build quality marked an all-time low; bits would just fall off them, and one of its biggest problems was that it had a tendency to overheat. Police mechanics invented an ingenious device to keep the engine cool when the car sat at idle for any length of time at an incident. To stop the car from boiling over, an additional water bottle attached to a small electric pump was fitted under the bonnet, then when the cooling fan was activated by the thermostat it would blow water all over the outside of the engine block, creating steam everywhere, in an effort to keep it cool! It wasn't the car's inability to stay in one piece or keep cool that gave it its legendary place in police history but its appalling handling and distinct lack of road holding. It would roll and lean over at some incredibly unnerving angles whilst negotiating the average bend or roundabout. Push it hard through a faster bend and the body roll, in conjunction with massive understeer, would ensure that the driver would be fighting the wheel for supremacy. Every now and then the car would win and the driver was looking at the possibility of a re-test! Officers involved in a serious accident have to do this.

Although most forces gradually abandoned the Marina, with few tears of regret being shed, it must be said, the Met had to buy British-manufactured cars so it purchased a fleet of the later Itals, which were basically Marinas with a slightly sloping nose and higher tail; a restyle done in-house by Harris Mann NOT Ital Design . . . BL were happy to perpetuate the myth that this had been done by Giorgetto Giugiaro at Ital, though, and even made a TV advert which hinted at that and was, amazingly, filmed on Ital Design's premises in Italy! Giugiaro, the master stylist who founded Ital Design, was less than keen and the car does not appear *anywhere* in the company's official history, despite it being the only car produced in large numbers that used their name. Ital Design had production engineered the facelift, so did have a major involvement in the project that they couldn't deny (no matter how much they wanted to!). In fact it was originally going to be called Marina-Ital but had its name changed just before the launch on the orders of Michael Edwards, who'd realised the Marina name was sullied. The Italians were apparently less than thrilled by this use of their name, but perhaps because they hoped for further work, they let BL get away with it. The Ital, like late-model Marinas, was fitted with BLs then reasonably new 2.0-litre and 1.7-litre OHC O-Series engines and the Met bought quite a large fleet of them as unmarked A to B general-purpose cars for CID and other tasks, which meant they often travelled with four burly coppers in them. In order to accommodate the taller OHC engine, BL had lowered the engine and gearbox in the frame but had not told the Met; within days of the fleet entering service most of them were off the road with a smashed automatic gearbox bell housing (all Met cars have been auto for many years because of their urban environment) as they would not clear speed humps at all. Engineering's solution to this was to fit the cars with a rally-inspired full-length steel sump guard that prevented the car's gearbox getting damaged but led to unmarked police cars being evacuated when 'beached' on speed bumps! They were left rocking back and forth like the coach at the end of *The Italian Job*, so had to be pushed off by said officers until they got to the next speed hump. When the Met phoned BL engineering liaison seeking a solution they were helpfully told, 'Oh we have a cure for that, Hong Kong Taxi specification rear leaf springs'. These were actually listed in the parts' book as 'Hong Kong Taxi', and when supplied they did indeed raise the car enough to substantially reduce the problem, if the car was only crewed by two officers. However, the problem remained when carrying four people (nothing is recorded about how CID dealt with felons being arrested and carried in the back when the Ital beached . . .), and the full-length sump guard tended to make the Borg Warner automatic gearbox over-heat and thus fail in service. With this issue unresolved the Met gave up on the Ital and sold off the cars much earlier than intended, only to jump from the frying pan into the fire by replacing them with . . . wait for it, automatic 1750cc Allegros. The EU fair competition laws had yet to come into effect, so the Met, unlike other forces, were still forced to buy British so had a limited choice.

Now it might sound like I'm indulging in the obvious tabloid-inspired sport of BL bashing here, but I'm not really. In fact I've owned various BL cars, and I especially adore the SD1 (a vehicle which well deserved its Car of the Year Award in 1977), the original Land Rover (which became a BL car for a while) and the original Mini. However, we are doing a book about police cars and it has to be said the police struggled with both the Marina and Allegro, no matter how fond today's classic car enthusiasts are of them. It's great that examples of both are restored and enjoyed by their owners but they do represent a low point in British car production and design engineering. They emerged from a compli-cated situation, beyond my remit here, which involved lack of development funding, overly militant unions, ill-advised mergers and many other factors. It should be said that there are numerous police cars that are universally loved, though – the Mini, Senator, Granada, 6/110 and Minor all fall into that category – but during the research for this book, which involved talking to many ex and current police officers, there were really only two that were universally disliked: the Marina/Ital and the Allegro.

The Allegro 1100 Auto was used as a panda car to replace the much-loved ADO16 BMC 1100/1300 range, which had in turn replaced Morris Minors in many forces, both of which were regarded as strong, well-built and reliable cars by most police forces, although of course the police never kept them longer than three years so they never had the rust issues that later owners faced. There were therefore high expectations of the Allegro as its bloodline was good. The first force to get them was the Met in 1973, just as the car came out, and engineer Richard French was working in the garage the day the first one came in for evaluation and for the servicing protocols to be established. The police's relationship with BLs new medium-sized car did not start well, as he explains. 'The Allegro was the biggest disaster in UK police car history. The first problem was that there was nowhere safe on the chassis or suspension to put a trolley jack in order to jack it up. You could only use the factory side jack, which is obviously not safe, or quick, in a garage and unusable on a ramp. So what we had to do was make a bar that fitted between the towing eyes on the front and was located by two pins, then we could jack it up on the bar. They then started to come in for their first services, so we used our bar and windscreens started to pop out! Even on the early pandas, which were two-door, you couldn't jack them up with the doors open because the body was so floppy it twisted and popped out the screen. We had to install a protocol to make sure the doors were firmly shut before one was jacked up. We ended up with 25 major faults and 50 minor faults being looked at by BL; the worst aspect of which was body cracks because they cracked spot-weld to spot-weld to the point where we had one come in where the police driver's head was at window height because the whole seat pan had collapsed! This is on a car that's less than a year old, remember. We also had a spate of rear wheels flying off and it turned out to be because an accountant had saved some money by fitting a rear bearing lock washer that was too small. We must have had 600 Allegros in the Met at this point and they were just awful, bits just kept falling off or going wrong. Yet they bought more of them!'

I can hear Allegro enthusiasts preparing insults for me at the next

NEC Classic Motor Show that I host, and I want to defend the Allegro a little because I can see the car's appeal. Its ride and handling were impressive in the day, and the styling, had it been made to Harris Mann's original designs, was up to the minute and almost rakish. The larger OHC engines and 5-speed gearbox were also novel then, with high-tech features that many competitors couldn't boast. However, I can't really defend it further than this, because although you might think Richard's memory exaggerates the awfulness of the Allegro, it is, sadly, the other way round. Research at the wonderful British Motor Museum archive in Gaydon (who hold the surviving records for BMC/BL and many other makes – it's well worth a visit) done by my man Lakey has turned up a document compiled by BL after liaison with many police forces, although led by the Met's engineering division. Its title is **METROPOLITIAN POLICE ALLEGRO COMPLAINTS**, which let's face it is not a good start, and it is 11 pages long! There are 24 major areas of complaint that have been investigated at the time when the report was written, in autumn 1974, and that are explained away in sometimes ludicrous terms, plus another 11 points that had been raised more recently and were yet to be fully investigated . . .

I was quite tempted to reproduce all 11 pages here, but the publisher intervened and made muttering noises about my anorak. He has let me put in some choice phrases and reproduce one page, though:

1. *Jacking:*

Limitations being imposed by the design are being considered a nuisance by the police.

BL Service Division's reply: It is envisaged that the facelift Allegro is being engineered to improve body rigidity.

2. *Water Ingress to Interior:*

A supply of sealers and applicators has been delivered to the Metropolitan police on a free of charge basis, enabling them to carry out their own rectifications.

3. *Front Seats:*

The back release lever mechanism is flimsy in relationship to the 10 SWG Wire and is as such prone to failure (10 failures per week are considered a reasonable average).

BL Service Division's reply: Not accepted as a general problem, the facelift Allegro could include seat modifications.

Note the 'could' – NOT the same as will!

Hampshire-based Sergeant Nick Carter's previous association with Stan Faulkner from Rover had developed into a very good, friendly relationship, and in late 1974 Faulkner turned up at Basingstoke Police Station and asked Nick if he would like to test drive a prototype Rover on that notorious stretch of the M3 known as Black Dam. Nick was driven to a quiet layby on the A30, where he saw a trailer with a rough-looking American pick-up truck on it. The Rover technicians then pushed the truck off the trailer, undid some wing nuts and removed the dummy front end. They then removed the rear canvas cover to reveal a very sleek-looking car, finished in matt black, with no name badges or identifying features on it except for the Jensen Interceptor-type rear window that was fitted. Inside, the car looked like a mobile laboratory, with dozens of gauges and dials, to test every component. This was the prototype of the Rover SD1. Nick thrashed it up and down the M3 to test the car's performance and handling characteristics in the strong side winds, whilst the technicians checked their instruments. The new car passed its test drive, but it would be another two years before the public caught a glimpse of the finished product. It's interesting to note that the police market was so important to Rover they were prepared to risk a sighting in order to get feedback from the Service. The SD1 was so alien to Rover's image, though, that perhaps few spotting it would have believed it emanated from Solihull's development shop.

As modern cars became more reliable, there was less need for such skilled intervention by police mechanics. However, certain tuning companies also stepped into the frame from the late 1960s to the mid-1970s and offered performance packages that the police took a long hard look at. For a start these companies could offer all the necessary mechanical items required straight from the shelf without the need for the police themselves to design, build and test them. The other bonus appeared to be that for relatively small money the police were obtaining the practicality of a standard car but with sports-car performance. There were several companies to choose from, dependent on what cars your force already used. At the same time, of course, cars like the Lotus Cortina and Mini Cooper S became available and a number of forces took them on because they offered power and performance in a saloon body; both hold almost legendary status in police car history today. However, specialist roles meant forces often looked at alternatives and even in the mid-1990s Strathclyde Police had a couple of Rover 800 saloons fitted with BMW 5 series engines during the BMW/Rover Group collaboration.

But for the purposes of this chapter we will be looking at the aftermarket bolt-on goodies that the police used.

Ruddspeed

When the Hampshire and Isle of Wight Constabulary took on the Volvo 121 Amazon estates in 1965 it caused a sensation politically. The force had dared to go foreign, something that was unthinkable back then. The force weathered the political storm but it had other concerns surrounding the cars themselves. The whole reason the force opted to buy the Volvo was because of the truly awful reliability issues surrounding the then current crop of Austin Westminsters or similar Wolseley 6/110s and, given what we have already discussed so far here, it was an obvious move on their part. The cars were designed in-force to become the new Accident and Emergency Tenders, a sort of early Range Rover-esque load carrier. The car was loaded up with a large wooden box in the rear that contained the latest sheet metal accident warning signs, plus the angle-iron (finger breaking!) collapsible stands they stood on. Then there was the very heavy, mobile petrol-driven generator and the two Mitralux floodlights (that weigh about 10 pounds each!) that were powered by the generator, plus the metal tripod stands they sat on, a large reel of electrical cable, first aid kit in another wooden box, 12 cones on heavy rubber bases, huge towing rope with metal shackles, blankets, shovel (metal, not plastic), broom and two of those blue repeater lamps that stand in a metal base, filled with concrete to prevent them from falling over. Oh, and let's not forget the very heavy VHF radio receiver that was bolted to the inside of the inner rear wing. When you look at period photos of the Hampshire Amazon it looks incredibly rear-end heavy, even with the rear seats folded flat to spread the weight. In fact, it's a wonder the cars moved at all! So the car's decent turn of speed was now seriously compromised.

The solution came from the Sussex-based company Ruddspeed Engineering. Ken Rudd was a racing driver and mechanical engineer who saw a niche in the market to produce top-quality mechanical solutions to enhance a car's performance potential. His early work from about 1962 involved race-tuning AC Ace and Austin Healey sports cars, and he even finished tenth in the Le Mans 24-hour race in a car he'd developed and built, so he clearly knew how to prepare an engine.

What isn't known is who approached whom. Did Hampshire approach Ken Rudd or did he come to them with an idea to help out? I rather suspect some knowledgeable Traffic officer or force mechanic who knew of Ruddspeed took a trip down to Worthing, made the necessary enquiries and brought the ideas back to Winchester. Whatever the case, the Amazons (the force had five in total) were soon receiving the Ken Rudd treatment and the difference was extraordinary. The engines had the head skimmed and ported, and a four-branch manifold exhaust was fitted, together with twin SU carbs. On the handling side of things Koni dampers and springs were fitted, together with Cinturato tyres. The Volvo 121 estate's standard performance gave the car a top speed of 98mph. With the full Ruddspeed conversion fitted, this went up to a staggering 115mph. However, by the time all that kit was put back into the car that top speed had been reduced to about 107mph. Still very acceptable for the era and a vast improvement over the somewhat bogged-down standard car, and the force now had the Accident Emergency Tender it so desired.

In 1968 Volvo replaced the Amazon with the all-new 144 range, in both saloon and estate. Hampshire, now deeply in love with the product, bought the new car by the bucket load. In fact, from 1968 to 1974 when the car was replaced by the 244, Hampshire purchased no fewer than 94 of them. But the early cars suffered the same way as the Amazon. The force bought the saloon rather than the estate, which meant that all the kit had to go in the boot, albeit much of it had either now been left out (like the generator) or replaced with lighter products, like the new canvas and vinyl roll-up accident signs to replace the metal ones. Much of the previous equipment was now being carried in a new type of accident tender: the LWB Mk1 Ford Transit van and, by the early 1970s, the Range Rover.

So the early 144 saloons also received Ruddspeed conversions, but these were of a slightly lesser specification than previously. The heads were skimmed and re-ported, plus the exhaust and carbs were replaced but the suspension and tyres were left standard. This obviously brought the car's performance back into line, but as Volvo brought out more powerful versions themselves so the need to upgrade them became unnecessary. The Volvo's role within the force was also changing as it got downgraded to an area car when the Rover 3500 V8 SD1 took on the mantle of Traffic car.

As many of you will know, the Hampshire Constabulary still owns one of the original Volvo 121 Amazon estates, which was rescued many years ago. Whilst doing the research on this car I discovered that a previous owner had removed the original Ruddspeed engine to put it into an Amazon rally car he was preparing. He replaced the police engine with a standard unit; genuine Volvo Ruddspeed engines are now incredibly rare.

Blydenstein

In December 1966, Team founder Bill Blydenstein, already known for his tuning prowess, persuaded Vauxhall dealership Shaw & Kilburn to sponsor a saloon race effort with promising and spectacular driver Gerry Marshall. He knew the still-secret belt-driven OHC 'slant-four' engine was to be dropped into the HB Viva to create the Viva GT, which was actually launched in March 1968. He moved into the former railway station at Shepreth with engineer Gerry Johnstone – the secret ingredient to their success – and from that start Dealer Team Vauxhall (DTV) was created. Marshall drove their immaculately prepared machines with such gusto that his efforts in cars such as Baby Bertha (a V8-powered Firenza droop snoot) made the team probably the biggest spectator draw in saloon-car racing. They achieved tremendous success in saloon racing throughout the 1970s.

Bill Blydenstein's reputation in racing meant his business was tuning road cars; Vauxhalls obviously, and in particular the FD and FE Victor-based VX2300, VX 4/90 or Ventora, which was a big bruiser of a car anyway and a sort of forerunner to the Senator with the same tail-happy rear end! His cars were reasonably successful, although they didn't seem to carry quite the same respect as the Ruddspeed conversions, which may have been a consequence of the era. However, that didn't stop the police becoming interested. Vauxhall produced a couple of special VX 4/90 FE saloons in black (no other VX4/90 or VX2300 was ever issued in black) and it then got the Blydenstein treatment with reworked camshafts, twin Webber carbs and other modifications. It increased the power of the 2.3-litre in-line-4 from 110bhp to 135bhp and from its standard 101mph to 115mph. It wasn't a huge difference but it was enough to make the police interested in taking a good look at it.

In 1974 that man Nick Carter also test drove one of the two Blydenstein Vauxhall VX4/90s to see if it was suitable for police work. It wasn't a successful trial as the handling was described as vague and woolly, with the three-speed automatic gearbox coming in for severe criticism – having the police use auto boxes outside of London was a few years off yet – so as a police car it failed to make the grade. One of these cars has survived today and has been very nicely restored by its current owner.

Broadspeed

Broadspeed Engineering, from Sparkbrook in Birmingham, was famous for race tuning Mini Coopers and of course for building the now very rare Broadspeed GT Coupe. Ralf Broad started out in 1962 as a Mini race enthusiast and was very successful at it. As his company and reputation grew, so other car enthusiasts wanted him to make improvements to their cars too. They started tinkering with Ford Capris, in particular the 3000 V6, and in 1973 added a turbo charger to one. Not any old turbo charger, you understand, but one from a diesel truck! It was fitted to the front of the engine, and although it looked a bit crude, it was very effective, pushing the standard 138bhp up to a massive 230bhp. This gave the car a top speed close to 140mph.

Broadspeed then looked at doing the same thing with a Consul GT, and this is where the police became interested. The 3000 Essex V6 motor was the same as used in the big Capris and so a truck turbo was fitted together with further modifications to the suspension and brakes. A front spoiler or splitter was fitted beneath the front valance, then Humberside Police took delivery of the car and were mightily impressed with it (and you thought their first performance car was a Subaru Impreza!). However, the turbo and the engine weren't made for each other and it wasn't long before reliability issues came to the fore. It was an interesting car nonetheless.

Savage

Even the name conjures up an image, doesn't it? And Cortina Savage just sounds cool. In the late 1960s Jeff Uren formed 'Race Proved', a company that increased the performance on Fords mainly. He'd won the British Saloon Car Championship in a Mk2 Zephyr in 1959; as a matter of interest he was only the second winner of the title, which continues today as the BTCC. However, they really hit the headlines when they shoehorned the Ford Essex 3-litre V6 motor into a Mk2 Cortina and created the Cortina Savage.

They didn't stop at the engine bay, though. Other modifications included fitting stronger cross members whilst the chassis members on the engine bay were seam welded. Bigger engine mounts were used, as was a 2000E gearbox, modified springs and shocks were fitted, as was a bigger alternator and an uprated exhaust, and the battery was moved to the boot to help spread some weight. Total output was said to be 136bhp, a big leap forward from the standard Cortina's 86bhp.

Over the years there have been all manner of stories about this car from older officers who all claim to have driven it when it was on test with their force. Some say it was fully marked, others dispute that and say it was red or black! Some even insist that it wasn't a Mk2 Cortina but a Mk3 (and Savage did make a Savage Mk3, which was based on a twin-light GXL). What is known is that a Cortina Savage was definitely trialled by the police, but to date no photos seem to exist. Can you help?

Prodrive

Prodrive was founded by 1980 World Rally Champion co-driver Dave Richards and ran Subaru's rally operation for 18 years, making World Champions of Colin McRae, Richard Burns and Petter Solberg in the process. They are based near Banbury in a large building that's impossible to miss when passing it on the M40, and have a large multifaceted operation building where they are developing both road cars and motorsport projects. They hit the police headlines in 1998 when they built two Subaru Imprezas for the Humberside Police that had coppers from other forces putting in transfer requests as soon as they saw it. When the Subaru Impreza Turbo was launched in Europe in 1994, it totally reinvented the performance-car rulebook. With room for four people, this conservatively styled saloon was faster point to point than any supercar, but cheaper than a top-of-the-range Mondeo! Curiously, the police didn't use Imprezas until 1998 when two Prodrive Performance-modified Vehicle Crime Unit cars entered service with the Humberside force. At this time Humberside had the worst vehicle crime figures in the UK, but nothing on the road could outrun an Impreza, as the local criminals soon found out.

The Humberside force worked with Prodrive before ordering the cars, picking and choosing upgrades for increased engine performance and reliability under consistent high-speed use, plus larger brakes, as they were often using the cars for pursuits in urban areas and housing estates, meaning brakes that could take consistent hard use without overheating and fading were a priority.

The improved performance figures are shown in the chart below:
Engine capacity: 1994cc flat-4 turbo.
Power: 237bhp@6000rpm (standard car 208bhp@6000rpm)
Torque: 259lb-ft@3500rpm (standard car 214lb-ft@3500rpm)
0-100mph: 14 seconds (standard car 15.8 seconds)

In the meantime, it seems that the era of the small-time car tuners is just about dead. Although those who do it these days 'remap' the engine's ECU via a laptop, of course! The days of the police using such cars is definitely over, because manufacturers now produce cars that are so good it's simply not needed. However, it's doubtful that it will ever stop certain members of the public from asking that age-old question, 'Ere, mister, is your car all souped up?'

What makes a police car?

I make no apology for devoting a whole section to the SD1 and its development into a police car. Partly because of my age, this is one of my favourite ever police cars and a car I've always loved; a Ferrari Daytona that I actually might be able to afford when I grow up! However, the story of how it became a policing legend – in fact, how it became a Met Police car at all – is fascinating in itself and centres mainly on the brakes. For more information we spoke to retired Metropolitan Police vehicle development engineer Richard French, a senior part of the team that made the SD1 fit for service.

The Metropolitan Police's working environment is very harsh on cars. The area they police is very densely populated with lots of high acceleration and high G braking as officers move from traffic light to traffic light, thus many manufacturers have struggled to meet the Met's criteria until recent times.

However, the late 1970s was a particularly interesting time because the febrile nature of politics in this era impacted on the Met's vehicle acquisition policies. Hampshire pioneered the purchasing of foreign police cars with their Volvo Amazons in 1968. However, because the Met were funded directly by the government they were absolutely forbidden from buying cars that were not built in the UK; a policy that would eventually become illegal under European single market competition law but was very much enforced in this era. That meant that when their two mainstay large area car and traffic patrol cars, the P6 V8 Rover and Triumph 2500, were being phased out by BL, they were faced with assessing a new vehicle for use. Evaluation tests were carried out on a number of vehicles, including BMWs, Volvos and other foreign makes, because they liked to understand the competition and keep up to date, even if at that time policy meant the car could not be purchased. Ford's Mk2 Granada was much favoured, but although Ford had substantial presence in the UK building Transits in Southampton and a variety of products at Dagenham, all Granadas were built in Cologne, in Germany, so they were not an option. The same was true of Vauxhall/ Opel's larger Senator/Carlton family of cars, although again Vauxhall had a substantial presence in the UK at Luton and Ellesmere Port, building smaller cars.

The Jaguar XJ range was disregarded as being too large, too expensive, too soft in dynamics terms and, perhaps most importantly, just not roomy enough to carry the equipment needed, especially by Traffic, although their greater power did make them suitable for armoured work. This left the Met with only one option: BL's then relatively new Rover SD1, which had been launched on 30 June 1976 to great applause from the motoring press, who voted it Car of the Year for 1977. The 4-door Ferrari Daytona styling and up-to-the-minute, almost sci-fi TV programme-interior style were much admired and, despite it reverting to a cheaper (although well located by a Watts Linkage) live axle: rather than the De Dion rear suspension used on the P6, it was much applauded as a driver's car as its subsequent wins on the race track proved. However, within weeks BL were fielding complaints about quality, both in electrics (which suffered from damp) and paintwork – some colours regularly needed total resprays in dealerships within months of being made. Never willing to make rushed decisions (a policy that turned out to be wise), the Met ordered a large supply of both Rover P6 V8s and Triumph 2500Ss in 1977, the last year of production for both as they were kept going while the behind-schedule 6-cylinder version of the SD1 was being developed. These cars could be gradually put into service as their predecessors reached their mileage or service life allowance, and BL even agreed to honour warranty from the date the car entered service rather than when they were purchased. This explains why news footage from as late as 1982 shows P6 Met Police cars on duty, something that's always puzzled me!

However, government contract rules insisted that, when evaluating cars for purchasing, at least two vehicles had to be compared and contrasted, and, Jaguar excepted, there were no other large British cars in production or planned. Rover, however, provided the solution by suggesting to the Met that they should look at the forthcoming 6-cylinder 2600cc SD1 as the second car, thus neatly circumventing the rules whilst still ensuring Rover got the contract work and the ensuing kudos from police use. Incidentally, another political consideration prevented the obvious in-house BL rival, the wedge Princess, being considered at this point for area car or Traffic duties. The unmarked cars used for transportation of senior officers had to be dignified and, importantly, had to be different from the marked patrol cars. The Princess was reserved for this role so could not even be considered for any other! However, it must be admitted that no one on the Met made any real attempt to circumnavigate this as they were not great fans of that particular car. A number of black examples were used in the senior transport role but the DPG also used a small number of Wedges in their distinctive bright-red colour scheme.

With the short-term supply of cars sorted, as the 'new' V8 P6s and 2500Ss mothballed stock came on stream as needed, the Met's engineering team were given a pre-production prototype 2600 6-cylinder SD1, which had started life as a production V8, to assess. It was widely known in the industry that 6-cylinder versions of the SD1 were on their way, and they were officially announced in late 1977.

Both V8 and 6-cylinder versions passed their initial assessment, the handling in particular being praised as a biddable driver's car, which, despite its bulk, could be hustled along far more quickly than the P6, which while comfortable had a tendency for the rear De Dion to tuck-in, limiting the car's performance, or the 2500S, which understeered in a predictable but frustrating manner. As we'll see, the SD1's ability in this area would come to haunt it when it went into service because, confident in the car's biddable handling, officers drove them harder than they ever had the P6s or the big Triumphs. The hatchback body was also considered useful for policing because it offered great and more versatile carrying capacity. However, the Met's engineering team recommended that the force only used the V8 as both an area and Traffic car because its smooth power was loved by all and because it was a known quantity; as long as the oil was changed regularly to prevent the hydraulic tappets getting bunged, it was a very reliable product. However, the bean counters (who are as frustrating in the police as they are in the TV industry!) insisted that Traffic would get the V8 (you can recognise this on old news clips because they had two blue roof lights) but the area cars (which had one roof light) would be the new, approximately £600 cheaper, 6-cylinder 2597cc version.

I'm getting ahead of the story of the SD1's development into a competent police car here, but let's finish this section on the Triumph-designed Rover's SOHC straight-six PE166 engine then go back to developing the cars as a whole. In service the smaller engine used at least as much fuel as the V8, if not a little more, negating one perceived accounting advantage, but the accountants should have listened to engineering who had grave, and it would seem well-placed, misgivings about BL's ability to design and deliver a new engine, because the 2600 engine proved to be a nightmare in service. That saving was spent tenfold on constant cylinder head repairs. The engine had an abnormally high incidence of camshaft and valve gear failure, resulting from blockages of the oil feed passage to the camshaft and valve rockers, which was designed to provide an intermittent oil feed to limit the amount of oil in the top end and thereby reduce oil consumption. However, sludge built up and could block the oil supply completely with disastrous results. What made it worse was that the shims needed to set up the cam were almost impossible to set up accurately. The mechanics were tearing their hair out trying to set them up and repaired cars would often be back for an identical repair in weeks. This was because when you do a top overhaul on the engine you put a dummy shim in, measure the shim then put the correct one in. However, when you bolt down the cam cover the shimming figure changes! Eventually very experienced mechanic Sid Taylor became the man in the police garage at Croydon who ended his career almost constantly doing nothing but 2600 SD1 cylinder heads because he had 'the touch'. Proof again that accountants should never be allowed to make decisions . . .

The SD1, in both V8 and 6-cylinder form, eventually went into service with the Met in 1980, but before it did so, they had to change the steering wheel. The 'Hendon shuffle' method of steering could not be used on the cars as they were because they had been launched with an Allegro-style 'Quartic steering wheel'; albeit with much less publicity about this feature than the Allegro received. This was also exacerbated by the fact that all Met police cars – SD1s and others – were at this time ordered without power steering in a misguided attempt to both save money and provide officers with consistent weighting and feel to different types of car. The after-market steering wheel fitted was much preferred by officers driving the cars.

Immediately the car entered use, however, it started having serious brake problems, especially in 6-cylinder area car form, which usually took the form of the rear brakes catching fire! Engineering were puzzled, as both versions of the car had passed what was considered a severe brake test. In actual use the front brakes faded out because the cast-iron calliper that was bolted to the large metal hub conducted the heat into the fluid and boiled it. This transferred the braking effort to the rear drums, which could not cope and just burst into flames! The Traffic, royalty and special escort teams used 3.5-litre V8s, and as they tended to be more mechanically sympathetic and were not answering emergency calls in quite the same way, they did not present with the braking problems as quickly or as often as the 6-cylinder area cars. The team were trying everything to reproduce the fault, often on quiet public roads, and French was testing one on the Caterham Bypass, putting big inputs into the brakes and trying to drive as though on a shout. Looking in his mirrors, he could see the rear brakes were on fire, billowing smoke and flames, so, knowing there was a fire station nearby, he got off the road and pulled in; Richard had to physically prevent them putting out the fire, shouting 'let it burn' until he'd got underneath to see where it was ablaze! The fire service were not amused.

After working out that the braking test was inadequate for the actual use that the area cars were being put to, Richard's engineering team created a new test which was based on cars accelerating to 50mph then stopping 50 times on the trot. This was tougher on the brakes of faster cars because the faster it accelerated the less time the brakes had to cool. Thus armed with a new test, they had a new standard to aim for and set about analysing the problems. The SD1 was, at this time, using cast-iron 2-pot callipers on solid front discs and, in the Met's specification, steel wheels. The first change was to Rover's own alloys, but they offered very little advantage. So with both engineers having a motorsport background they approached Minilite and sourced a wheel that used the same tyre width, as they wanted progressive controllable breakaway rather than ultimate grip; today, of course, cars have anti-skid control so that philosophy has changed. It did improve matters, as they offered a great deal more airflow, but delayed it rather than solved it. The next move was logical, but politically sensitive. The Mk2 Granada was evaluated by the Met and well liked; it even passed the new tougher brake test, but it couldn't be purchased because it wasn't considered British, so they grafted a set of Granada 4-pot callipers and vented discs on to the front of an SD1! Rover were shown the car and were very impressed until they were told, after the evaluation, where the Met engineers had sourced the brakes from!

The rather shocked Rover engineers took the car back and re-engineered the braking system using the same Granada vented disc casting (which was an AP part, not Ford) but a new 4-pot calliper made from joining a Ford Transit and Fiat calliper together from the AP parts book, which was actually a neater 4-pot system than the Granada's and had a great swept area. The revised system also used a TR7 brake bias valve (bringing more brake balance to the now-stronger front brakes), a different rear shoe actuated by a different wheel cylinder, a thicker rear brake drum to prevent warping when hot, plus a master cylinder and servo unit. This, combined with the Minilite wheels, more or less cured the problem, and was actually listed on the BL Microfiche parts book as 'Met Police Brake spec only', and changed everything except the car's brake pipe. This later formed the basis of the braking system used on the Vitesse model 'go faster' SD1 which did so well in competition.

However, even the new specification created a knock-on issue. Officers were soon reporting that they would brake hard into a roundabout then bleed off the brakes (remember, SD1s were never fitted with the then-emerging technology of ABS brakes), and the brakes would stay locked on, leaving the crew sailing towards an obstacle with both front wheels locked! Pumping the pedal hard and taking your foot off it usually made it spring off, but not before new underwear had been ordered . . . This happened numerous times so it was obviously a facet of the new system, and whenever it did happen engineering were given the car so they could reproduce the issue and sort out a fix. But they couldn't seem to reproduce the problem, until eventually one car from Croydon garage did do it, consistently. French drove this car to Lockheed in Leamington Spa and demonstrated to some very nervous engineers that the brakes were locked on fully at high speed while his foot was not on the pedal . . . They took the car apart that day and eventually found that the operating fork in the servo, which is stamped out of metal and is smooth one side but has a burr on the other, was assembled the wrong way round and would just catch and thus lock when the foot comes off the pedal. Lockheed's test rig for every built servo only measured the pressure going up, not going down, so the manufacturing fault had not been picked up. Lockheed's manufacturing process was altered to prevent this, as was their testing procedure, and that weekend over 200 SD1 Met Police cars were fitted with newly checked Lockheed brake servos by hard-pressed police mechanics.

The SD1 had one further brake-related foible, which, but for a little luck, could have gone unsolved for years or even precipitated the removal of the car from the whole Met Police fleet. We all know that brake fluid is hydroscopic, meaning it absorbs water, which is obviously not good for brake systems with steel or cast-iron components, as they will start to rust. Once the brake problems had been solved the garages found that they had to change brake fluid more often than expected. Because brake fluid boiling was still marginal, even with the new brake kit, the force went from DOT3 to DOT4 when it was announced because it has a higher boiling point, although it drops off more quickly, but they were still getting drivers losing the pedal even when changing fluid regularly. Tests showed the fluid on these cars had deteriorated to the point where the boiling point would be below 150 degrees Celsius – very low – meaning that it had absorbed too much water through the flexible hoses. In the short term brake fluid service intervals were reduced to nine months or less and regular tests were instigated on all SD1s to check the condition of the brake fluid. The hoses were sent away for analysis and various other investigations were done, but to no avail. Drivers were still losing pedals because the fluid had absorbed water to the point of being unfit for purpose.

At the time, French was running tests with Mobil on their new long-life oil, Mobil1, and went to the garage first thing on a cold damp winter's morning to take some oil samples off a car that had been left parked overnight. He lifted the bonnet and watched the water run off the back, down the seam on the inside and straight into the master cylinder! The Lockheed master cylinder cap had a little hole for the fluid level plunger and it would fill up with water whenever a damp bonnet was lifted up, then pump the water into the brake system as it was agitated by driving along. Almost a year of technical investigations had been solved by luck and good observation. Lockheed came up with a rubber cap for the system that sorted the problem completely, and this was fitted to all SD1s thereafter, so for the handful of SD1 owners reading this book, go and check your rubber cap! Amazingly, this story has a sequel, because Richard French noticed this problem many years later on a Mk4 Astra panda car after Vauxhall had phoned asking if they had any brake problems with that model. A bonnet vent hole was dripping water and washer fluid onto the master cylinder, causing the brake reservoir to become overfull and burn paint off the bulkhead. Vauxhall had been scratching their heads for weeks wondering why the brake reservoir was flooding and leaking what they had assumed was brake fluid, not realising it was actually increasing in volume because of the water ingress. They used the same SD1 cap to solve the issue!

Once the brake problems had been properly sorted and the cam problems accepted as an issue (but because BL were now in crisis, never cured), the SD1 had a remarkably successful and long career. It's worth noting two other problems, though. The first in-service problem manifested itself initially as a damp carpet, and this turned out to be the heater matrix, which leaked from its joints constantly. Rover never addressed the problem in any serious way and the sodden underfelt and carpet, followed by low coolant level, meant a really complex and difficult job for the mechanics, although they did eventually work out that the job could be done, with difficulty, by levering the dash upwards rather than removing the whole dash structure. The second issue was the threaded end of the anti-roll bars, which was so small that emergency U-turns done over kerbs snapped them off in a way that never happened with previous cars. This was quickly cured by the adoption of a much heavier-duty component to do the same job.

Covertly armoured cars – a brief history

The first covertly armoured car ever used by a government anywhere in the world is, allegedly, an armoured Cadillac that the US government captured from gangster Al Capone and then used as presidential transportation! However, if you actually interrogate the story, no photos match Capone's car and many now believe this to be a myth created by former Secret Service agent Michael Reilly in order to sell his book, *Reilly of the White House*. It's certainly true that Capone and other gangsters pioneered covertly armoured civilian cars, as opposed to armoured military vehicles, but this type of vehicle really started to be used more widely after the tragic assassination of President Kennedy in 1963. Today world leaders routinely use convoys of armoured vehicles because, as we all sadly know, the threat of terrorism is ever present. This threat has led to the somewhat ironic sight of world leaders, often professing green credentials, arriving at events and conferences to debate climate change in large V8- or V12-powered limos that do 15mpg, on a good day . . . This is because both the power and actual structure of smaller, more fuel-efficient cars are incapable of supporting or pulling along the 3 or 4 tonnes of armour needed for full ballistic protection. The loathsome terrorist attacks suffered in recent years worldwide have made this an even bigger priority and thus nothing is going to change; an effect of terrorism that is not often discussed but is very real nonetheless.

This has also led to an interesting purchasing dilemma in these modern times. Armoured cars built outside of the manufacturer's original facility by an after-market armouring company are not actually type-approved or crash-tested and, perhaps obviously, lack the energy-absorbing crumple zones designed into cars from the 1970s onwards. As the Met at this time took the view that officers were far more likely to die in a road crash than get shot, they were not considered for marked work in this era. Modern armouring technology is somewhat lighter and some manufacturers now offer ballistic defence vehicles which are factory built and thus type approved; ensuring no organisation can be sued if an operative dies or is injured in a normal non-terrorist-related car crash while driving an armoured bullet-resistant vehicle.

As an aside, never use the phrase 'bulletproof'; even the best are merely bullet resistant and are designed to stay intact for *just* long enough for the vehicle to move away from the line of fire; they have defence systems such as run-flat tyres and armoured fuel systems that should ensure they continue to be drivable whilst under attack. They cannot sit there and be shot at for ages, a mistake TV and film often make . . .

The Metropolitan Police High-Performance SD1

Once the SD1 V8 had entered service on Traffic duties, Richard French was asked to develop a faster version because, although the V8 produced plenty of lazy torque, it was sometimes left breathless when high-speed work was needed. His team started by fitting high-compression P6 V8S sourced pistons and a 4-barrel Holley carburettor and an uprated cam, which did offer an improvement, but the car was still not as fast as requested. So the force's engineering team approached Janspeed, who built a twin turbo intercooled unit blowing *into* a 4-barrel Weber carburettor in order to reduce turbo lag compared with the firm's previous Rover V8 turbos, which had sucked through SU carbs. The engine performance was extremely impressive, probably well over 350bhp, and once the car was fitted with Bilstein-developed suspension and a lower-ratio limited slip differential it absolutely flew. The car accelerated so violently that test drivers reported the ATA calibrated police speedo looked more like a rev counter. Unusually for the Met Police, it was a five-speed manual car and on a test track it was measured at over 150mph in fourth gear! No one ever ran it flat in fifth . . .

The car did three months of operational service and was then withdrawn and put back to standard specification because it was realised that other motorists, unconsciously, did not expect an SD1 to be *that* fast, which made the car dangerous. This was an early realisation that sometimes an overtly high-profile performance police car, such as a Sierra Cosworth or Subaru Impreza, does have a role, as the public expect a level of acceleration from it that is both a deterrent to criminals and a contributory factor to the car's operational safety.

The infamous Metropolitan Police brake test

The National Standard for brake testing a model of vehicle prior to fleet acquisition during the 1970s was 20 stops from 60mph with a 45-second interval. So the driver accelerated from standstill to 60mph when the stopwatch started then braked very firmly to a halt after 45 seconds. A brake pedal strain gauge was fitted to ensure consistency, and braking very firmly did not mean doing a full emergency stop, as that could not be repeated consistently or measured. The driver then put the car into neutral and the passenger started the watch to repeat the process – and yes, the drivers who did this much preferred to work with passengers who had a reputation for holding on to their lunch . . . This was repeated 20 times or until the car had no brakes at all, whichever came first. At the end of this a heat-soak test was done: the car was stopped with the footbrake held on hard for a minute then released for a minute, a cycle that was repeated three times. It was surprising how often at the end of this the pedal would just go to the floor because the heat soak had boiled the fluid. This was an important test because it replicated what was quite a common occurrence when driving at high speed in this era. For instance, when escorting ambulances cars would be held at junctions on the footbrake (handbrakes were not used for this because after hard driving they were so hot they would either not apply or not come off!), or when dealing with an incident a car may have driven fast then be parked across a road as a temporary traffic stop before cones, etc., can be put into place. Needless to say, this was always done on test tracks or closed roads; mainly at Hendon, MIRA, Chobham or Millbrook.

However, London is absolutely unique in police vehicle service times, worldwide. They have dedicated, very skilled drivers with pretty good equipment who are keen to get from A to B safely but quickly in an area that's just not designed for vehicle usage, let alone fast-response vehicle use. For instance, in the 1970s one area, Marylebone, had 73 sets of traffic lights from one side of the patch to the other, so officers were braking and accelerating constantly when responding to emergency calls. Sensors showed it was not unusual for brakes to reach 900 degrees Celsius in this situation, something more usually associated with racing at Le Mans than 45mph stops in urban environments . . .

The problems with the SD1 that have been detailed earlier in this book actually caused the test to be modified in 1979 into an even more severe test: 50 stops from 50mph but with no time gap in between. The car was given a 0.5G stop (gauges had improved enough to monitor this by then) every time it reached 50mph from zero, then immediately accelerated up to 50mph again, then stopped at 0.5G again. This meant the faster the car the more often it was braking and thus the more likely its brakes were to overheat. A beautifully elegant solution to the problem of cars getting faster and brake technology not keeping up with their improved performance. This is still the basis of the standard brake test used today, although vehicle manufacturers now know about this and it's very rare that a police spec demonstrator under evaluation actually fails it. Amusingly, when it was first introduced, the Met were still buying a few vehicles that this test was not appropriate for (armoured Land Rovers being one such case), because they took so long to get to 50mph that the brakes were completely cool before the second stop!

A case in point is the early Mk2 1.8 Vauxhall Cavalier, which came out in 1981 and had a master cylinder and servo on the left-hand side as the conversion to right-hand-drive was poor – although this was later rectified. The police refused to take them on after testing because under really hard braking the bar pivoting across the engine bay actually just twisted, acting like a torsion-bar spring! The bulkhead flexed as well because the twisting bar tended to make policemen not getting any brakes press even harder! Once the car had been re-designed and entered service the brakes still proved troublesome, as they ate brake pads and therefore the car would not pass the brake test. The Met called in Opel in Germany, who had designed the car, and they sent an engineer over with some pads, all of which had been tried before and had failed! They eventually produced another 'super-strong' pad and their engineer accompanied Richard French on the brake test and was amazed to see the brakes glowing cherry red, as it was by this time reaching dusk. On inspection the pads were removed and the friction material just fell off the packing plate. The engineer returned to Germany and a few days letter a letter arrived from Opel suggesting, respectfully, that the Metropolitan Police refrained from buying the Cavalier because it was not fit for their purpose.

The problem appeared again on the Mk3 Cavalier, but this time they persevered to a solution, and in order to get that far they had to understand why the brakes were getting and staying so hot. The answer lay in what was then an emerging science in vehicle design: aerodynamics. To achieve both high-speed stability and good cruising fuel consumption, the Cavalier/Ascona (like many cars of their generation and beyond) had negative pressure inside the wheel-arch at speed; this of course translates to no cooling air whatsoever when the car is being driven hard. The solution was to design an air scoop for each front wheel that bolted to the bottom arm and fed cooling air directly to the front wheels, thereby undoing all that careful wind-tunnel work! However, GM actually adopted the design almost unchanged for their later high-performance version of the same basic car. The Met Police garage then spent some years regularly refitting these brake-cooling air scoops that got knocked off by the speed bumps that were starting to appear at around the same time. Officers on a shout, when time is of the essence in the apprehending of a criminal or some other emergency situation, did not slow down for speed bumps – another hidden cost to the tax payer of horrible speed bumps!

The Met were by that time used to working with different brake-pad and brake-mechanism manufacturers such as Ferodo, Mintex and Textar: often using bespoke materials which were similar to racing pads but with a softer element as well so they worked from cold (unlike racing pads, which have to be warmed up to provide any serious retardation). Some cars reached legendary status in the police for being poor rather than good, and one such is the MkIV 2.3 V6 Cortina, which was known for getting only 900 miles out of a set of brake pads! The team tried all sorts of solutions, but without a total re-design they were just not good enough for the high-speed urban use that Met Police area cars receive. The aim was to make the brakes reach a 6000-mile service life – less than a third of the normal expected life, but good for a Met Police car.

The Met and the automatic gearbox

The Met's unique service conditions meant they adopted the automatic gearbox before most of the public, let alone other police forces. There was a dual logic in this; clutch changes were very frequent in an era when clutches were abused and the logic of automatic gearboxes was partly that they made driving safer, because the driver could concentrate on steering the car safely in tricky conditions, and partly to reduce service intervals because clutches did not need to be changed. This meant they effectively became

Borg Warner's testing ground; they actually had monthly meetings with the automatic gearbox manufacturer for many years and had their own automatic gearbox rebuild specialist facility at the Finchley workshop. In the 1960s, 1970s and early 1980s, autos would often last only 12,000 miles, which is 4–6 months or so of use in the Met's terms. Electronic control of the gearboxes has now massively improved this.

CHAPTER TWELVE

BUYING CARS – POLICE VEHICLE PROCUREMENT AND DEMONSTRATION VEHICLES

Just how do the police choose their cars? What makes a good police car? Is it all about speed? Acceleration? Fuel economy, maybe? Reliability? Value for money? Ease of maintenance? Budget? Resale value? Safety? Politics?

In a nutshell, yes, to all of those points! A police vehicle has to score well in all of those departments in order for it to be considered worthy. In the past, local politics often took an overriding role; for instance, Longbridge-built cars proliferated in Birmingham, Fords in Essex and Vauxhalls in Luton. This meant each force bought their own cars and as you drove the length and breadth of Britain police cars would differ widely from Mini vans being used as pandas in rural areas, where the ability to transport a loose pig or other livestock might actually be quite useful, to Ford Escorts fulfilling the same role in south-east London, where transporting pigs is less likely but collecting lost children who will be safe in the back seat of a two-door car is more so.

Vehicle types are deliberately chosen to undertake specific roles, and thus the criteria for that vehicle will differ from that of another used for a different purpose. However, who initially decides which vehicle to look at and ultimately makes the decision to purchase a number of them for force-wide use? In twenty-first-century Britain it's a complex jigsaw puzzle, but this hasn't always been the case.

In years gone by where the UK's police forces were much more localised than they are today, with large numbers of independent city and borough forces of perhaps no more than a couple of hundred officers, virtually all policing decisions, including those of vehicle purchase and use, came down to one man: the Chief Constable. He may have taken advice from certain staff or followed the recommendations of others, but in the end it would be the Chief who signed on the dotted line and ordered the latest motor cars into his force. And when they were delivered it was such a big issue that the force photographer and often the local media would be invited along to record the occasion. Police car history is littered with such photographs.

But there was a substantial flaw in this process, and this concerns the personality of the Chief Constable himself. If he (and it certainly in this period was a he; by the time we got to female Chief Constables the vehicle-buying decisions had been removed from their remit) had some personal knowledge and enthusiasm for motor cars then his involvement and possibly his final decision would, or at least could, be quite sound, because he'd want his force to have the best available at that time. If, however, he had absolutely no interest in things automotive and to him a car was just a box on four wheels to get you from A to B, it's very likely he would merely opt for whatever was the cheapest or most locally beneficial after he'd played golf with the local car factory MD and the constituency MP, who of course was keen to protect jobs there, and his chaps would just have to put up with it. Police car history is littered with such decisions!

As an example, do you remember Arthur Charles West? He was the Chief Constable of Portsmouth City Police from 1940 to 1958. He was a proper petrol head; he loved his cars. In fact, so much so that he encouraged his officers – actually that's not true, he ordered his officers – to enter their patrol cars into Concours d'elegance competitions both locally and in neighbouring areas and was delighted when they often won them. Because he liked cars, West took a very personal interest in selecting the vehicles required to police. The Portsmouth Watch Committee weren't happy that West's beloved Riley RMs cars would put a penny on the rates, and without his passion for cars it's doubtful that the Riley RMB and later RMF would have patrolled the streets of Portsmouth even though they were without doubt superb quality cars.

By the late 1950s the major manufacturers could see the worth in promoting their products as police cars, and once they had successfully done so, they sold to fleet managers and public alike on the basis that if the police use our cars it's good enough for you, because the police only use the best. What better advertising can you get than to have your car being used by the local police? So we started to see glossy adverts in magazines and on posters with catchy slogans like 'When Quick is the Word' and 'On All The Evidence' in an effort to tempt police forces and then the public to buy their products. The almost subliminal mobile advertising boards and TV news featuring these cars became a must watch across a national rather than a local audience.

By this time it was also becoming obvious that a 'police specification' vehicle was needed, so manufacturers responded by making cars specifically for police work. Typically these were a standard showroom car with lowest-spec trim (the government have always been reluctant to pay for officers to travel in luxury, so, depending on the era, lovely wooden dashboards were replaced by tinplate ones, electric windows with wind-up versions, headrest-equipped leather-trimmed seats by low-back vinyl ones, air con was deleted, as was power steering, etc.), but did include the highest-spec engine brakes or suspension. If you ever find a Wolseley 6/110 with a grey-painted metal dash or 2.3V6 Cortina Mk4 with vinyl seats and steel wheels rather than alloys, chances are it will be ex-police. This meant that manufacturers produced brochures for police specification cars, and this has opened up a whole new world to me – I could quite happily collect police car brochures! Just don't tell anyone . . . During my day at Gaydon driving cars owned by members of the PCUK Club (an owners' club for ex-police cars who have been invaluable in the research for this book), I had a look at a few, and the period ones with paintings on the cover are superb.

The Humber Hawk and Peugeot 405 may be separated by many years but they of course basically come from the same factory and company, Peugeot having taken over the Rootes Group from Chrysler.

As police forces grew, much of the autonomous decision-making by chief constables was devolved to other people within the force who had responsibilities for certain aspects of the job, and this included a new post of Transport Manager. Successful candidates were selected because of their background knowledge of the motor industry and how it worked. Most, if not all, of them were (and still are) car enthusiasts and really know their business. As the years went by they became quite powerful and influential people within their respective forces, with many holding almost legendary status. These people have shaped the progress of police vehicles. In 1973 the Wiltshire Constabulary gathered a number of transport managers together for the first time to discuss transport problems with a view to collective pressure being applied to the manufacturers to solve certain issues with their vehicles. This Annual Police Fleet Engineers Seminar ran successfully for 13 years, before evolving into the National Association of Police Fleet Managers (NAPFM) Conference and Exhibition in 1986. It is essentially a two-day police car motor show where the car and equipment manufacturers showcase their latest products; everything from the latest cars to blue lights, from sirens to storage solutions, from armoured vehicles to the latest fuel-saving initiatives and everything in between. It's a multi-million-pound business that both sides take very seriously.

In 2003 the NAPFM and the major manufacturers introduced a National Framework Agreement that categorised approved vehicles into a number of groups. This meant that the vehicles had already passed certain criteria like braking and handling tests, spares back-up, warranty and costs. This framework removed much of the legwork required when considering certain vehicles, as most of it had been done in advance. All the force needed to do was choose which particular brand they preferred and submit their order. With the recession that began in 2008 and the austerity measures that followed for the next decade and beyond, that framework agreement was adjusted somewhat. Now called the NAPFM Blue Light Framework (Government Contract ID RM 1070, which, because it's a public government contract, is available to read at http://ccs-agreements.cabinetoffice.gov.uk/contracts/rm1070 if you ever *REALLY* can't sleep), it has formed two large buying groups: the Northern Buying Group and the South Central Buying Group, which allows far greater discounts with large-scale vehicle procurement, all in the interests of saving money.

The Northern Buying Group consists of the following police forces: British Transport Police, Cheshire, Cleveland, Derbyshire, Durham, Dyfed-Powys, Greater Manchester, Gwent, Humberside, Kent, Essex, Lancashire, Merseyside, Norfolk, North Wales, North Yorkshire, Police Scotland, South Wales, South Yorkshire, Suffolk and West Yorkshire.

The South Central Buying Group consists of the following forces: Avon and Somerset, Bedfordshire, British Transport Police, Cambridgeshire, City of London, Civil Nuclear Constabulary, Devon and Cornwall, Dorset, Kent, Essex, Gloucestershire, Hampshire, Hertfordshire, Leicestershire, Lincolnshire, Norfolk, Northamptonshire, Northumbria, Nottinghamshire, Suffolk, Surrey, Sussex, Thames Valley, Warwickshire, West Mercia, West Midlands and Wiltshire.

Geographically speaking, one or two of these don't make much sense and the likes of Kent and Essex have two bites at the cherry, having signed up to both groups! To add even more buying power to the groups, a number of Fire and Rescue Services have also jumped on board. The vehicles themselves have been separated into three groups called lots. In 2017 the lot for cars and four-wheel-drive vehicles had the following approved manufacturers: BMW, Ford,

General Motors, Honda, Hyundai, Mercedes-Benz, Peugeot, Audi, VW, Skoda, Seat and Volvo. The lot for vans had the following approved manufacturers: Ford, General Motors, Mercedes-Benz, Peugeot and VW. The lot for motorcycles had the following approved manufacturers: BMW, Honda, Kawasaki and Yamaha. You may notice that Mercedes are in the commercials group but not cars; this is because in the late 1980s the Mercedes UK boss took the decision that having his clients pulled over in a Mercedes was actually *bad* advertising, and they withdrew from the procurement process for *marked* police cars, although they are still very active in the commercial four-wheel-drive and Ambulance/Fire/Rescue vehicle area and some unmarked Mercedes are used, especially in VIP protection roles. Mercedes passenger cars were too expensive to be widely used anyway, although some forces had used them as motorway cars. A shame. I'd have enjoyed having a Merc to use in my time!

Within those lots are categories of vehicles like Low Performance Segment, Intermediate Performance Segment, High Performance Segment and a whole host of others too numerous to list here. So now much of the research work that forces once needed to do has been removed and the manufacturers take the industry so seriously that they produce what is commonly referred to within the trade as 'Turnkey' vehicles. These cars, vans and motorcycles are taken off the standard production lines, stripped down again and rebuilt with everything the police would need. This includes the huge amount of additional electrics to power today's modern technology requirements, including ANPR, digital radio systems, telecommunications, black box flight recorders, data tracking and CCTV, as well as other requirements like prisoner cells, dog cages, emergency equipment storage, multi-function blue light systems and sirens, with most now supplying the vehicles fully liveried in Battenburg and ready to roll. Gone are the days when each individual force had to do all of this work themselves.

In some respects this is rather sad because it has taken away each force's individuality and its previous identity. It has also removed the different choice parameters that each force had initiated – whether they be political, whimsical or just down to the marque loyalty of those in charge who might like Fords because they were a fan of rallying Escorts, or BMWs because they liked smooth 6-cylinder engines, and nobody really did that better than BMW.

It has to be said that when it comes to the vehicles the police use, and their skill or otherwise in choosing them, some of their choices are bizarre. This testing of police specification demonstration vehicles would vary a great deal. Manufacturers who thought they had a vehicle with promise in a certain role would prepare one or two demonstration versions and then tout them around the various different forces. Some were only tested by the force driving school, others out on trial at stations, some for only one day, and some more serious contenders being put through their paces for a month or more – and in rare cases a year! So let's take a look at a few of the more interesting vehicles to have made an attempt to gain a permanent blue light. Some made it in small numbers, most didn't, which is what makes them interesting cars to ponder! No records exist as to why certain cars failed to make the grade, as each force kept their own records, and many have now been destroyed. I'll make no attempt to second-guess why some failed, but anecdotal evidence or research by owner enthusiasts does shed light on some; others are just patently daft! However, the sheer variety of vehicles tested has to be seen to be believed and just goes to show how important that market is in both outright cash sales and marketing prestige.

1973 MGB GT V8 Police Evaluation and Demonstration Car

We have not seen the police driving sports cars since the mid-1970s (although I'd rejoin if they used 124 Abarths!) so it might come as a surprise that MG, one of my favourite marques because my first car was a Midget 1500, have a great tradition of working with the police service. In today's world police cars have to carry a huge amount of equipment, be reasonably fast and nimble in both motorway and urban situations, and carry extra passengers when needed as well as the two-person crew. However, before the Mini Cooper was launched in 1961, saloon cars tended to be cheapish and slow, or fast and expensive, so for the police forces buying on a budget, MGs made sense as they were fastish (for the period) but reasonably priced. The MG factory also had a reputation for looking after customers, making modifications when needed and for making stylish, well-built and reliable machines. Their long-term marketing slogan, 'Safety Fast', could easily have been the motto of the Hendon police driving school!

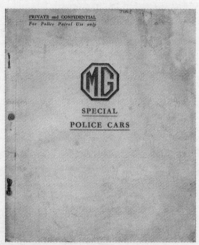

You will find references to MGs in other parts of this book but it's a mark of just how important this marque is to the history of the police car that as far as I know the only book written about a specific make of car and how the police used them is the excellent *MGs on Patrol* by Andrea Green, a book that inspired this tome to some extent and that is well worth reading. MG took this market very seriously and the oldest 'Police Special' specification brochure that we have found whilst researching this book is for their 1939 range – both the TB Midget and larger 1.5-litre VA saloon. Interestingly, the police demonstrator photographed, BRX 266, was later purchased and put into service by the Kent Constabulary.

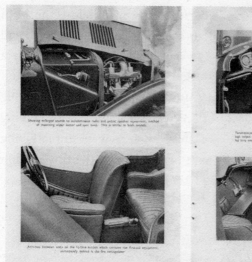

This 1939 MG brochure shows how seriously the firm took the police market;
they were right to do so because the British police used almost 650 MGs
from the late 1920s until 1973, a small but still significant proportion of the
firm's production. The greater benefit to a small company with a moderate
marketing budget, however, was the public seeing the police in MGs and
buying one after reasoning, 'well, if it's good enough for the police
it's good enough for me'.

1970 MGB GTs set off on patrol from the Edinburgh police garage.

The 1962 MGB carried on this tradition and was widely used by the
police in both open and GT forms. Because of Andrea Green's research
we know that the UK force used 79 MGBs between 1962 and 1967,

while the GT was even more popular, with forces ordering 129 between 1966 (the GT was not launched until 1965, three years after the roadster) and, amazingly, 1975, long after the roadster had gone out of favour.

The Lancashire Constabulary even had a synchronised driving display team using MGBs. The 3-litre straight-six MGC found favour initially as well, with 31 roadsters and 21 GTs being ordered by various forces in 1969 and 1970. The police were critical of its nose-heavy handling, though, as were the public and press, so this, combined with its greater expense and very short production life, ensured that it was never going to be used in the numbers the MGB had been.

It's ironic, however, that perhaps the MG that is most perfect for police work never really found its feet in the police world in the way that previous models had because it was just a little bit too late. That vehicle was the MGB GT V8. The V8 was announced on 15 August 1973 and really was the car that the MGB GT should always have been. That wonderfully mellifluous aluminium 3.5-litre Rover V8 (one of my favourite engines) was slightly lighter than the normal 1800cc B-series and offered almost twice the torque from 2500rpm, giving the car a relaxed gait that disguised real pace. It had the feel of a miniature, budget Aston Martin and was none the worse for that. It should have been a massive hit, both in America and Europe, but sadly politics killed the project engine supply before it had even got going, and only 2,591 GTs and no roadsters (although a roadster V8 had initially been planned) were produced, making genuine examples of a rare car in the classic car world. Many an SD1 has been ransacked to build unofficial MGB V8s!

In 1970 racer and tuner Ken Costello had shown that it was possible to put Rover's compact Buick-derived V8 into the MGB, roadster or GT, and he created a successful business doing conversions for customers who received a car with E-Type performance for considerably less money. MG were aware of this and realised the car was a feasible production option because, since the creation of British Leyland by the government encouraged the merger of British Motor Holdings (the old

BMC plus Jaguar and body builder Pressed Steel) and the Leyland Group (Rover, Triumph and Alvis) in 1968, they had been in the same family of companies.

However, political infighting within BL led to muddled decision-making, and the dominant company, Triumph, backed the production of a new 3-litre OHC V8 for the Stag, which proved disastrously unreliable, cost a fortune in warranty claims, and lost the Stag a promising position in the US market. Ever-changing American safety legislation also meant design engineers were constantly on the back foot in the late 1960s and early 1970s, trying to keep their existing cars' type approved for the then biggest consumer market in the world, which also delayed the MGB GT V8 programme.

Consequently BL never expanded the production of the Rover V8 engine, and neither Range Rover nor MGB V8 were federalised to meet US-type approval (although Range Rover finally got there in 1986 and was an instant hit) as was originally intended: profitable export sales of MGB V8s and Range Rovers might just have been the cash cow that BL needed to bring enough money in and thus invest in new models. Hindsight is a wonderful thing, but had they tripled Rover V8 production capacity, fitted it to Stag and produced an American MGB V8, history may have been very different . . . However, MG could never get enough of the V8 engines because the Range Rover and Rover P6 had first call on the limited production capacity. When the oil crisis began in October 1973, thirsty V8-engined cars seemed doomed, so BL had yet more justification not to expand Rover V8 production.

As the MGB GT V8 was developed in early 1973, MG produced two prototypes for evaluation by the police, including the GOF 88L, which was completed on 30 January 1973 and registered new in Birmingham just two days later. It was painted in 'Police White' and initially supplied to Birmingham Police for evaluation. During the car's time with the police service, it was fitted with a triangular roof sign, two-tone siren, a radio and other police equipment. The car spent time with the Sussex, Staffordshire, Thames Valley and West Mercia police forces as well, and was praised by all. An enthusiastic in-depth report in the magazine *Police Review* stated, 'It is so perfect for police work that it could have been designed specifically for the job'. Sadly, however, the MGB GT V8 never had the prolific police career that this promising start may have indicated; it was a wonderful and characterful car out of its time. By 1973 saloon cars had become both cheap *and* fast. For instance, a V6 3-litre Ford Granada could be as quick point to point as the MGB BGT V8 but also held considerably more equipment and had back seats to put bad guys or drunk drivers in when needed. The days of a sports car doing mainstream police work were numbered as soon as high-performance saloons such as the Mini Cooper and Lotus Cortina came out in the 1960s, and by the 1970s the Escort RS2000, Granada 3 litre, Triumph 2500 and Rover P6B V8 all offered brisk performance, reasonable handling and space aplenty for the increasing amount of kit needed, such as signs, cones, first aid equipment and radios.

The MGB GT V8 may be the greatest potential police car MG ever made, but it was also their last hurrah. Only three were sold, all to Thames Valley, who used them as unmarked motorway Q cars, the only clue to their ownership being a cord-operated police stop sign mounted in the load area, which rose up into view in the tailgate window, illuminating as it did so. The only people who saw the sign were those being stopped for speeding or other offences! Despite the two marked demonstrators being universally praised by enthusiastic officers, none was ever purchased as a marked Traffic car. Unless you count two MG Montegos and an MG Maestro, the octagon's 40-year-plus association with the UK police ended with the BGT V8.

Specifications
Engine: 3528cc OHV V8 in low-compression Range Rover form
Power: 137bhp@5000rpm
Torque: 193lb.ft@2500rpm
0–60mph: 8.6 seconds
Max speed: 124mph
Weight: 2,385lbs (without police equipment)

VW Beetle 1303

At the height of the panda car era, in the early 1970s, when the cars were all painted pale blue and white, Volkswagen made an effort to get into that market by pushing the ubiquitous Beetle onto the police. The 'people's car' fitted the criteria in some respects because it was a low-performance, low-budget car that had been used very successfully in continental Europe where the likes of Germany (obviously), Belgium, Holland, Switzerland, Austria, Sweden, Finland, Spain and Greece had all used the car in significant numbers. It was also well built and of proven reliability.

A Beetle 1303, complete with suitable registration number of AYM 999H, was passed from force to force in an attempt to push it onto the panda fleets, but it really didn't inspire anyone within the UK's police. The car was finished in Bermuda blue with white doors (the most common panda scheme) and came with an illuminated police roof box and a second police/stop box positioned above the rear registration plate and front spots. Although the records do not show how long the car was tested for, they do suggest that it was trialled at Basingstoke on 21 June 1970. The car obviously created a lot of interest and became the subject of official force photographs. Apart from in Hampshire, the car was also trialled in Staffordshire and Leicestershire and may have gone to other forces as well.

It's quite an interesting one to reflect on. Remember, the Beetle was by then a 30-plus-year-old design. It may have been super reliable (Woody Allen successfully starting a 2000-year-old VW in the film *Sleeper* shows just how etched into the world's consciousness the reliability of Beetles was in the 70s) but it was slow and noisy, had a cramped interior, poor brakes, and because of the basic inefficiency of an air-cooled engine, used a lot more fuel than, say, the Morris 1100 or Ford Escort Mk1 that was competing for panda car orders; both were much more modern designs. I'm not dissing Beetles, I like them, but what's appealing in a classic car now was not necessarily what would have appealed to officers working long shifts in the winter months or police accountants paying fuel bills.

The VW Beetle demo car when being trialled by the Hampshire force in 1970 photographed at Bramshill Police College.

VW Vento VR6

Rather amusingly in retrospect, VW spent a lot of money advertising their new Vento saloon as an entirely new model in late 1991, going to great lengths to explain that it wasn't just a Golf with a boot, despite it being, clearly, precisely that! The new Mk3 Golf had come out in August 1991 and it was entirely logical and predictable that VW would launch a saloon version, although they dropped the previous Jetta tag in favour of Vento. However, VW had launched their new narrow-angle 2.8-litre VR6 engine in 1991. With only 15 degrees between the two cylinder banks, it was almost a compacted straight-six, with the cylinders nestling together in each other's 'corners', so it was very short. It had taken many years to develop and was specifically designed for transverse FWD installation, and whilst only an SOHC design it was quite advanced for its time, with a single cylinder head for both banks. Oddly from an industrial perspective, it shared no parts with the similar-capacity Audi V6 of the same era, but when fitted to various Golf-platform-derived cars it produced a characterful pocket rocket of a car which looked promising for police work. In 1993 VW produced a test Vento VR6 that really did get the pulses of police traffic drivers racing. This compact car came with a big 2.8i V6 engine, giving it a top speed close to

140mph, with acceleration to match and the most amazing road holding. The test car was striped in yellow and even sported a VW Police Test Car crest on the rear pillar. A full-width Optimax light bar from Premier Hazard was fitted and it became one of those rare cars that everyone wanted to drive. It got rave reviews from all who got the chance to drive it and looked set to become a staple. However, in actual fact it wasn't quite big enough to be a mainstream Traffic car and never really caught on, although Hampshire ordered one as an unmarked video Traffic car. That example, finished in a sort of dark blackberry colour, proved to be one of the most successful unmarked patrol cars ever procured and it holds the record for the lowest 'down time' of any car ever used before or since.

VW Vento demo car photographed by a private officer who was assessing the car.

Ford Cortina Mk2 1300 De Luxe 2-door
Ford demonstration panda car for Manchester and Salford Police, and others

As police forces modernised in the early 1960s, the UBP (Urban Beat Policing) cars required by the Home Office's plan for future policing became known as pandas because they were often initially in a black, then mainly pale blue, and white livery. Manufacturers soon realised the huge sales potential that this new policy promised, as the total UK police vehicle fleet was likely to increase tenfold or more if the policy worked and was rolled out UK-wide. Accordingly, most manufacturers prepared police specification 1100cc or 1300cc saloons as 'panda demonstration cars' that were then loaned to regional forces for evaluation. A Cortina Mk2 1300 2-door was developed as a panda car by Ford, painted Nordic blue with white doors, then loaned to various forces for evaluation in the months after the model was launched in October 1966; it was photographed while being used by the Manchester and Salford Police in the course of their loan period.

Ford had a perceived gap in their range in this period as the Mk2 Cortina was significantly bigger *looking* than the Mk1 because it was so much flatter and lower. It *was* 61mm wider (much less than it appeared) than the Mk1 but, and you won't believe this because it's such a convincing optical illusion, it was actually, at 4267mm long, 7mm shorter! It's a good job I never play car trivia games for money as I'd have bet my shirt on the Mk2 being both wider and longer . . . This of course suddenly meant the Anglia was perceived as being a lot smaller than the Cortina, a fact that was remedied when the Escort came out in 1968. Panda cars were usually in near base model trim (unfathomably that was called De Luxe in the mad world of 1960s Ford Cortina Mk2 nomenclature) and used smaller engines as the majority of police drivers using cars in this role were only qualified to standard level and not allowed to drive more powerful cars. Despite Ford's attempt to gain panda car sales with this 1.3-litre pre-crossflow-engined demonstrator they were not successful as none was ordered, forces choosing smaller 2-door cars for this role instead, such as Ford Anglias, Morris Minors or ADO 16 1100s and, from 1968, Ford Escorts. You'd have thought Ford might have learned their lesson after experiencing similar rejection a couple of years earlier when they tried the same trick with a Mk1 Cortina estate, which was also finished in light blue with white doors and a narrow white band over the front part of the roof. Again, it was too big to be used as a panda car and underpowered for area car use.

This is one of those strange quirks of police procurement. The Cortina was just a little bit more expensive (but surely much more useful?) than an Anglia or Minor. The fact it was so up to date also counted against it because the police were reluctant to buy fleets of cars for high mileage roles until a small number had proved the type reliable and it had been in production a year or so. Police procurement often got itself into trouble because of this policy during this era; once a certain force had decided they were using car X for role Y they just carried on blindly buying that same car, no matter how out of date it had become, which is why the Morris Minor was still being purchased in large amounts as a panda car up until the day production ceased. Individual forces even sometimes bought up the last stocks of cars as production ended, usually at a seriously discounted price, and stored them for a year or two before using them. This led to some almost-antiquated cars pulling over much more modern machinery – for instance, in some areas the big 6-cylinder BMC Farinas were still being used in the early 1970s. On the other hand, when production did cease they were suddenly in a panic because they had to choose a new car for that role, which could take some time! This changed as procurement became more nationally organised and it's now much more of an ongoing assessment.

The Mk2 Cortina was, however, very widely used by the police in a variety of other roles, including marked traffic patrol, area cars and CID investigation – who used discreet unmarked examples. Many police forces also used the Lotus-engined Mk2 (which had three different names during its three-year life) as high-speed pursuit cars and Mk2 Estates as load- or dog-carrying vehicles, so it's perhaps odd that the Mk2 was never purchased as a panda car despite Ford's best efforts. The accountants, of course, won that argument.

Specifications
Engine: 1298cc 4IL OHV
Power: 53.5bhp@5000rpm
Torque: 71lb.ft.@2500rpm
0–60mph: 21.4 seconds
Maximum speed: 81mph
Standing 1/4 mile: 22.2 seconds

Renault 30 TS and 25 V6

In 1979 Renault jumped on the UK police bandwagon and offered the Renault 30 TS as a Traffic car. The 2664cc PRV V6 had a good reputation for durability when used in Volvos, and Renault made a serious effort to tempt the police with a complete package car. It came with a single blue light, twin roof-mounted spot lamps, a police/stop box and two additional flashing rear fog lamps, and on the front grille the car sported twin blue repeater lights. Officers who used it particularly praised the super-comfortable seats. Renault tried again in 1986 with the Renault 25, which used an improved 2849cc PRV V6 engine with 153bhp, powerful in its day and smooth, too. Renault thought they had a winner on their hands and gave it to the police, rather bravely saying 'break it if you can'. It has to be said that certain officers did try, but they all failed! Renault had delivered a nice package to the police and the car came ready striped and with a Premier Hazard Olympic light bar and a police/stop box on the rear tailgate. Although it seemed promising, the Renault 25 never quite got the backing of the UK public or police, despite that memorable TV ad campaign featuring well-known stars Malcolm Stoddard and Rosalyn, 'You're not being a bit . . . hasty about this, are you?' Landor relying on an R25 at key points in their life, which was really the first continuing returning soap-opera-style ad for any product, coffee included. Sadly, the ads are likely to be this under-rated car's biggest contribution to motoring history, in the UK at least: *'It's time to go it alone, John's with me, Ian's with me and we've got the backing.'*

Renault launched the overweight Safrane to replace the 25. It was a sublimely comfortable and quiet car with superb NVH that cost a fortune to develop but failed to sell well anywhere, even France. They didn't even bother to try to tempt the UK police away from their Granadas, Rovers or Volvos with that one and left the conventional executive car market when the Safrane ceased production. They have never returned to that segment under Renault badging, preferring to use their Nissan offshoot, Infinity.

Renault 25 (top) under evaluation in a police garage – eagle-eyed readers can just see the Hampshire BMW poking out behind. The Renault 30 (bottom) was liked but failed to win any orders.

Saab 900 Turbo and 9000CDi

In 1980 the police tested a rather unusual car that is still remembered with fondness today by those officers lucky enough to drive it. The Saab 900 Turbo was one of those rare cars that got instant cult status among police officers and public alike, partly due to its excellent performance but also because of its sheer Swedishness. It was very quick, with excellent handling, and had a reliability record that some manufacturers would have killed for that had been built up over many years of Scandinavian testing and rallying – a sport Saab have excelled at. By the time the police trialled the 900 Turbo it was actually quite an old, basic design, having started life as the 99 in 1968 and been progressively developed into the larger 900, which came out in 1978 and was easily identifiable by its longer

American safety-regs-inspired nose. Amazingly, the engine in the Saab started off as a joint development between Saab and UK maker Triumph. As I found out while making the Triumph Stag episode of *For the Love of Cars*, it's basically half a Stag V8 and shares its basic design with the engine in the Dolomite 1850 or TR7; the Sprint, of course, used a 16-valve version. Saab developed the engine independently and produced a car that was many years ahead of its time and had thoughtful touches like the entre-console-mounted ignition which locked the car in reverse. Both the car and the engine were tough enough for the job, I'm sure, but the thought of expensive turbo chargers going wrong put the police off and Saab withdrew the car. The test car came equipped with twin blue lights mounted on a roof rack bar, and in between these were two spot lights and the siren and on the front grille were twin blue repeater lights. It was one of the most modern-looking police cars of its era.

In 1996 Saab decided to rejoin the police vehicle circus and offered the Saab 9000 CDi, a car that was, like the 900, derived from a joint venture, but this time with the Fiat group, who spun the resulting 'Type Four' large front-wheel-drive platform into three cars: the Fiat Croma, Lancia Thema (and who doesn't secretly lust after a Thema 8.32? Google it if you don't know what it is as it's a bonkers car story; if you do know, just nod in a sage-like manner and carry on reading about police cars) and Alfa Romeo 164, all of which were released in the mid-1980s. Saab, of course, made great play of the fact that they had redesigned the car extensively to the point where there were only seven interchangeable components or pressings, to make it much safer than the Italian ones! Fiat group were understandably unhappy about this, but by this time GM had bought a substantial stake in Saab, so any further collaboration was then out of the question. Saab repeated this trick by substantially redesigning the GM Vectra platform that their 1994 900 was based on; perhaps we can see why they never quite made the money GM hoped and, very sadly, eventually went under in these two little stories . . .

The demo police car was almost as popular as the old 900 Turbo with the officers who evaluated it. It was quick, comfortable, with a brilliant seat, and pleasant to drive. It came equipped with a Premier

Hazard light bar and Saab's own livery; it even had a Saab police crest, which was a nice touch. A good number of forces trialled the car and the likes of Kent, Strathclyde and the City of London Police purchased them, although perhaps not in as large numbers as Saab had initially hoped. Some were also purchased for unmarked duties, so the demonstration probably earned its keep.

Fiat

In 1982 Fiat arrived with the 131 Super Mirafiori, a 2.0-litre saloon car with a fantastic engine but very poor build quality. Tested by a number of Traffic Departments, it was said to be great around town with excellent acceleration and good handling but it wasn't up to long-term motorway speeds. This may be due to the fact that it was 'under-geared' in the Italian manner; although it had a 5-speed gearbox, it really needed a sixth 'overdriven' ratio for sustained high-speed work.

At this time corrosion was still seen as an issue with Italian cars – a story with massive political overtones that's not for this book and involving the supply of Russian steel for the Italian auto industry in exchange for building the Tolyatti-based Lada plant. This undoubtedly put off UK police buyers as well.

Jaguar

Jaguar have been supplying cars to the UK's police for decades with varying degrees of success, and in 1984 it came up with one of the best-looking police cars of all time, the 3.6-litre XJS Police Special 5-speed manual which used a Getrag gearbox also used on certain BMWs. The XJS (initially XJ-S) had been in production for over 10 years by this time but only as a V12, which was basically auto only, although a very small number of 4-speed manual V12s were made early in the car's production life. However, the development of their new all-aluminium AJ6 engine for the forthcoming (although much delayed) XJ40 range of saloons was trialled in the XJS. Jaguar has a long-established practice of debuting new power units in lower-volume sports cars, ironing out the bugs, then announcing them in the higher-volume saloon range. The improved economy offered by the 6-cylinder, plus the manual gearbox option and the general rise in the quality of the product, prompted Jaguar to offer the car as a motorway traffic patrol vehicle. Truth be told, this sleek Jaguar didn't really fit the criteria as a patrol car, but nonetheless it looked beautiful with its stripes and full-width light bar. The demo car, registration number A329 KHP, was quite possibly one of the widest-travelled and most famous of police demo cars ever produced, with forces including Kent, Warwickshire, Hampshire, Greater Manchester, Lincolnshire and the Met, to name but a few, all having the car on long-term evaluation. The Met Police in particular used it as a motorway patrol car where it was at its best, obviously, but in terms of practicality it was never really a serious contender. It had a cramped interior (but was still a big car) with very little boot room and the running costs were high. The Met also spent a day at MIRA doing brake tests and managed to melt the wheel bearings into the hubs by overheating the brakes, losing the grease, then running without the grease until it all fell apart. Their motto was to finish the test unless the car ceased to function entirely; destruction-testing of the most direct sort! Jaguar were apparently less than happy when the car was returned to them with an arc of wheel-bearing grease sprayed up each flank like mud . . .

For Jaguar it was a priceless exercise in PR and product development that they were very happy with. In police hands the car clocked up thousands of miles in a relatively short time and they took it back in order to strip it down to examine how the component parts had stood up to the rigours of police life, which, Met Police brake test apart, was apparently quite good; maybe Jeremy Clarkson's successful use of an XJS in the *Top Gear India* special owed something to police destruction testing. Thereafter the car was rebuilt and loaned to the Greater Manchester Police Museum before it went back to Jaguar as part of their Heritage Collection, where it remains on display to this day.

Prior to the XJS making an appearance, the firm had even put that most famous of Jaguars on police trial with the Buckinghamshire Constabulary getting the use of an E-Type. Being a traffic cop in that force must have been amazing! How many other young men got to drive an E-Type flat out on the new M4 motorway and got paid for the privilege? It seems incredible now, but only one photo ever seems to have been taken of the car. You'd have thought there'd have been many more.

The XJS on trial in the Met Police garage; note the SD1 Traffic car and Mk1 Senator plus the Defender behind.

The XJS outside Jaguar Museum at Gaydon, Warwickshire.

Lotus Carlton

The Lotus Carlton 1991 is remembered in police automotive circles as the car that no one was allowed to drive! Unless, that is, you were an instructor at the force driving school. The driving schools deemed it too fast to be driven by mere mortals. Based on the Vauxhall Carlton saloon (although GM's engineers had originally started the project on the longer Senator B, which used the same platform), Lotus didn't compromise when it came to playing around with the mechanicals, and they came up with a car to beat the Cosworth Sierra as a headline-grabbing super saloon. Lotus were owned by GM between 1986 and 1993 and became a sort of unofficial skunk works for GM, working on various projects, although the Carlton was the most high-profile. A 3.6-litre straight-six engine, based on Opel's famous CIH block but fitted with a new double overhead camshaft 24-valve cylinder head with twin Garrett T25 turbo chargers running water-cooled intercoolers, gave the car an incredible 377 bhp@5500rpm and an even more incredible 419lb.ft@4200rpm. It came with a 6-speed manual gearbox, taken from the Lotus-developed Chevrolet Corvette ZR1, and power was laid down by some of the largest tyres ever to roll on British roads, with 265/40x17s on the rear and 235/40x17s at the front. The 0–60 times were around 5 seconds, depending on how much wheel spin you got, and the top speed was 176mph! The press rounded on GM for having the temerity to make the car, and they found themselves in the position of defending its very existence to uneducated scandalmongers from tabloid newspapers, even wheeling out Chief Engineer Paul Tosch to defend the car. How things change! Vauxhall's legendary Technical Director, Laurence Pomeroy, had been feted almost as a hero by the press when he produced the fastest sports car of its era, the Prince Henry, in 1910, which had a top speed of 65mph . . .

The car *was* awesome, and was tested by a number of force driving schools, but despite some desperate pleas from certain Traffic officers, that's where it stayed. Finished in Imperial Green, as all 300 British-registered Lotus Carltons were, the car was just not practical in terms of police vehicle use. Shame!

Suzuki Vitara

A rather strange vehicle arrived for testing in 1991 and, like one or two others, it was difficult to see just where it would fit into the force. The Suzuki Vitara JLX 1.6i four-wheel drive was one of a new breed of small, four-wheel-drive cars, aimed at the younger 'upwardly mobile set' with an active lifestyle. The car was great fun to drive and very responsive. It did suffer from a rather plastic interior that wasn't really considered to be 'policeman proof', and it probably wouldn't have lasted the course in terms of police use. It came striped in the usual way and had a full-width light bar and twin repeater lights placed on top of the front bumper. The only force to have purchased the Vitara following their evaluation was the Wiltshire Constabulary, who bought a small number for their rural patrols where it was eminently suitable, doing the majority of work previously done by Land Rover products but in a vehicle that cost far less to buy and fuel.

Ford Zephyr MkIV Ferguson Formula four-wheel drive

The blue oval has been responsible for producing some of the very best patrol cars ever used by the UK's police over many, many years. Most have proved to be very popular, but there have been a couple that, although extremely innovative, just haven't been accepted.

The Ford Zephyr, from its inception as the Mk1 in 1951, through to the very successful Mk2 in both saloon and Farnham-bodied estate, and of course the *Z Car*, the Mk3, was universally accepted as being one of the greats. Then Ford ruined it all by giving us the MkIV, the car for which the phrase Dagenham Dustbin could have been born. It was a big, wallowy barge that was instantly criticised by the press and police evaluation team alike. Ford were stunned. It *should* have been brilliant! Although Ford rather wonderfully built three pilot-build MkIVs in late 1965 in the run-up to the Christmas break, when the first proper production MkIV, or Project Panda as it was known internally, rolled off the line in early 1966 it carried a development bill of £28 million on its enormous shoulders. (Apparently only just over £2 million of that figure was for the car, the rest was for the new range of Essex V4 and V6 engines that had been developed separately and would go on to give Ford UK sterling service for many years.) The car's chief engineer, flamboyant American Harley Copp, had insisted it be substantially larger overall and have a much larger visual bonnet-to-cabin ratio than any competitor, so much so that Canadian-born stylist Roy Brown (who had been sent to the UK from Detroit after playing his part in the Edsel disaster by producing a grille which was likened to a lady's private parts) was forced to house the spare wheel in the nose in front of the engine! Ford UK's Chief Passenger Car Engineer defended the idea strongly in an interview with *The Autocar* in April 1966, although it's hard not to wonder if his tongue was planted firmly in his cheek at this point because he said, 'We say here the big car man has power under the bonnet, and the longer the bonnet the stronger his image!'

Strong image or not, it produced a very nose-heavy car when actually the shorter V engines should have enabled more compact packaging – as they would in the replacement Granada. When this was added to Ford's new independent suspension (this was Ford UK's first IRS car and they made sure everyone knew that during the launch), which tended to veer left on its inside rear wheel and cause an unseemly lurch during high-speed cornering, the damage was done. This brave all-new design lasted only until 1971, selling a mere 148,347 – about half what the Mk3 had sold in two fewer years on the market. I have to say, as a kid, the MkIV did have one thing going for it, those mad hubcaps with a red, white and blue central rim surrounding five gold stars on a black background. Brilliant, when you are seven years old; I wonder if they were inspired by the Docker Daimlers?

In 1968 the Home Office, at the request of various police forces, undertook to test and evaluate the advantages of a high-performance car fitted with the new Ferguson Formula four-wheel drive (as used on the Jensen Interceptor FF of 1966) and Dunlop Maxaret anti-lock braking system. The police required a fast four-wheel-drive car with good load-carrying capacity that could be used on the growing number of motorways being built around the UK. Several manufacturers were approached with a view to having their cars converted, but Ford was the only company willing to adapt and still warranty their cars, so the Zephyr was chosen. However, its rather oversized character helped here because Ferguson had realised packaging the four-wheel-drive system into some of the larger 6-cylinder cars available would have been difficult. The Ford was also the most up-to-date large car being built in the UK at that time, with BMC and Rootes both pensioning off long-serving models and looking to launch new cars that were not quite ready. The car chosen had to be British because the cars were bought by the Ministry of Technology (of which Tony Benn was the then current minister) who actually purchased 50 MkIV Zephyrs to identical police spec, 25 of which were converted and 25 remained standard; the idea being that a force would run one of each on similar duties so they could be compared directly. The cost of the conversions, around £1000 a car, was also borne by the Ministry, but they were aware that this could be reduced to around £400 a car if a batch of 5,000 were to be made – and that was seriously being considered. As an aside, at least three more were produced: an estate which FF used for many years as a management car, and two saloons, one for the Ministry of Defence (although no one knows why they wanted it or what they did with it) and one for BSM's high-speed driving course which was based at Brands Hatch! It's worth noting here that this was not just a technical exercise to equip the police, although that was important. From a Motech perspective it was a sort of under-the-counter, government aid programme design to keep our motor industry at the forefront of technology and thus hopefully provide jobs and prosperity into the future. Tony Benn had been a pilot during World War II, and although he is best remembered now for his outspoken views on programmes like *Any Questions?* he was genuinely interested in engineering and technology. It's doubtful BL would have been created had Tony Benn not been in this job, but that's another story . . .

As well as having the 'Ferguson conversion', the cars were fitted with the high-compression 3-litre V6 Essex engine (usually only fitted to the upmarket Zodiac range, the standard Zephyr 6 making do with a 2.5-litre engine) along with a high-pressure oil pump system, again, not normally fitted. Other upgrades included heavy-duty suspension, reversing lights and a 'police spec' wiring loom to allow for 'blues and twos', plus the power supply for radios and other ancillary equipment. To cope with this, a larger-capacity alternator was fitted, this being a positive earth type, as police equipment in those days was still positive earth.

The Zephyrs left Ford for the Ferguson works in Coventry during November 1968 and were then comprehensively re-engineered over the next five months. To fit the four-wheel-drive system the complete front suspension assembly was removed and the inner wing structure was modified to allow double wishbone and coil springs to be fitted, along with a modified trailing arm and anti-roll bar. The top wishbone, coil springs and shock absorbers were standard Ford Mustang units supplied through Ford of America.

A new engine support was fabricated to allow the engine to be raised and moved across to the left to allow the fitting of the front axle differential unit, housed within a modified aluminium sump. This allows the left-hand drive shaft to pass through the sump, with the differential unit on the right.

The only surviving Ferguson Zephyr, restored. It is now resident in a museum in the USA.

The transmission tunnel was also enlarged to accommodate the four-wheel-drive transfer box, which was coupled to a standard Ford C4 automatic gearbox. The Dunlop Maxaret unit, which senses and operates the ABS, was attached to the rear of the transfer box. The brakes were also uprated with a dual-circuit system with larger front discs and a more powerful double-acting servo allowing the operation of the ABS (the brake-master cylinder being the same as fitted to a Ferrari 250 GTO!). The whole package ensured that the wheels would not spin or lose their grip in icy or slippery conditions, but it did add around 240 pounds to the weight of the Zephyr. However, this was to some extent concentrated on the front wheels, and as all 50 cars had power steering, the drivers did not notice this weight.

Once the work had been completed to the satisfaction of the Ferguson engineers, one of the cars, registration number WUL 365G, was dispatched to TRRL (Transport Road Research Laboratories) in Crowthorne, Berkshire, on 2 April 1969 where it was subject to a proving process to evaluate the four-wheel-drive and ABS systems under various conditions on the test track and skid pan. Several modifications were incorporated by TRRL to improve the systems, including extra vacuum tanks to supplement the ABS and electronic counters to record the operation of the ABS. They also added specially developed four-way hazard flasher warning lights to the car, making WUL 365G the first car in the world to receive this now-mandatory safety feature.

On 29 April 1969, the fully sorted WUL 365G was put through its paces at TRRL to demonstrate what it was capable of in front of representatives from various police forces who had been allocated the other cars for evaluation. After the display the other 21 Ferguson-modified Zephyrs were ceremonially handed over to their respective operators by former racing driver (winner of the 1953 Le Mans 24-hour) and war hero, Tony Rolt MC, Managing Director of Ferguson Research.

After the trials were completed, WUL 365G was used for a limited time on general police duties. Unfortunately for the Zephyr development team, while all their hard work was going on, rival engineers at Rover's Solihull plant were beavering away on the revolutionary new Range Rover, which made its debut in 1970. With its commanding driving position, huge load area and unparalleled blend of on-road civility and off-road ability, as far as the police were concerned it rendered the four-wheel-drive Zephyr obsolete overnight. WUL 365G was restored by an enthusiast over many years and in 2010 was sold at auction; it currently resides in the Transport Museum in Tampa, Florida.

Whilst the development of the Zephyr four-wheel drive is fairly well documented, what isn't that well known is that Ford and Ferguson also produced a couple of Mk1 Capri 3000 GTs. One, if not both, of these cars was not only trialled by Thames Valley Police from 1972 onwards but they kept them in service, too. Ferguson also later produced a number of four-wheel-drive Mk1 Opel Senators for military use, mainly around the Berlin Wall in the early 1980s.

Ford Granada concept car

By now you probably understand that it's a rare and expensive experiment when a major motor manufacturer decides to rewrite the rule book when it comes to designing a police car. In late 1986 Ford were given a brief by the police to design a purpose-built car based upon their all-new Mk3 Granada 2.8i four-wheel drive (now a common fit to a number of Fords!), which also came with anti-lock brakes as standard. It had to be built to police specification, which included uprated rear shocks to carry the extra load of around 160lbs of kit. But Ford Engineering's John Willis disagreed with this, saying, 'It's a requirement of theirs, but a bone of contention with us because 160lbs is only equivalent to two-up plus luggage. But the police are the customers so they get their way.'

The gearing was also altered to make fifth gear on the manual gearbox a proper top rather than an overdrive because the police were more interested in speed and response over economy (how times have changed!). The car was fitted with RS alloy wheels and a full body kit to help with the aerodynamics. But the feature that caught everyone's attention was the aerodynamic roof-mounted pod. Representatives from Ford Engineering had met several fleet managers at the 1986 NAPFM exhibition and one of their biggest gripes was that roof-mounted light bars slowed down most cars by almost 10mph and drastically reduced fuel consumption because of the drag. So John Willis took his team and several light bars to a wind tunnel in Cologne to conduct a number of experiments and eventually built this roof pod in conjunction with blue light technicians from Woodway Engineering. Thus Ford's Concept Granada was born. The pod reduced drag by a massive 28 per cent and came with a blue strobe light on every corner, faired in with blue Perspex and illuminated 'police stop' signs front and rear. The top of the pod also hinged upwards to reveal signage to be used whilst static at an incident.

To help complete this futuristic-looking package it also had to be practical. Two blue repeater lights were set into the tailgate so that when it was opened and obscured the roof lights at least the officers were afforded some protection. Inside, the Concept Granada got the very latest kit, including a Marconi RC690 radio system to enable the crew to talk to any force in the country, a Hico-Neas VASCAR system that printed out an offending car's speed from the box mounted on the dash, and a second Plessey computer that could 'talk' to other computers using code words on things like HAZCHEM codes and symbols, or even direct PNC checks on registration numbers or person checks. This item didn't get the go ahead and it's only now in the twenty-first century that such technology is finally being used, but the idea was sound.

In conclusion, John Willis is quoted as saying, 'What we are trying to do is provide the police with the closest they can come to an ideal car, so they'll buy our car rather than someone else's'. So did they? Did it work? Before answering that question, just for the record there were actually two versions built over a three-year period. The first version had a lip on the rear of the roof pod and incorporated the word 'police' along the sides of its then-futuristic-looking blue-and-yellow diagonal graphics (most police cars of this era had plain orange stripes). The second version had a revised roof pod without the lip, and clear lens lights in each corner. The side graphics were revised and the 'police' wording was dropped.

It's not clear how many Concept Granadas were actually built, although the numbers were almost certainly in single figures. Plenty of forces trialled the cars, and the Met Police eventually bought two of them on the cheap once Ford had finished with them. Wiltshire Constabulary bought one of the later versions and kept it for many years, only disposing of it in 2010. The experiment was just that and nothing else. The roof pod, brilliant in design and practice, was hugely expensive to make, and despite the cost savings it made in fuel it was never repeated by either Ford or Woodway.

But the car had a huge impact on those who saw it when it was unveiled. It was light years ahead of anything else we had ever seen. Police car expert, historian Steve Woodward, who has been central to the putting together of this book, told me an anecdote which, whilst a little grisly, explains the importance of this car: 'I attended a murder scene in 1987 when my force was trialling the car and the Traffic boys turned up to help out at a road check the next evening. The car drew crowds like a moth to a flame. Nobody was interested in the murder, only the Concept Granada! The Inspector had to tell them to bugger off in the end. Surely that says it all?' It does indeed, and it seems a great shame that the ideas embodied in this car took so long to filter through and in some cases still have not done so. A classic case of accountants ruining the world by placing costs into boxes then minimising that individual cost without any thought, or knowledge, of what that will do to the overall life cost of operating that car; but that does not matter because it comes out of a different budget. I'm sure the roof boxes would have paid for themselves in reduced fuel consumption over time and would have reduced emissions a little as well. Madness! The Met Police's 2.6-litre SD1 area car debacle detailed elsewhere is another case in point of this phenomenon.

Proof that light bars can have a massively detrimental effect on both fuel economy and performance comes from a lovely anecdote involving the West Midlands force. One warm summer night, the coppers were moving a human organ for transplant from Birmingham to Northampton using a 5.3-litre Jaguar XJ12 Se1 with a full light bar on. Over a distance of around 65 miles, the car, flat out with its blue lights on for most of the motorway route, as time was of the essence, used well over 20 gallons of fuel! A fact the crew didn't realise until they spotted that the previously full tank (which held 24 gallons) was empty as they were setting off to come home with the transplant organ safely delivered. That's less than 3mpg . . . Even allowing for the Jag V12's famously thirsty character, that's pretty bad!

Ford Galaxy 2.8 VR6

And now for something completely different! The Multi-Purpose Vehicle (MPV) or people carrier, as they became known, really took off in the mid-1990s. Although Renault's Espace had been around for several years, it was some time before the rest of the industry caught up. When Ford released the Galaxy as a joint venture with the VW group (as such, it was often called the Sharlaxy, because the VW version was called the Sharon), few would have even dreamed of using it as a police car. However, Ford did just that and came up with a 2.8i VR6 motor taken from its tie-in with VW (as in the Vento discussed earlier) and issued it as a full police specification Traffic car, with a multipurpose role. It could be used as a people carrier, a motorway back-up unit or a combination of both; it could even be had with four-wheel drive. The rear seats had been removed and in their place was a wooden racking system capable of holding a huge amount of equipment, and there was still room for five adults! It was very quick and had excellent road holding for a vehicle of this size. In short, it was far better than the Land Rover Discovery currently in use at that time as a motorway back-up unit, and there were those who were dismayed when it didn't make the grade, although its maximum loaded luggage weight was over 450lbs less than the Discovery, which did worry some forces. It did eventually find favour with a number of forces who utilised the car for Accident Investigation Units.

The Ford Ecostar van

On 13 March 1995 the Hampshire Constabulary held a press conference to announce that it was about to commence a 30-month trial with Britain's first ever electric police vehicle. The Ford Ecostar van was to be used as a section vehicle at Fleet, where it would be directly compared to a Ford Escort 1.6 petrol van based at Hartley Wintney and a Ford Escort 1.8 D van at Odiham. Comparisons would be made on the basis of performance and environmental impact. The idea of using a silent vehicle to approach a crime in progress in the middle of the night captured the media's imagination and the vehicle made national headlines as well as local ones, featuring in several TV programmes and car magazines.

The vehicle was built around a Mk4 Escort van shell. It incorporated unique battery technology and some of the most sophisticated electronics ever installed in a vehicle designed for normal day-to-day use. Some of the Ecostar's key features included a sodium/sulphur battery, providing up to three times the power of a conventional lead/acid automotive battery. It gave a driving range of 120 miles between charges. A high-speed, three-phase AC drive motor delivered the power to a single-speed, direct-coupled transaxle. A system called regenerative braking recharged the battery whilst the vehicle was decelerating.

Its top speed was restricted to 70mph, with a 0–50 time of 12 seconds. Acceleration from a standing start to 30mph was quicker than the comparable petrol and diesel vehicles, due to the instant high torque at low speed. At the end of each day the car would quite simply be plugged into a special socket located in the rear yard at Fleet Police Station and recharged.

Other differences included a unique front grille with a series of different-sized holes in it, unique wheel trims and colour-coded bumpers. A Britax light bar, together with blue-and-white chequered side stripes and all the necessary police signs and force crest, were also fitted.

Generally speaking, the trial was a success. Those officers who drove the Ecostar were impressed with its performance, even if driving a silent vehicle took some getting used to! There are no records to state how many criminals it surprised by creeping up on them. At the end of the day it was just a trial for the force and for Ford. It was used on a daily basis for two and a half years and is now a force museum piece; in fact, it is the only vehicle the Hampshire Constabulary has ever kept. Former Transport Manager, John Bradley MBE, managed to purchase the vehicle from Ford for a mere £100, not bad when you consider that the estimated value of the car when it was first built was a cool £250,000!

Ford Fiesta

There is one more product from Ford that we simply cannot ignore, in part because it's doubtful that most people will have ever heard of a two-stroke Ford Fiesta. Go on, admit it; you hadn't either until just now. The two-stroke engine is generally thought of as a cheap motorcycle engine reminiscent of 1970s Japanese machines that left plumes of blue smoke in their wake, or powered weird, communist block,

Wartburgs and Trabants. But in 1992 Ford developed an advanced two-stroke motor as an experiment and at least one of them was trialled by Surrey Police. The press release read as follows:

A pilot run of 25 two-stroke-engined Fiestas has been built at Ford's Dagenham plant in the UK. They will be used by key component suppliers and selected fleet operators. A further run of 35 prototype cars will be produced during 1992. All 60 test vehicles are being built to gain early service experience of this advanced engine concept as part of the run-up to low-volume production of two-stroke cars, planned by Ford for the second half of the decade. About half the test cars have been built with left-hand drive for operation in Germany while the other half have right-hand drive for use in the UK.

The new Ford two-stroke engine has been developed over the past four years in cooperation with the Orbital Engine Company of Western Australia. Its use of the two-stroke combustion cycle allows the valvetrain to be eliminated and provides double the number of power strokes. This makes the engine more compact, reduces engine weight, increases simplicity and power density and significantly enhances performance feel.

A further advantage that cuts directly against the current two-stroke image is that the new Ford engine also has very low exhaust emissions. As a result of a technological breakthrough that overcomes the traditional disadvantages of two-stroke units, the new Ford engine has already proved it can meet the latest US Clean Air Act emission standards.

Ford in Europe has been working closely with the Orbital Engine Company on the development of two-stroke engines since signing a licensing agreement with the Australian-based company in the middle of 1988. The engine design is aimed primarily at improving vehicle efficiency, increasing running refinement and lowering exhaust emissions.

The Ford two-stroke is an in-line 3-cylinder engine with a displace-

*ment of 1200cc. It develops 80PS (60kw) at 5,500rpm (about 10 per
cent more than an equivalent four-stroke design) with a peak torque
of 125 Nm at 4,000rpm (about 14 per cent more than the torque of
a 1.2-litre four-stroke). The fuel economy improvement over an equivalent four-stroke engine is between 10 per cent and 12 per cent.*

The press release continues but then gets very technical, so it is perhaps
better that we take more notice of someone who actually drove it to
see what he thought about it. Trevor Thompson was a patrol sergeant
with Surrey Police, stationed at Reigate in 1993 when the two-stroke
Fiesta first arrived there on trial.

'I wasn't there when it arrived but heard a couple of the PCs talking
about it when I returned to work. The feelings were mixed; some liked
it, with others were saying it had no guts.

'The keys were on the vehicle board as the PCs working that area
were two who were not interested. So a call came in and I grabbed the
keys and dashed out. It was a priority job so I leapt into the Fiesta. I
started the thing and it barked into life. (I later found that the exhaust
pipe was very much like the drainpipes found on the modified boy
racer cars of today.) My first impression was that it sounded very tinny.
I revved it a couple of times and boy did the rev counter fly up and
down. I selected first gear and found that you had to have more revs
on to set off than you do in a four-stroke car. Not a problem, being
used to motorcycles.

'I cleared the front barrier at the front of the nick, turned right to
pass in front of the station towards Redhill and I was off; the bloody
thing revved and revved like a demented hornet. I was later told that
the senior management team were having a meeting at the front of the
building and they got up to find out what was going on!

'So long as you kept the revs on the engine above about 2500 the
thing flew; there was no engine braking, but who cared, it was a different
way of applying the 'Police Driving System'. You had to think about the
engine and power band, but once you got used to it, boy was it fun.
None of the Astras or Escorts could get anywhere near it.

'After that when I went into work there was no argument, the keys came off the board and were in my pocket. When I got back to the station and got a chance I had a good look under the bonnet. The engine was so small it just sat looking at you, narrower and set lower than I thought and with a big two-stroke oil tank tucked away.

'The looks the car got especially when I stuck my boot down! The boy racers knew to beware and the word was that we'd had it specially modified to deal with them. It turned heads everywhere. And there was no typical blue smoke from it either.

'We did not have to bother with two-stroke oil when refuelling, we only had to fill up with petrol as with other police cars . . . the two-stroke oil was only examined and refilled when the car went for service. So no mistakes to be made by numpty coppers . . . it was "Copper Proof", if ever there was such a thing.

'I was sad to see it go and I gave it glowing reports, pity the EU put the death knell on two-strokes; it had everything going for it and I believe it was less polluting than most other cars.

'The strange thing about this tale is that the 100 or so engines produced for the project were built for Ford by a company in Perth, Western Australia, just down the road from where I now live. I must find the time to write to them with a view to finding someone who worked on the engine to let them know what I thought of their work.

'It was the best engine I have ever come across for its size and I have driven some crazy stuff in my life, including Works rally cars as well as the stuff the police got.'

Sadly, no photos of this police version of Ford's two-stroke car seem to exist. However, if you wish to go and see the engineering, the excellent Haynes Museum in Sparkford, Somerset, actually have one on display with the bonnet open for examination.

Ford Mustang

In 2016, about two weeks before the NAPFM exhibition, rumours abounded that Ford would debut the Mustang 5.0 V8 on their stand. The Mustang was every boy's dream car but it had never been officially imported into the UK in right-hand-drive form, so definitely no police versions then. But now the company had finally bowed to public pressure and built the car with Europe very much in mind. It got good, if not rave, reviews from the motoring press, but the public lapped it up and the cars sold like hot cakes. To have a fully marked-up police version really was hot news and, let's be honest, it does look the business. The staff on the Ford stand were quite insistent that the car was definitely a serious effort and not just a showpiece celebrity item to draw in the crowds, but in reality that's exactly what it was. No police force was going to buy a two-door coupe with a 5.0 V8 under the bonnet in this day and age. Since the exhibition, the car has gone out on loan to a couple of forces but was only used as a promotional item at public events.

Audi 100 and 80

In 1986 Bedfordshire Police tested a rather nice-looking Audi 100 quattro. In the mid-1980s Audi weren't exactly the household name that they are today, but their big advertising campaign of the period included the strapline 'Vorsprung durch Technik' ('Advancement through Technology'), which seemed to strike a chord with the public, especially those adverts which were voiced over by the lugubrious Geoffrey Palmer explaining how the Reinhardt family get to their holiday villa in time. The quattro 4x4 system had turned the rallying world on its head in the early 1980s and really started to build the foundations of Audi becoming a competitor to BMW and Mercedes. The 100 was a first quattro (and before you write in, Audi insist it's a small q) on a mainstream saloon car, and its credentials seemed to be ideal for police use. The car was aimed at the traffic fleets and in Bedfordshire was up against established cars like the Mk2 Ford Granada and the Mk1 Vauxhall Senator. Audi's big sales pitch majored on its aerodynamic efficiency, which was a big deal in the 1980s, and they quoted a drag coefficient of just 0.30, thanks in part to its revolutionary flush-fitting windows, which also added to its sleek looks. The 2144cc, fuel-injected, inline 5-cylinder engine pumped out 136bhp@5700rpm, giving the car a top end of 125mph.

Bedfordshire Police obviously had it on long-term evaluation because the car got the full Bedfordshire livery treatment complete with the force crest on the doors. On paper the car looks like a winner, but for reasons unknown it wasn't taken on, although it may well have come down to cost on things like technician training, parts supply and initial price. The Audi 100 was the first Audi ever trialled by the UK police, and a couple of years later the slightly smaller Audi 80 also got passed around a number of forces but again failed to make the grade. It wasn't until the second decade of the twenty-first century that the German car finally got the green light, and a number of forces currently use the A4 and A6 variants.

Alfa Romeo

At exactly the same time as Bedfordshire were trialling the Audi they also took a long hard look at the Alfa Romeo 75 Twin Spark 2.0i. This 4-cylinder version used Alfa's classic all-alloy double-overhead-camshaft engine fitted with two spark plugs per cylinder (hence the name) and produced 148bhp@5800rpm, giving it a top speed of 126mph. So it was a quick car and its transaxle design ensured good handling, which magazines raved about, but where would it fit in? It was physically the right size for area car use but probably was of too high a performance, and had too complex a maintenance schedule. However, it wasn't quite capacious enough for traffic patrol so it fell somewhere in the middle, like the Vento in a way. As with the Audi, overall cost figures probably played a part in it being declined, although Alfa's woeful resale values and reputation for fragility during this era probably also played a part.

Mind you, this wasn't the first time the Italian firm had made an approach to the UK police market, in 1979/80 they put together an Alfa Romeo Alfasud and an Alfetta in full police livery. The Alfasud had a 1490cc engine and front-wheel drive but was slightly too big, and certainly not economical enough, for panda car use and was too small as an area car.

The Alfetta 2000L had been rather successfully used by the Polizia in Italy as its mainstream patrol car, so it probably made sense to try their luck over here. But it was a short-lived trial. Like the Alfasud, it just didn't fit in anywhere, and if that was the case the rest of the trial was pretty irrelevant. No doubt the performance was pretty good, given Alfa's tradition of producing good drivers' cars, but build quality and reliability have always been a bit suspect with those cars, and that is a massive part of the policing criteria.

SsangYong Musso GX220

One of the strangest-looking vehicles to be tested was the SsangYong Musso GX220 four-wheel drive. This Korean-made off-roader was fitted with a 3.2i Mercedes-Benz engine that gave it excellent performance, and although it actually looked a little lumpen it had been styled by Briton Ken Greenley, who had previously worked on both the Aston Martin Virage and the Bentley Continental. It was quick, handled like a car and had as much room as the Range Rover. It did suffer from a rather poorly made interior and bits did tend either to break or just fall off.

The car was tested in 1997 over a period of four weeks, mainly at Whitehill Traffic Section. It was given full Hampshire graphics because of its extended test and even received a fleet number. But the car failed its overall test for unknown reasons. It was later tested by Humberside Police, still in Hampshire's livery! Perhaps it's good it never made the grade as a police car, Musso apparently means 'water buffalo' in Korean, and you can imagine the jokes, can't you . . .?

Reliant Regal

No, not the 3.0 V6 Scimitar but the three-wheeled Reliant Regal. During the 1970s oil crisis police forces were quite literally forced to save fuel and cars were only allowed to go out if absolutely necessary. One force therefore trialled Reliant's frugal 'plastic rat' as a panda car; they had it painted blue and white and invited officers to give it a go. With its Austin 7-based engine of 701cc and scintillating 29bhp it didn't entice, and the three-wheeled stability need not be discussed in length here. For some reason they all refused! The trial was probably the shortest in police car history.

Chrysler 180/2 litre

Chrysler UK Ltd was formed in June 1970 when they finally bought out the old Rootes Group totally, having first purchased a stake in 1964 and effectively taken control by January 1967. Slowly the old Rootes brands – Humber, Hillman Sunbeam and Singer – disappeared and they even rebranded existing products so that the later Hillman Hunters became Chrysler Hunters. Their first product, the Avenger, was a simple Ford Escort competitor that had been developed by the old Rootes team. It did well in police use, since it was slightly roomier than the Escort, and it had a good reliability record. However, Chrysler's next car, the code-named 'C-Car' project, was something of a debacle. Chrysler had purchased Simca in France around the same time and effectively amal-gamated the large car programmes both companies were developing. The end result used the basic engineering and styling of the UK car but with a Simca-styled nose and interior. Most importantly, the new 3-litre V6 engined version, which would have been badged as a Humber, built in Coventry and fitted with a more traditional wood and leather interior, was cancelled at the very last moment. The tooling and development of the new engine had been completed *and* installed in the factory in Coventry, at a combined cost of £31 million! At the last minute Chrysler got cold feet, though, and pulled the UK assembly of the car, meaning it would only be available as a 1.6, 1.8 or later a 2 litre, and only be built in France. The UK tooling was ripped out of Humber, and Rootes'/Chrysler's long-developed, up-to-date and apparently very good 3-litre V6 was never produced. Thus ended a chapter in UK policing: no more large Rootes-designed police cars from Coventry. A sad end for a glorious record and a fascinating insight into what might have been if the V6 engined 5-speed version of the rear-wheel-drive Chrysler 180 had been produced; it would surely have been a very good traffic patrol car. As good as a Granada? Sadly, we'll never know.

Chrysler did prepare a 180 1.8-litre police demo car, which is pictured on the next page, and, as you can see, it does look alarming, like a sort of big Avenger, but I suppose that's no surprise as the basic style came from the same styling studio, led at the time by designer Roy Axe. It was touted around a number of forces and met with a lukewarm response that typifies the car – it *wasn't bad*, but it wasn't really desirable either and no one actually really wanted one! Sadly, that applied to civilian customers as well as the police because it sold appallingly in both the UK and France and as such is now super rare as a classic car. However, it was used in smallish numbers by at least three forces – Lincolnshire, Sussex and Warwickshire – mainly as a Traffic car. Lincolnshire had the largest number. These days it's a forgotten car, duly remembered for its small role in British police history.

The Chrysler 180 demonstration car when just finished and photographed for their publicity material.

Talbot Tagora

Every now and then a car comes along that gains instant legendary status in police circles for all the wrong reasons; think Morris Marina or Austin Allegro. The Talbot Tagora is such; a car that induced genuine fear amongst the few officers unfortunate enough to drive it. It followed on from the Chrysler 180 but was a bigger car, aimed squarely at the executive market which was growing at this point, post energy crisis. Originally called Chrysler C9, its styling was developed in the UK but its technical development took place in France at what had been Simca. Sensibly, they decided they wanted a V6 engine to compete in this market, but having bizarrely cancelled a good one only a year or two previously, they did not have one to use! They tried to buy the Peugeot, Renault, Volvo (PRV) joint venture unit, but were rebuffed on the grounds that it would be fitted into a competitor car. They were still wrangling about this when Peugeot bought the company in 1978, for $1! Chrysler were really keen to get out . . . PSA delayed production of the C9 in an attempt to fit more PSA group parts – although at least the engine argument was now sorted . . . The re-engineering left the rear track too narrow, giving the whole thing a slightly pinched look, but it emerged at the Paris show in 1980 and sank without trace in every marketplace in the world! PSA group, of course, already had the successful and elegant Peugeot 604 and the avant-garde Citroen CX in this market segment, so adding a Tagora, which again was only made in France, seems illogical. However, it was too far down the road to cancel when they took over the company.

Peugeot UK, as they had become by 1981, produced a police version in the hope of creating some positive feeling around the car because, after all, if the Old Bill are using your cars you must be doing something about right. But as we know, coppers are bloody good at moaning and the Tagora set new records in the amount and the volume of their outbursts. They hated the car. It was slow, handled badly, rolled from side to side, had poor brakes and had a build quality that made Leyland products look like they'd been built by Rolls-Royce. The demo car was passed around a number of forces, mainly in the north, including

Glasgow in 1983. For some inexplicable reason it managed to secure a couple of orders from Cumbria Constabulary, Northumbria Police and the West Yorkshire Police. The Talbot-badged arm of PSA expected more; their archives show they hoped to sell 100 cars a year, and they had settled on a police spec of grey plastic seats and specially made TRX steel wheels. The car also came equipped with three Weber 2 barrel carburretors and apparently had almost race-car-like intake roar!

Rover

In February 1994 BMW took over Rover and the British-owned mass-production car industry died. It wasn't a marriage made in heaven and BMW sold what they had by then termed 'The English Patient' in 2000. The takeover came during the Rover 800s' reign as a pretty successful upper segment car both in the public's eyes and with the police. The larger-engined 827 V6 used Honda's 2.7-litre engine, which won a lot of praise. However, with the BMW takeover, a new engine was required for the series 2 Rover 800 launched in 1996, and a new in-house engine, the KV6, was developed. So what's all this got to do with this chapter on demonstrator cars and missed opportunities? Well, somewhere along the line BMW popped its own multi-valve 2.5i straight-six engine into a Mk2 Rover 800 and gave it to Strathclyde Police for them to evaluate over the next three years. The car was based at Motherwell Traffic Garage and the workshop manager at the time, Alex Lachlan, confirms that the car was given three years' hard running before being returned to Rover. Rumours abounded at the time that the car was subsequently crushed, and while this cannot be confirmed, it's unlikely that it was ever re-sold to the public. Nevertheless, it was an interesting idea and a rather clever method of testing an engine whilst hiding it in a car nobody would take much notice of. It may also have been a way of evaluating a front-wheel-drive installation for an engine that hitherto had only been used in rear-wheel-drive cars.

Daihatsu

A new manufacturer to British police forces in the late 1990s was Daihatsu, who arrived with the Mk1 Fourtrak four-wheel drive. This robust and reliable diesel-powered, four-wheel-drive car looked like the ideal vehicle to be used as a rural area car. The trial was a long one and proved to be successful. However, a decision on purchase was delayed because Daihatsu were about to release a Mk2 model, which was slightly bigger than the original. Once the police were happy that the new one was basically as sound as the original, a number of forces, including Wiltshire, Hampshire and Dorset Police, went ahead and purchased them. It proved the value of trialling the cars first.

The prize for the strangest car of them all has to go to the Daihatsu Move. It remains one of the weirdest police cars of all time, looking like an upturned wardrobe on castors, but it actually proved to be a very practical tool. It was on trial for two months prior to purchase by the Hampshire Constabulary. The diminutive Daihatsu was a mere ten feet in length, but it still managed to have five doors and five seats. The 850cc, 3-cylinder engine was good for 80mph and returned an average 47mpg. Why was it ever used as a police car? Following a successful trial in the New Forest, it was purchased and used by rural beat officer PC Colin Gordon at Ashurst. Like most other rural beat officers, he had the use of a lightweight motorcycle to go about his business. The Daihatsu had several advantages over a motorcycle. First and most importantly in PC Gordon's eyes was the fact that he remained dry; previously, if he had been on an enquiry somewhere and the weather was bad, he would spend the first ten minutes on arrival removing wet-weather clothing and apologising for dripping all over someone's carpet! His paperwork remained dry as well. He was also able to take passengers without the need to call up another four-wheeled unit, and as the rear seats folded down, the Daihatsu was also capable of carrying small loads. With typical Japanese reliability, it was actually cheaper to run than a small motorcycle.

A Premier Hazard light bar was fitted by Daihatsu prior to delivery and Hampshire decorated the car in its old traffic livery, complete with a small amount of rear striping in orange and yellow, and, believe it or not, it also came fitted with a siren! The car drew comments from the public wherever it went. Some laughed, some scorned, some shook their heads in disbelief, but all of them talked to the officer who was driving it and it really helped break down some barriers, in particular with adolescent kids who couldn't quite believe what they were seeing. In that respect alone the car has to rate as a success!

Riley Riviera and Silhouette

We did mention the BMC 4-cylinder Farinas in the area car chapter, but there was also a private enterprise attempt to revive the Riley marque in police use based on the Riley 4/68 version of that family of cars.

BMC was formed on 31 March 1952 (a day too early?) and, although announced as a merger of equals, was a takeover of Nuffield/Morris by Austin; something Austin boss Sir Leonard Lord had sworn to do since falling out with his then employer, Morris, before World War II, and swearing to take Cowley apart 'brick by bloody brick'. By 1958 Lord's plans to rationalise BMC into 'a first division car maker' (no premier league then) were taking shape. He had to keep the Nuffield makes – Morris, Wolseley, MG and Riley – because most UK towns had competing dealers with a loyal customer base who were selling only one or two of these marques. Lord could see that the solution was to make one car with many names: enter, in November 1958, the BMC Farinas. All were based on a re-skin of Austin's 4-cylinder Cambridge and 6-cylinder Westminster by Italian stylist Pininfarina and featured unique front and rear styling plus different interiors, created, ironically, at Cowley by stylist Sid Goble's team. However, the two more sporting models, the Riley 4/68 and MG Magnette MkIII, were roundly criticised (in a polite way; we are in 1958) as disappointing, lacking the luxury, elegance and easy power of the Riley RME or the sporting élan of the Z-type Magnette. It was into this obvious market gap for mid-sized sporting saloons that Mr John Goodfellow, the MD of Wessex Motors in Salisbury, wished to place his brainchild, the Riley Riviera; a heavily modified version of the 1489cc Riley 4/68 featuring an MGA engine, Healey 3000 front disc brakes and modified styling.

Goodfellow, a larger-than-life ex-military man with a severe limp, known as G by the staff, was always accompanied by a glamorous secretary who, unsurprisingly, every red-blooded young Wessex mechanic remembers more fondly than the cars! Goodfellow lived in Stockbridge and often roared into work in a bright red, supercharged, pre-war Alfa Romeo that Wessex had restored FOC by booking the hours to accident repair jobs! Wessex was a large company, bought out by Henlys in 1956 but left to run autonomously. They occupied all of New Street, Salisbury, and employed at least 50 people on site, but also had satellite garages in Andover and Winchester holding franchises for Rolls-Royce, Rover and Guy trucks, as well as BMC (not Austin, which was handled by a another local dealer).

The modifications for the Riviera were planned by Goodfellow, Daniel Richmond of legendary tuners Downton Engineering, who was almost part of the business, and their mutual friend Eric Fenning, another local BMC dealer and a successful racer. Although all Rivieras were built to order, Wessex published a price list of three conversion levels and chose to launch the car in December 1960, perhaps not realising that many of their changes would be rendered less appealing by BMC's own 1622cc update of the whole Farina range, which was to be announced only a year later.

Wessex's body man, Stan Gordon, cut the fins off to a down-ward-curving line drawn on the side of a car by Goodfellow, while their engineer Maurice Townsend raided the parts bin to work out how he could fit disc brakes and a bigger engine. The standard Riley 4/68 was the most expensive of the Farina range and was listed at £1,028 including purchase tax, while the base A55 Mk2 Cambridge cost only £802. Wessex offered three levels of 'improvement' for their Riviera, the cheapest of which, at £100, included an MGA 1588cc engine, offering a 15bhp increase in power, recessed fog and spot lights, a handbrake warning light, altered rear fins/doors, plus, bizarrely, Wolseley 15/60 front seats instead of the Riley's more sporting buckets. Even more oddly, the standard Riley rear seats were left in so the front and rear seats didn't match! Maybe Mr Goodfellow only managed to 'get' front seats out of the back door of BMC but didn't have a contact for the rears . . . The front disc brake and servo option was in the second level of modifications, which cost a further £100 and also included 15-inch wire wheels. These two modifications had to be done together, otherwise the cars would have had mismatched wheels front to rear. Those customers wanting to go for the full conversion, which cost an extra £71 on top, also received high-compression pistons and related engine tuning, an HMV radio with an electric aerial (surely a novelty in 1960?), extra sound insulation, a Servais straight-through silencer and, most weirdly of all, a rear window de-mister which was not connected to any hot air supply and so literally just blew cold air from inside the boot (where the petrol tank vent is . . .) at the rear window; this made a great deal of noise but had little or no effect on clearing a misted-up rear window beyond giving any villains travelling in the back reason to complain about a weird humming noise and a mysterious draught . . .

Wessex launched the Riviera locally with an event from 1–12 December 1960 and claimed to have loaned the car to various police forces around the south of England for evaluation as a high-speed car. Whilst the engine and braking mods could have interested the police, I'm sure budgets even then would have been looked at very hard before cut-down fins or hand-made brass badges saying the word 'Riviera' were being paid for!

However, this all became immaterial because, unbelievable as it may seem, Leonard Lord and BMC's hierarchy only found out about the Riley Riviera project when *Autocar* published a review of the car in its 3 March 1961 edition. They were not amused! After that, the project faltered and it's believed around 12 to 16 were made before BMC pulled the plug. Goodfellow eventually negotiated with BMC to have another go, and he made three Riley Silhouettes using the same basic bodyshell but with the later 1622cc Riley 4/72 as a base, and he fitted an MGB 1800 engine. Again, these were loaned to the local police for evaluation and apparently were quite well liked, but cost was a massive issue when compared with new cars such as the Cortina GT, and none was purchased.

Only one running Riviera is known to exist and this has been restored by an enthusiast, Mick Holehouse, who also researched the history of the project; much of the information I've shared here would not be known if it were not for his work.

Mick Holehouse driving his Riviera, with its cut-down fins and unique badge, plus that wonderfully pointless rear window de-mister.

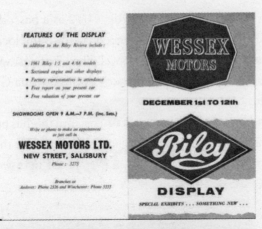

Wessex Motors' original sales material for the car.

CHAPTER THIRTEEN

FANTASY CARS

Policing is all about communication; no matter what the role, a copper's job is all about having the ability to communicate with people on any level and on any subject. Whether it is a foot patrol officer listening to the gripes of an old lady plagued by young hooligans on her street, the traffic officer about to knock on someone's door to deliver the worst message in the world, or the detective interviewing a rape victim, every officer must have the ability to listen, empathise and talk appropriately. Gathering evidence doesn't just mean dusting for fingerprints or crawling on your hands and knees looking for clues, it also means listening to the public, usually at grass-roots level, and noting their concerns over a neighbour or some slightly suspicious incident. On the face of it these comments may not mean much, but any one of them could be that piece in the jigsaw that your police colleagues have been looking for.

Whenever things go wrong in society the police often get the blame for being oppressive or overzealous, and those with an axe to grind are all too happy to face the camera and sound off about it. When things have calmed down a bit it's usually the local beat bobbies who go back in and start the softly softly approach by taking the time to talk – especially to the younger element, who are often the most volatile. It's a difficult task, but breaking down those barriers is absolutely vital and the key to good policing.

When the police do it well, they usually do it very well. They are good organisers and are forever coming up with innovative ideas to bring the community together, and to let everyone know that it's okay to talk to the Old Bill because they actually have everyone's interests at heart. They attend local fetes and other community events, they'll referee a local charity football match or even put together a team themselves to play against the local youth (always an interesting experience! You know how much I love football, and I played some unlikely games in the name of police public relations). Or they will organise a 999 show and gather the Fire and Rescue Services, the Ambulance Service, Coastguard and a whole host of other organisations in one place then invite the public to come and meet them and talk to them. Mounted police and police dog handlers play a part in this because the British public are an animal-loving people.

Sound advice from such professionals is gladly given out to adults on subjects like first aid response, fire safety, crime prevention and a thousand other subjects, whilst the kids usually go home with a badge, some new pens, a plastic police helmet and a huge smile! In the aftermath of the recent terrorist attacks in Manchester and London there were a lot more armed police officers on our streets who were actively encouraged to talk to people; the media did their bit by suggesting that the public should approach these officers and have a chat with them because, even though they were armed to the teeth, they were still approachable. It was heartening to see a huge number of photos in the papers and on social media of officers posing with the public, all of them with big smiles on their faces. Barriers broken down, fear reduced.

So what's all this got to do with police cars? Apart from being the practical tool to aid officers to carry out their work, police cars are often the very subject that members of the public will approach an officer to talk about. Young lads in particular want to know how fast your Volvo T5 goes, or they'll brag that their mate was driving a stolen car and managed to outrun the cops in their BMW 330d 'cos he's a better driver than you lot'. Whilst at times it can be a bit irritating, especially if you are right in the middle of a job, it does indicate that the cars the police drive can and do act as an ice breaker to enable communication to flow, even if it's only a brief encounter, with a bit of banter thrown in, so that youth will leave with a much more positive attitude towards the police.

When the police support a local event like a school fete or a community day they will more often than not take one or two patrol cars along, purely because the public love looking at them. They want to know how fast it goes, what it's like to drive at 140mph on the motorway, how the VASCAR and ANPR work, what other kit it carries, and then that bloke with the elbow patches and the dodgy-looking haircut appears again and asks that age-old question: 'Is it souped up, mister?' Meanwhile, all the little kids want to get inside it, turn on the blue lights and hit the siren button. It can be a long and very noisy day. But the car is a tool that assists the police in building those vital bridges with the public.

By now you hopefully understand what the police car or bike is all about, apart from the obvious. The combination of car, police officer and public is a proven recipe for good communication, so why not go one step further and introduce some exotica, some eye candy, some cars that stop the public dead in their tracks as they let out a 'whoah, I didn't know you guys had those', and instantly the ice is broken. Police officers are superb blaggers of stuff. If it's shiny, they want it, and more often than not they'll get it. And if there's a new supercar around they'll go to extraordinary lengths to grab one, get it marked up as a police car, then use it – not as a genuine patrol car, but as a publicity tool at that local event we've talked about. More often than not the media gets involved, quite often at the request of those same officers who are using the car, and the publicity it generates encourages the public to come to that event. It works brilliantly; except for one recurring theme, and that's those who miss the point completely.

Almost without exception, as soon as the celebrity car is put into the papers or on social media all the moaners come out of the wood-work like flies descending on a dead carcass. It's almost like they've all been issued with the same little red book and they choose one of half a dozen derisory quotes and, like all good keyboard warriors, they just can't help themselves:

'What a waste of money.'

'As a taxpayer I object to the police driving around in such luxurious cars.'

'Why can't they just use diesel Astras like everyone else?'

'I'll be writing to my MP in the strongest possible terms about this.'

'I thought the police budget had been slashed? I don't think so.'

'How can they justify using such cars?'

Pardon me whilst I yawn! Dave Gorman bursts the pomposity of this type of comment superbly on his show *Modern Life is Goodish* quite regularly, so I won't go any further down this road other than to say that I would have thought, obviously, these cars don't cost the taxpayer money! Just as manufacturers are keen to produce demon-strator cars for police assessment, and sometimes even end up altering road cars once the police have broken their lovely demo car, so they are keen to get publicity of these and thus foot the bill for the Bill.

One of the first celebrity cars that I can recall came in around 1992 when Grampian Police got hold of a Jaguar XJ220 supercar, popped a light bar on the roof and the Grampian Police crest on the doors, then took a photo and turned it into an enamelled pin badge which they sold on behalf of their in-house charity, the Grampian Police Diced Cap Appeal, which still runs today. The pins sold like hot cakes and the photo appeared in newspapers and on the internet. It had an appropriate registration plate of 999 JAG attached to it as well.

A year or two later, a yellow Lotus Esprit S4 suddenly made an appearance in the motoring press, the dailies and even on *ITV News*. It looked fantastic with its Federal Signals V Bar on the roof, but for once it wasn't the idea of the police but Lotus themselves, although it wasn't long before Norfolk Police and then Warwickshire Police cottoned onto the idea and borrowed it for all manner of local events.

The Head of PR at Lotus Cars from 1991 to 1996, Patrick Peal, now takes up the story.

'We came up with the idea of the Lotus police car to gain some positive coverage and improve community relations at a time when there were many dark clouds around the company and not much other good news. We arranged with Norfolk Constabulary to make a car available for display at community and local events such as the Royal Norfolk Show, school fetes and so on.

'We got help from 3M with the decals (specially produced to be not too sticky so we could return the car to press car duty when not being a police car) and from the top UK supplier of light bars (mounted on a spare removable roof panel which we had to reinforce), and the rest was done by the Lotus skunk works ably led by Bob Anderson and Colin Moy.

'It's Colin in the car, with another Lotus employee being the local youth on a bike. The setting is the turnoff from the A11 dual carriageway to East Carleston and, yes, we had police in attendance to make sure we didn't get into the role of police patrol for real ... even so, the number of cars that hit the brakes as they saw us was most rewarding!

'We also got our five minutes of fame on *ITN News at 10* (in the good old days). Another middle-England police force (Warwickshire) was so struck by the car they asked if they could borrow it for a high-profile road safety campaign, which attracted national TV coverage as well. It led to all sorts of letters to the editor which ranged from "sign me up for the police if they're using Esprits" to "I'm hanging up my car keys if the police are using Esprits".'

Steve Woodward: In 1999 I organised a huge 999 Emergency Services Show with classic police, fire and ambulance vehicles attending from all over the UK and Europe, together with all the in-service professionals giving out that aforementioned good advice together with the hats and badges. The event was staged to educate the public and to raise funds for the RUC Dependants Trust. Amongst the many interesting vehicles at my disposal was a genuine Ford Crown Victoria

Police Interceptor from the Orange County Sheriff's Department in Florida. I discovered that the car was often used by the Northumbria Police as a publicity tool and I eventually managed to entice the owner down south; we used the car to gain as much publicity for the event as possible, with photos and a write-up in the local media followed by a ten-minute slot on BBC South Today where we took it to a local shopping area, filmed the car and interviewed a few members of the public, most of whom gave it the thumbs up. Then this really annoying bloke with elbow patches and a dodgy-looking haircut turned up and ... well, you know the rest! But from the point of view of the interest the car generated it was certainly worth all the hassle involved in getting it down there. But as a work tool to drive? Erm ... no thanks, it was a truly awful car; imagine driving a Range Rover with four flat tyres and that just about sums up the Ford Crown Vic.

Some police forces are better at this sort of thing than others, and one of those is the Durham Constabulary who have used a number of high-profile cars, dressed them up and thrust them onto the red carpet. In 1996 they produced one of their best ever with an Aston Martin DB7. This is surely the stuff of dreams for all of us with four star running through our veins. A fully marked-up police spec Aston Martin. What's not to like? And that was the point. It drew crowds like bears around a honey pot, with everyone wanting their photo taken with it or whilst sitting in it. The registration on this one was RV1, which if memory serves me correctly did in fact belong to the proprietor of the local Aston Martin dealership, which would suggest they probably supplied the car for the event whilst Durham Police dressed it up.

It wasn't the first time an Aston Martin had been used. That accolade goes to the Lincolnshire Constabulary who managed to grab an Aston Martin Lagonda, of all things, back in 1977. This huge car featured Aston's famous hand-built 5.3-litre quad-cam V8 mated to an autobox, Space Age looks, cutting-edge display electronics and a fuel consump-

tion of around 10mpg . . . A natural police car! It was 'dressed' in
Lincolnshire's livery for an exhibition of policing through the ages at
the Lincolnshire County Show and was used on the road as a police
car, if only for a few hours to do some pictures. It created a massive
amount of PR, which of course was the point, as it was, bizarrely if
you think about, the Lagonda's first public showing since it had stolen
the headlines at the 1976 motorshow. The car's designer, William Towns,
would surely have been proud!

One of the most celebrated of these publicity cars was the Ford RS200,
a 4x4 Group B rally weapon that never really got the chance to make
the history it surely deserved because the Group B Formula was
withdrawn before the car was properly developed, after a number of
tragic accidents involving spectators and drivers made the FISA realise
that doing 160mph, on ice, in a forest, was not sensible. To anyone
who saw them or was involved, the Group B cars are legends today,
and deservedly so. In order to homologate the mid-engined four-wheel
RS200 for motorsport, Ford had to build 200 of them, and they did
so – or rather, well-known fibreglass and three-wheeler experts Reliant
did on Ford's behalf in their Tamworth factory. Even in 'standard'
trim they offered a 0–60 time of 6 seconds or less and utilised a
246bhp@6500rpm transversely mounted twincam engine 4-valve per
cylinder 1803cc turbo-charged unit. In competition form this easily
made 650bhp, and some of the rallycross specials later went much
higher than that. With an on-the-road weight of 1,285 kilograms
because of the large amount of cutting-edge (for 1985) composites
and glass fibre used in the construction, this bespoke motorsport car
had handling and performance undreamt of by most, although the
top speed was just 140mph because of the low rally gearing. With the
car banned from its main reason for existing, the World Rally
Championship, Ford had a problem and decided to recoup at least
some of their investment by getting coachbuilders Tickford (owned
at this point by Aston Martin, who were in turn owned by Ford) to
trim them to a Ghia specification and sell them as a luxury sports

car. As a driving machine they were apparently sublime, but as a car to actually use they were still coarse, harsh racing machines that suffered from a lack of decent heating and ventilation and a very noisy transmission system, although apparently their ride was surprisingly supple.

So what better way to generate a bit of publicity than to tell everyone that the police were now using them? The rumour mill went into overdrive when Ford's publicity department had a couple of photos of it splashed all over the papers and motoring magazines. Being a Ford, the car was liveried up as an Essex Police car, with full graphics and light bar. Never before (nor since) has one car spawned so many stories. Here are just a few of them:

Essex Police have two of these cars now.

They don't have the RS200 engine, just a bog standard 1.6-litre Mondeo engine.

Two Essex Police officers were killed when they wrapped the car around a tree!

An Essex copper died when he lost it and hit a bridge parapet.

This car was tested by most of Britain's police forces but was too fast for them to handle.

The sneaky buggers are using an unmarked RS200 in Essex.

Ford is selling these RS200s to the police for the same price as a Granada.

But the bottom line is that the car was most definitely a Ford publicity stunt. Essex and possibly Suffolk Police only got to drive it on a test track at Boreham. And thankfully no police officers were killed in the process!

No doubt there will always be those who have convinced themselves or others that they were pulled over by Essex Police in an RS200 or they saw it flat out on the A12 at 155mph or whatever, but without doubt this car was never, ever used by the police as a patrol car or anything else. How can we be so sure? Because coppers like posing for photos with such cars to prove that they did in fact drive it, and suffice to say there are no such photos in existence.

In 2016, those good people in Ford's publicity department struck on a new idea for a retro theme. At the exact same spot where the police RS200 had stopped that Cosworth Sierra, they placed a civilian model RS200 having been stopped by a fully marked-up police model Ford Focus RS. Genius.

Jaguar has produced some of the very best-looking cars ever made and a good number of them have become successful police vehicles throughout the UK. But in early 1998 a photo of a burgundy-coloured Jaguar XK8 in full police livery appeared in a January edition of *Auto Express* with only the briefest of explanations. An approach was made to Jaguar Cars for an explanation and this was the response:

'In 1997 Jaguar Special Vehicle Operations in Coventry prepared this car, based on the XK8 coupe. The 150mph 2+2 4.0-litre car, dubbed the Swift Pursuit, was loaned to the Bow Bells (Metropolitan Police) Motor Club for entry in the European Police Motor Club Rally in Tramin, Italy. Marked up in current Metropolitan Police livery, the car was crewed by Superintendent Peter Dowse and PC Steve McCabe and, once over the Channel, it linked up with over 200 police crews from all corners of Europe.'

Almost every red-blooded male hankers after the thought of driving a Lamborghini. How many of us had a picture of a Countach or Miura on our bedroom walls as a kid? They are top of many people's shopping list if they ever win the big one on the lottery. So what better car to draw the public in, dressed up as a police car? Not once, but twice, and both times by the Metropolitan Police. In 2006 the Italian Polizia hit the world's headlines when they were given a Lamborghini Gallardo as a gift to patrol the streets with. Photos of the fully liveried car were splashed all over social media and appeared in every motoring magazine known to man, with some journalists asking why the boys in blue over here had to make do with a diesel Astra whilst their Italian colleagues were being treated to something rather more special.

So someone in the Met hit on the idea of doing the same as the

Italians. Well, almost. The Metropolitan Police had an initiative called Road Safety Week coming up, with lots of media focus on promoting all manner of road safety issues, so what better way to promote the entire thing than a bit of eye candy to drool over? They managed to acquire a yellow Lamborghini Gallardo from a local dealership, decorated it in Battenburg graphics, a blue light and Met crests and sent it out to work. The press had a field day and teased the public by saying that it would be patrolling the Met's section of the M1 and M25 in an effort to reduce motorists' speed and bad driving habits. Within seconds those members of the public armed with one of those little red books were there again with their 'waste of public money' quotes.

A year later the Met followed it up with a silver Lamborghini Murcielago, giving it the full Met Police graphics treatment. They invited the press in, who obliged by splashing it all over their respective papers and magazines and, hey presto, instant publicity to promote this year's Road Safety Week and not one penny of tax payers' money was used in the process, other than maybe a few gallons of high-grade petrol to quench the beast's thirst.

The Lotus Esprit that the company dressed up in the 1990s wasn't the only Lotus to get the police treatment, with at least two more models dragged in to raise awareness of certain public safety issues.

In 2007 Sussex Police took delivery of a Lotus Exige, fully marked-up in their livery, and it was initially used by the Roads Policing Unit based in Hastings. The £35,000 Lotus had 220bhp on tap with a top speed of just under 150mph and a 0–60 time of 4.1 seconds. Sergeant Paul Masterson, who worked on the force's Casualty Reduction Team, said the Lotus was the fastest police car that he knew of, although standard police cars can reach similar speeds.

'As far as I know it's probably the fastest quoted speed of any police car I know of, certainly in the south-east region,' he said. 'Its top speed is just shy of 150mph, which is what most of our bigger patrol cars can do anyway.'

Sergeant Masterson described the car as impressive. He added that the car was not an operational police car and was unable to get involved in high-speed chases. 'The idea is that drivers will come to us and have a chat,' he said. 'We generally turn up at events where youngsters hang out, and with this vehicle they actually come over and want to talk to us. Then we introduce the road safety message.'

Then in 2011 a Lotus Evora appeared, and this one caused quite a stir when it hit the national papers and even the TV news. The Central Motorway Police Group (CMPG) is a joint motorway unit comprising officers from West Midlands, West Mercia and Staffordshire Police, and with the National Exhibition Centre on their patch they are quite often called upon by the organisers of motoring exhibitions to put on some kind of a static display. Apart from the obvious day-to-day patrol cars, they felt it might be a good idea to bring in some eye candy, some of which we will feature here.

The Lotus Evora had a 3.5-litre V6 lump pumping out 280bhp, giving it a top speed of 168mph and a 0–60 time of 4.9 seconds. Cost was a mere £50,000. The car sported full Battenburg graphics and the CMPG logo and hit the headlines a week or two before the exhibition. The publicity it generated was huge and it goes without saying that members of the 'moaners and gripers' club immediately set about raising the roof over this latest waste of tax payers' money by the police. It has to be said that part of the propaganda quotes given to the media didn't exactly help the situation, but maybe that was part of the plan? The following two paragraphs are taken from the *Daily Mail*.

It is on loan for a two-week trial to a team of officers who will patrol a patch covering 450 miles of highways in the West Midlands. Complete with its police livery, badges and flashing lights, the white supercar will be used on major routes including the M5, M6 and M42 motorways and the M6 toll road.

PC Angus Nairn said: 'It's a very quick car and we hope it will prove an effective deterrent to anyone thinking of speeding or trying to outrun us,' he is quoted as saying, and then continued with 'It has incredible performance yet still does around 30 miles to the gallon. It will attract a lot of attention on the motorways but that is the whole idea – it will remind drivers of the need to keep to the speed limits at all times.'

Conversely, right at the end of the article another officer is quoted saying all the right things about why they had the car.

PC Steven Rounds, from the Central Motorway Police Group, added: 'The Lotus is a visually stunning machine which offers us the opportunity to engage with the public and reinforce and promote the life-saving messages of road safety.'

In 2014 the CMPG needed a new publicity tool to display at the Autosports Show at the NEC and managed to persuade McLaren to lend them one of their MP4-12C supercars. At a cost of £240,000 it was one of their more expensive celebrity cars, and that fact, coupled with its performance figures, which include the astonishing 616bhp it gets out of its 3.8-litre V8 motor and a top end of some 207mph, was guaranteed to get it some attention. Needless to say, it worked, although by now the 'moaners and gripers' club were positively apoplectic in their disdain at this latest move by the CMPG and didn't they realise just how much damage they were inflicting on the planet?

The *Daily Mail* was once again keen to report on the issue and rolled out a number of well-worn clichés in their article:

Criminals might have to think again about attempting a getaway after police revealed their newest and fastest police car – a 207mph McLaren worth £240,000. The specially adapted two-seater supercar – which can hit 60mph in three seconds – will be unveiled at a car show in Birmingham this weekend. Although it is not practical as an everyday patrol vehicle, officers from the Central Motorway Police Group in the Midlands hope it will help to break down barriers between police and motorists.

PC Angus Nairn – currently starring in *Motorway Cops*, the real-life TV policing show – said, 'Everyone who sees the McLaren wants to come and talk about it; it is an excellent way to get people to have a conversation with the Police.' He continued, 'The car breaks the ice and gives us the opportunity to get the safety message across, especially about the dangers of excess speed. The McLaren is a very powerful car with a top speed of more than 200mph but it wouldn't be very practical as an everyday police vehicle – there's not a great deal of luggage space and it has gull-wing doors that open outwards and upwards so you have to leave plenty of space when you park.'

The car was supplied by local McLaren dealers, and sales director David Tibbett said, 'We were delighted to help the police and their work'.

Other cars that the CMPG have displayed include the only Jaguar XF-S Shooting Brake (estate) ever seen in police livery, a BMW i3, a rather nice-looking Dodge Charger in black-and-white American police livery, a Caparo T1 RRV, and in 2017 they even showed off a Polaris Slingshot with police graphics and an array of blue lights. The three-wheeled vehicle looked like something out of a *Batman* movie, and it goes without saying that it drew in the crowds for all the right reasons.

We haven't mentioned Ferrari yet, have we? Surely the police can't have used a car from the prancing horse stable? Well, they did, and this story is somewhat different in that it was Ferrari who came up with the idea, asking the police to drive it on their behalf. The following press release has all the details:

612 SCAGLIETTI HGTS FOR BRITISH POLICE ON FERRARI 60 RELAY

18 April 2007 – Ferrari has created a special version of the flagship V12 four-seater 612 Scaglietti liveried in police colours to be used when Ferrari's 60th Anniversary relay starts the UK leg on 18 April. The relay is part of the celebrations marking the 60th anniversary of when the first Ferrari car was produced in 1947 by Enzo Ferrari. The relay began on 29 January 2007 in Abu Dhabi and has already visited the Middle East, China, Japan, North America, Spain, Portugal and Holland and will involve over 10,000 Ferrari owners and their cars of all ages. The cars are symbolic 'bearers' of a specially commissioned relay baton adorned with 60 badges symbolising the most extraordinary events in the Marque's 60-year history, which is travelling across the world.

Ferrari has provided a dedicated 612 Scaglietti HGTS liveried in police colours, which will be driven by police officers at the front of the UK tour throughout the 1500 miles from Belfast to London. All of the police officers driving the 612 are volunteers from five forces across the UK, and are driving in their free time (not as part of their normal duty) to support the tour.

Representatives from police forces across the UK will also visit the various handover and meeting points throughout the UK, promoting the government-backed 'Think!' campaign on road safety.

Massimo Fedeli, Managing Director of Ferrari GB, said: 'Ferrari is the world's ultimate high-performance luxury car but we are very focused on ensuring that clients enjoy that performance in a safe and responsible way. We offer driver training courses for all customers through our Pilota programme, and also in the UK a track-driving experience called Fiorano Ferrari.

Our 60th anniversary tour is the perfect opportunity to provide this special 612 Scaglietti HGTS for the police service of England, Ireland and Wales to drive. This reinforces Ferrari's commitment to responsible driving and promoting road safety. The 612 Scaglietti HGTS is the ideal car for this activity, with four full-seats and the Ferrari Grand Touring heritage as the flagship of our range.

PC Paul Atkin, ACPO Liaison Officer for Roads Policing, said: 'Ferrari have worked closely with ACPO to ensure the relay has been organised responsibly and we are very impressed with their approach to the issue of road safety and their considerations of the effect the relay would have on other road users. The tour provides a perfect

platform for the police forces across the UK to promote the 'Think!'
campaign.

The car provided to the British Police for the tour is in Tour de
France colours with a beige interior. The car features the HGTS
(Handling GTS package) and also has Daytona-style seats, person-
alised stitching and leather details, yellow rev counter, yellow brake
callipers and Scuderia Ferrari shields. The price of the car as featured
is £200,411, including four-year warranty. Ferrari GB would also
like to thank Woodway Engineering for their assistance with the
emergency lighting bar.

Well, it's a tough job but somebody had to drive it, I suppose!

Meanwhile, back in Sussex their Casualty Reduction Team were on
the lookout for a new project after their Lotus Exige went back to the
dealer they blagged it from. What they found was something so unusual
that it was bound to have the right effect. They managed to obtain
two Can-Am Spyders, a three-wheeled vehicle that bore absolutely no
resemblance to Del Boy's three-wheeled Reliant. The trikes were built
by a Canadian company and powered by a Rotax 991 V-twin motor
of 998cc mounted behind the front two wheels. Top speed was limited
to 125mph with a 0–60 time of just 4.5 seconds. Average price was
less than £20,000. With the vehicles all dressed up in Battenburg
graphics, it looked pretty awesome. The *Daily Mail* thought so and
once again was keen to report on it:

As it whizzes by, you might think someone had given their ride-on
lawnmower a Space Age makeover. But this super-trike could become
the police's latest weapon in the fight against crime. With a top speed
of 125mph and a 0–60mph acceleration of just 4.5 seconds, the same
as a Porsche 911 Carrera, it could keep pace with the fastest of
getaway cars.

At the moment, however, it has a less racy role – enticing young
drivers and motorcyclists to talk about road safety. Two of the three-
wheeled Can-Am Spyder Roadster touring trikes have been loaned

to the police in Sussex as part of a month-long trial. Officers are displaying the trikes at road safety events in an effort to cut the number of deaths and serious injuries among young people on roads in the force's area.

Made in Canada by a firm which specialises in snowmobiles and sports boats, the Spyder is currently sold through a handful of dealers across Britain. The Spyder comes with speed sensors on each wheel that feed the trike's on-board computer with information to reduce the chance of the rider losing control during acceleration, turning and braking. There is also an intercom feature that allows the rider and passenger to communicate.

Adrian Yeomans, manager of Collision Recreational Limited, which supplied the two Spyders to Sussex Police for the trial, said these vehicles appealed to officers because they were much safer than traditional motorbikes and could be ridden with a car driving licence. 'With the unusual design of two wheels at the front, they are just far superior to ordinary motorbikes for safety,' he said. 'In terms of stability, when you go around a corner on one of these it is virtually impossible to flip them or fall off.' He added, 'The trikes certainly turn heads – everyone from six to 60 stops to look when the police drive down the street on them.'

Sergeant Tony Crisp of Sussex Police said he hoped the trikes would help the force get its safety message across to young drivers and motorcyclists, the two groups most likely to be involved in road smashes. 'They help police officers engage with motorcyclists and young drivers. They ask questions about them and we talk to them about road safety issues.'

Along similar lines, the Avon and Somerset Police went one step further in 2014 and got hold of an Ariel Atom, once the fastest car ever driven by The Stig around the *Top Gear* test track. The Atom 3.5R has 350bhp on tap with a top speed of 155mph but can accelerate from 0–60 in a neck-breaking 2.5 seconds! The press went wild, with just about all the daily papers and every glossy car magazine reporting on it, whilst social media had dozens of photos of it. Meanwhile, a small mushroom cloud was seen just above the 'moaners and gripers' club as they all imploded in despair.

Amongst the plethora of reports came this one from *The Independent*:

In the endless arms race between police and rogue drivers, the vehicles keep getting quicker, lighter and more powerful. At first glance then, it would seem as though Avon and Somerset Constabulary has just taken the upper hand. The force's latest recruit is a brand new custom-built Ariel Atom 3.5R – which they have affectionately nicknamed the PL1 – that has been brought in to 'deal with' the region's motorcyclists.

Boasting a supercharged 350bhp Honda engine as well as the latest aerodynamic Hella pursuit lights and striking law-enforcement livery, it can go from 0–60mph at a time of 2.5 seconds and beat just about anything in the world on the track. Sadly, however, that's the only place the PL1 is going to be used. Rather than getting involved in (very) high-speed chases, the car has been provided by Ariel as part of a safety campaign to get motorcyclists thinking about the way they ride.

Despite only making up one per cent of total road traffic, motorcyclists account for a staggering 20 per cent of all road user deaths, said Avon and Somerset Roads Policing Sergeant Andy Parsons.

He said, 'Modern motorcycles are capable of extraordinary performance, with supercar levels of acceleration, braking and cornering. Unfortunately with this level of performance comes an increased risk of death or serious injury as a result of inexperience, lack of training and rider error.'

Ariel have joined forces with the Police for the Safer Rider campaign, as part of which the PL1 will be making its debut appearance at a motorbike awareness and training day in Somerset.

Simon Saunders, the company's director, said that they wanted to promote a message of road safety for bike users ahead of the launch of a new Ariel motorcycle.

He said, 'Our business is about going fast, very fast, but there is

*a time and a place. The Atom is designed to be driven to a race
track, where you can drive to your limits in safety and in a professional environment designed for the purpose. The road really isn't
the place to explore your or your vehicle's limits.'*

We have mentioned the National Association of Police Fleet Managers
(NAPFM) exhibition in a previous chapter but didn't mention BMW
and what has now become a bit of an exhibition tradition for them.
Over the last few years they have gone to extraordinary lengths to
produce the most eye-catching car of the show, and more often than
not they pull it off. The cars are usually the very latest models and not
those you would ever consider as a mainstream patrol car, being
designed very much as showstoppers. Bearing in mind the audience at
this two-day police car show, virtually all of the delegates and attendees
are either in the police or more likely to be fleet managers and other
vehicle professionals who are more than used to seeing a variety of cars
with a blue light on the roof. But BMW's theory and target follows the
same principles as any other celebrity car: it makes people stop, stare
and then engage. It's a great marketing ploy and it works. In fact, it's
so good that many of the show's visitors head straight for the BMW
stand just to see what crazy thing they've dreamt up this year!

Over the last few years they have exhibited police versions of the
BMW Z4M, the Mini Countryman four-wheel drive, the i3, a 430i
Coupe, M135, and, at the 2015 show, by far the best one ever – the
awesome BMW i8 complete with its scissor doors, labelled as a Police
Hyperceptor! You could hardly get close enough with your camera
because there were so many onlookers.

The police in Norfolk have been given a Porsche Cayman GT4, which
will be used to break down barriers with young drivers and help officers
deliver life-saving road safety advice. The vehicle has been donated to
the Constabulary by the Lind Trust, a charity that supports the development of young people in Norwich and Norfolk. The GT4 has in fact
been loaned to the force for just two years for the purpose of engaging
with motorists around road safety – young drivers in particular.

Complete with police livery and blue lights, it will be at the forefront of police engagement and education with young drivers under the hashtag #Porsche999.

Young drivers are disproportionately represented in fatal and serious collision statistics; in Norfolk over the last five years 61 young drivers have lost their lives, while 504 suffered serious injury.

Chief Constable Simon Bailey said, 'Only this week we have seen the sad loss of three young lives in a road traffic collision in South Norfolk and my thoughts are with the families involved. Engagement and education are vital elements of the work on road safety undertaken by roads-policing officers. I'm confident the use of the Porsche will help break down barriers and enable officers to speak with young drivers and provide road safety advice.'

He added, 'The GT4 will certainly attract a lot of attention – but that's the whole idea. We hope the car will act as a conversation starter, which gives us the opportunity to engage with people, but more importantly those hard-to-reach groups like young drivers, and offer practical advice as well as describing what can happen when things go wrong.'

The GT4 will be used at events such as the Royal Norfolk Show and taken to areas where car enthusiasts are known to gather. The car will also be taken to schools and colleges throughout the county, as well as events and areas that are likely to draw crowds of young people.

Graham Dacre from the Lind Trust said, 'This is an important initiative to be involved in. The Porsche GT4 is every young driver's dream and it will be a talking point in their world; I can see them responding well to this initiative.'

So there you have it, unequivocal proof that the police aren't wasting tax payers' money on buying luxury supercars or bombing around in bonkers high-speed motors just for the fun of it. They are trying to deliver a serious message that cars and the people who drive them can and do kill if they are not used in a responsible manner. And if that message gets across it's certainly not a waste of anyone's money.

As we have already seen, those Sussex boys do like their exotica to entice the crowds to them at various events, so can you imagine what happened when they collaborated with this country's most famous car maker? In 2017 they took delivery of a Rolls-Royce Ghost Black Edition, marked it up in full Sussex Police livery, put a very fancy light bar on the roof and took it to a number of events in the county.

The car has a 6.6-litre twin turbo V12 motor with 603bhp on tap, an electronically limited top speed of 185mph and a 0–60 time of just 4 seconds! The cars are built just up the road at Goodwood, which is only five miles away from Chichester Police Station, so there was plenty of local interest. Everywhere the Police Roller went, so did a couple of 'minders' from the factory, who were on hand to answer any questions from the public, including all the usual questions. You know the ones . . .! You'd have thought that, given its price tag of just under £260,000, and with a cream leather interior, the car would have been fenced off with barbed wire and snarling dogs, but far from it. In fact, kids with ice cream dripping down their arms were allowed inside it to play with all the switches and to bury their muddy feet in the Wilton carpets!

When it comes to mad PR police cars, though, it's fitting that we leave the final car to the Isle of Man Constabulary, although sadly this is the only mention that this quite separate force will get as, technically, although they volunteer themselves to Her Majesty's Inspectorate of Constabulary for inspection, they are responsible to the IOM Ministry for Home Affairs. However, I just can't ignore this; they have a BAC Mono! Yes, a single-seater supercar with no roof that needs its driver to wear a crash helmet to operate it . . . as a police car! It's not as crazy as it sounds. The IOM is a temple of speed, as we all know, and BAC are based in Liverpool, just a short ferry ride away from the island. BAC have built up a relationship with the government and police there, doing their now quite well-known 'Festival of Mono' driving days for customers in which sections of the world-famous TT course and other roads are closed off for periods to allow BAC owners to enjoy their toys to the full, and what a bit of kit it is! The Mono comes with a 4-cylinder 305bhp 2.5-litre Mountune engine (based on a Ford unit but heavily modified) which takes the 580kg car from 0–60 in just 2.8 seconds! They recently launched a carbon hybrid wheel and announced the use of super material graphene, which is up to 20 per cent lighter than carbon fibre and as much as 200 times stronger than steel. It's basically a race car for the road, and this is emphasised by the F1-like 'waisted' centre and central driving position. BAC are adamant that although this is a publicity operation the car will be used as a police car, and PC Andy Greaves, who is an advanced driver, police motorcyclist and collision investigator, is the force's dedicated Mono driver, who was full of excitement when discussing his new steed:

'It's a remarkable vehicle, and clearly will be of huge interest to bikers and car drivers alike. We are here to promote safe use of the road, and it's amazing how many people want to talk to you when this is your transport. Every opportunity we get, we stop and talk to visitors. Having a showstopper like a BAC Mono police car on the fleet makes that so much easier.'

The PR spin is that, by driving the Mono, Greaves will be able to spread the good word on road safety and keep drivers both interested and entertained; that's great, but it is just THE coolest police car I have ever seen!

CHAPTER FOURTEEN

POLICE DRIVING

It may surprise many to learn that police officers are not exempt from road traffic laws and that they can, and do, face criminal proceedings if things go wrong. Police driver training, therefore, is tough, really tough. If you don't reach the required standard you don't get a driving permit. To drive basic cars, the panda cars of old, now commonly referred to as section cars, an officer will undergo an Improver Driving Course. This is basically a check and a refresh of a driver's standards, just to ensure they are safe to drive a police vehicle of that class. For those intending to do what are now called response car duties, you are looking at a three-week intensive course where you learn all aspects of the 'system'. It's a theory and practical course, with written exams plus a final drive examination which includes fast response on 'blues and twos'. For those officers wishing to go on to the Traffic Department, now called the Roads Policing Unit, on top of that they get another four-week course that takes them to the very top level and the coveted Class 1 Advanced ticket. There are separate courses for driving personnel carriers, towing trailers, TPAC (Tactical Pursuit and Containment) training and various others.

So getting your driving permit, no matter what level you are at, is no easy task, but losing it can be, which is why police driving standards are high; no one wants to end up wearing a big hat and walking the high street for the rest of their service! So if you get it wrong you might face being re-tested by the police driving school. If in their opinion you don't come up to standard, they either suspend your permit completely or put you on another course.

So it is only right that we look at the history and content of driver training in more detail, as it's an essential part of the UK police car story. We haven't got room to cover everything in detail, so this is an overview of the broad themes and history of a large and complex subject.

The human component – the nut behind the wheel; training the drivers

'It is clearly of the greatest importance that the standard of driving of police cars should be the highest possible. The public look to the police, especially the traffic patrols, to set an example of skilful and considerate driving, correct signalling, and the punctilious observance of the traffic code, and the courtesies of the road.

Also in the work of the kind which has to be done on occasions by the Flying Squad, area wireless cars and "Q" cars, the safety of the public and the lives of police officers themselves sometimes depend on the skill of the police driver.'

Lord Trenchard, 1934
(*Taken from a framed notice on the wall, Hendon Police Driving School; a principle as true today as it was then*)

Right from the earliest days of the police using cars it is plainly obvious – but so easy to overlook – that the cars needed drivers. From the very beginning, police cars were early cars, so there is an inevitable question about how those first traffic cops learned to drive. The detail that far back is pretty vague, but there are some specific events that appear in records; in 1913 the Chief Constable of Devon had a driver whose training had cost five pounds and ten shillings (probably to train someone who had hardly seen a car, let alone driven one).

It became obvious that the speed of response the police had always needed was actually possible through a combination of the police box scheme and the increasing use of the telephone. The 999 scheme would come later and would build on this new demand on policing to an even greater extent because of human nature; once people (and officers are people, remember) know that a disaster or a crime is in progress they will speed as fast as they can to it. Training was the only way; telling officers to slow down might have worked for a normal journey, but not for an emergency call when adrenaline takes over. The military's approach of train and train until a set of skills and protocols become utterly ingrained can't be followed in the police service, but they do try to make 'the system' become second nature and encourage police drivers to take pride in their craft.

The police box or, as it's better known today, the TARDIS (*Time And Relative Dimension in Space*)

The name TARDIS was invented for the *Doctor Who* programmes in 1963, when police boxes were still normal and common – so Doctor Who landing in his box would blend in, apart from that wonderfully weird noise, of course . . .

The boxes were to become obsolete only a few years later with the introduction of walkie-talkie radios and the national 999 system. They were known simply as 'police boxes' and were essentially phone boxes that only connected to the police station, to make it easier for the public to contact them. On top of the box was a light; if it was lit, patrolling officers knew the station was trying to get in contact, so they went to the box and phoned in, and thus could answer a call for assistance.

The box also provided a safety net for officers – they were expected to phone in from them regularly, sometimes at set times, and a missed call into the station could raise the alarm and trigger a search for a missing officer who might be dealing with an incident. The box also acted as a miniature police station where beat constables could shelter, process paperwork or even hold a prisoner for short periods; sadly, they are not bigger inside than out.

The first police boxes in the UK were in Glasgow in 1891, although the idea itself originated in America; the first one ever installed was in New York in 1877. The UK design that later became the most well known (the one that the TARDIS was actually based on) was created by the Met's own surveyor and architect, Gilbert MacKenzie Trench, in 1929. The main box was of pre-cast concrete with a door made from teak. Although this did become the standard UK design, some local authorities built their own, either to save costs or to employ local labour.

I wonder how many *Doctor Who* fans of today even know, or care, what a police box is, or was?

In the 1920s the Flying Squad (*The Sweeney* of TV series fame in later decades) acquired faster cars to deal with the criminals of the post-World War I era (initially Lea-Francis, or Leafs, as they were known), and drivers were trained at the then centre for UK performance driving, aviation and technical engineering progress, Brooklands, on how to drive and handle them. They gained a reputation for both safety and speed, and the police started to take driver training seriously, although each force approached it in different ways. The most experienced of the Met's trained drivers, George 'Jack' Frost, became not only an expert high-speed driver but also trained and advised both the Met and other forces on driving techniques and practices. As the Met's initial Leafs began to wear out and become outdated, in performance terms, they looked around at other vehicles and, perhaps through Jack Frost's contacts at Brooklands, consulted well-known racing driver and engineer Captain (later Sir) Noel Macklin. He recommended, perhaps unsurprisingly, the car he'd designed and produced himself, the Invicta 4.5 litre. However, it's a mark of how seriously driving capability in real use was now being taken by both the Met in particular and the police in general that a test of the Invicta was arranged at Brooklands (the example used being Invicta's Le Mans racer!), during which a team led by Frost put the car through its paces and pronounced themselves happy. The Met, now confident that their elite drivers could handle a performance car of this type, placed an order for some Invictas and were given a substantial discount because of the PR value. They soon also had other high-performance cars such as Bentleys and Lagondas on their fleet, and other forces followed suit.

By 1930 there were general concerns about road safety. Driving tests and licences were introduced in the early 1930s, along with the 30mph speed limit in towns, the Road Traffic Act 1930 being a significant point in the history of driving regulation. At the same time there was concern about the accident rate of police cars, and it was also in the 1930s that the 999 telephone system was first trialled, as well as the introduction of the police box and the use of radios in some police cars, creating more pressure on the drivers.

In London another, much more famous, racing driver, Sir Malcolm Campbell, had been appointed to test some of the drivers the police were using in the 1920s, to answer criticism in the press and parliament, and the apparent feedback was that they were very good. This seems to have effectively silenced the criticism, but it was decided to improve things further, and the now-famous police driving school for the Met at Hendon was set up in 1934.

The 999 service meant the police had yet another reason to get to places quickly

Britain can be rightly proud that we invented the concept of an emergency telephone service with one memorable number, 999, and without any shadow of a doubt its very existence has saved countless lives and banged up a few villains. Like many innovations in policing, it was launched in London, on 30 June 1937. Londoners were introduced to the idea via *The Evening News*, advising its readers thus:

Only dial 999 if the matter is urgent; if, for instance, the man in the flat next to yours is murdering his wife or you have seen a heavily masked cat burglar peering round the stack pipe of the local bank building. If the matter is less urgent, if you have merely lost little Towser or a lorry has come to rest in your front garden, just call up the local police.

Sadly, in a perhaps typically British way that would come to the forefront and be much needed only two years later, this whimsical light-hearted approach masked a sad truth; the pioneering system was developed in response to a tragedy. Prior to 1937 the emergency services in the capital could be contacted by calling Whitehall 1212 for the Information Room at Scotland Yard, or by operating a street post that sounded a bell in the nearest fire station. The instruments in the MacKenzie Trench police boxes that appeared in London in 1929 could also be used by the general public and, as with domestic or office phones, the GPO recommended dialling '0' – 'When the operator answers, ask for the service you require'. The problem with this procedure was that switchboards were unable to distinguish between priority calls and routine ones, resulting in delays. On 10 November 1935, a fire broke out in the surgery of Dr Philip Franklin at 27 Wimpole Street and his neighbour Norman MacDonald attempted to sound the alarm, only to be placed in a queue to speak with the Welbeck exchange. When he was finally connected, appliances were already on the scene but five women had perished. Had the call been connected earlier, the brigade might have been able to rescue them.

MacDonald wrote to the press about his experience, which stirred up interest, and the government established the Belgrave Committee to investigate the best form of emergency number and how to put the scheme in place. They recommended the digit '9' as being simple to remember, adaptable by existing telecommunication equipment and because it did not interfere with any existing number. As it was only one place removed from the stop on a rotary dial it was also comparatively easy to use in conditions of darkness or thick smoke, although paradoxically that meant it was also the slowest number that could have been chosen (excepting 0) as callers had to sweep around the dial and wait for it to return each time. There are videos on YouTube of how to dial a rotary phone if younger readers don't understand what I'm talking about . . .

To ensure that 999 calls would be promptly answered, the switchboard staff were alerted by a flashing beacon and a klaxon. The new code initially covered only 91 exchanges within a 12-mile radius of Oxford Circus, but there were plans to use it nationwide, with Glasgow adopting 999 in 1938, followed by Birmingham, Bristol, Edinburgh, Liverpool, Manchester and Newcastle in 1946; ironically World War II, which created huge domestic chaos from bombing, actually prevented the scheme launching nationwide as quickly as it would have done in peace time. The 999 code did not function on the manual exchanges that could still be found in many areas; amazingly, it would not be available across the entire UK until 1976 when Portree, in Scotland, finally gained an automated set-up. Yes, men had gone to the Moon before everyone could dial 999.

The Post Office issued advice on how best to use the code, and the service was promoted in cinemas by Pathé and Movietone. A wonderful 1948 newsreel illustrates how a police Humber can respond within minutes to a control-room message; it's called *999 – Scotland Yard* (1948) and is on Pathé's website – it's well worth a watch.

The GPO commissioned its own instructional films and the Communications Museum Trust have put some online; the best one is entitled '999 Public Information' and it gives advice on how to use a telephone kiosk to report an accident. From a 2017 perspective, the scenario may at first seem impossibly quaint, from the bell-clanging Wolseley 6/90 to telling viewers to 'ask for the service you require', but the message was absolutely serious. Even in 1969 less than 50 per cent of houses in the UK had a telephone and it was not unknown for Britons to have rarely or even never previously used the device. Guidelines such as 'give your exchange and phone number' were both simple to recall and essential when giving details to the operator. It's easy to forget just how easily available knowledge is now.

When 999 debuted, the capital had a network of 700 police boxes and constables were advised to keep them within sight as much as possible as the rooftop flashing light and gong were their signals to contact HQ, but the arrival of the 999 service also marked a key stage in the development of Britain's police, fire and ambulance communications. In 1922 the Flying Squad employed a pair of Crossley Tenders that could receive messages in Morse, in addition to a wireless van that was able to receive voice communications and transmit messages when it was stationary. By the 1930s the Met was using 'Area Wireless Cars' crewed by CID officers, and trained drivers and operators who could quickly respond to an incident, but in rural areas motor patrols might arrive at a phone box at a fixed time or be summoned by a disc being displayed from the station's window. In 1942 the government issued constabularies with short-wave radios, but these were only for the duration of hostilities. By the end of the 1940s, the Home Office recommended that all forces use vehicles with VHF wireless, and car fleets began to expand. Radios also started to be fitted to ambulances and fire appliances in the 1950s, in part to cope with the expansion of private vehicle ownership, and in 1963 some 3,000 police cars were equipped with wireless.

The demand for driver training was ever greater, as instant messages meant people wanted instant responses. It's a good job the police had started to look at driver training in earnest.

Mainstream police driver training began at the same time as the other developments around road safety, and, along with Hendon in north- west London, other police driving schools also opened, at Chelmsford in Essex and Preston in Lancashire. The Met appointed the Earl of Cottenham as a civilian advisor on driving and driver training, and he was also engaged by the Home Office to spread his methods.

Of course, every police force was entitled to do things its own way, but most employed common practices and themes. Courses were initially for new drivers, but the advanced course – the one that police driver training is best known for – has its origins in the 1930s too, starting quite soon after driving schools opened.

The whole thing was a huge success virtually straight away. Figures are available for the Met's fleet, and in 1934 there was one accident for about every 8,000 miles, reduced by 1937 to one every 27,000 – a massive change in a very short time, especially when you consider that when war broke out in 1939 there were over 2 million cars on UK roads.

Sir Mark Everard Pepys, 6th Earl of Cottenham, 1903–1943

Cottenham, who was educated at Charterhouse and University College London, where he studied Engineering, was a great motoring enthusiast who had made a name for himself racing at venues like Brooklands in the 1920s, and he was a team driver for Alvis and Sunbeam, among others. He was also the first motoring broadcaster in the UK, maybe even the world, as he had a regular radio programme about cars and driving on the BBC in the 1930s. He was appointed an advisor to the Met and the Home Office in 1937, and although he only had a short period of tenure his influence was huge, his training of police instructors leaving a permanent legacy. He is associated with strong basic concepts, such as being able to stop in the distance you can see to be clear, and with the creation of the System of Car Control, as it became known.

Cottenham had a philanthropic side when it came to road safety and police driving in particular, and he presented a Lagonda to the driving school at Hutton Hall, outside Preston, which was intended for use on the skidpan, and it's believed he gave cars to other schools at Hendon and Chelmsford as well. The Preston driving school's Lagonda was retained by the school until the early twenty-first century and in later years was treated as a reward; drivers who reached a certain standard were allowed to have a go in the Lagonda.

In 1928 he wrote *Motoring Without Fears*, in which he sets out a system of Advanced Driving including the block changing of gears. In his third book, 1932's *Steering Wheel Papers*, he set out his 'ten commandments of motoring', which continued to appear in *Roadcraft*, up to the 1977 edition.

The 1930s was a decade of change in many other ways, and the outbreak of war saw police driving schools closed, with the obvious pressures on manpower, vehicles and, of course, fuel supply. Policing was still important through the hostilities, though, and the police continued using cars for operational purposes, with untrained drivers being used in some forces.

By the outbreak of war, the usefulness of the motorcycle had also been identified, and some training was started for motorcyclists. Whilst not a theme of this book, the similarity of approach between police driving and police riding is worthy of note, along with a comment made by many police driving instructors that students who have already learned to ride a motorcycle generally make better car drivers.

Roadcraft

Originally published in 1955, still in publication, and now also as an electronic version. Originally published by HMSO and written by police instructors, there was a major change to its publication in 1994, when ownership of the copyright was passed to the Police Foundation (a charity) with TSO still providing the publishing support, and emergency service driving instructors setting the content. First printed at just 91 pages in 1955, it is now over 250 pages long.

> It's worth any driver reading it, as it gives great advice on how to lower your risks in any given situation, and we could all benefit from that.

After the war things gradually returned to normal and training restarted. Traffic levels were rising (as they have done pretty much ever since) and police activity in dealing with the traffic itself – already well established in the 1930s – became more specialised.

IAM and the League of Safe Drivers

In the mid-1950s there was (again) concern about road casualties, and both the Institute of Advanced Motorists (IAM) and the League of Safe Drivers were born. Both promulgated safe driving based on an advanced driving test conducted by police advanced drivers (mainly in their off-duty time), and aiming to achieve a similar standard to that of the police themselves. Both organisations continue today – the IAM as IAM RoadSmart, with over 90,000 current members who have passed their test, and Rospa Advanced Drivers and Riders, with about 10,000 who have passed theirs. The tests are very similar, and the two organisations operate using (often the same) volunteers, both as examiners and in helping drivers prepare for their test. If you want to improve your driving, take up their challenge . . . .

In the 1950s, after rationing had finished and with recovery well under way, road safety again became a subject of general concern. Part of the training for officers learning to drive police cars had included gaining some understanding of the machinery itself, and some theoretical learning about how to drive – not just the Highway Code (another of those 1930s developments), but also material about how to make decisions when driving, how to position the car on the road, how to identify hazards, and many other things. Students laboriously wrote out their own notes on this material, and eventually the instructors' notes were published by HMSO in a book called *Roadcraft* in 1955, a version having been published the year before by an instructor under the title *Attention all Drivers*. *Roadcraft* – still the manual used by the service today, albeit somewhat updated and hugely enlarged – was a publishing success. At about the same time comments made in parliament by a transport minister to the effect that it was a shame there wasn't an advanced driving test for the public like there is for the police led directly to the founding of the Institute of Advanced Motorists (now IAM RoadSmart).

Class 1 drivers

Police driving schools have traditionally graded the results of their training, and on advanced courses officers were encouraged by trying to gain the accolade of becoming a Class 1 driver rather than a Class 2. In practice, this is simply a case of the driver getting a higher mark on the course, but the result could have an impact on more than pride – some career options, such as joining the Flying Squad as a driver, could be affected by that result, being only open to Class 1s. The distinction was theoretically abolished in the late 1980s, with drivers being graded simply as 'advanced', but the practice did carry on in many forces until much later, with many police drivers very proud of achieving their Class 1. The phrase is often used wrongly to cover all drivers who have passed the police advanced driving course – although they are all very good!

Until the mid-1950s police drivers were essentially at one or other of two levels of training – they had either learnt to drive with a police driving school, and were what is generally known as a Standard police driver, or they had done the additional advanced driving course (generally known as Advanced drivers). Standard drivers were expected to drive as they had been taught, and to set an example on the road, but they were not expected to drive at high speeds, or under the pressure of answering emergency calls. Patrolling was still largely done by officers on foot, and those driving at this level were mainly doing so either for specific purposes, or to travel around. Advanced drivers were trained to develop the same skills to a far greater extent, and expected to cope with the pressures of driving in emergency situations, and at high speeds. The Met introduced an intermediate course in the mid-1950s, to overcome a problem of the big step in the performance of the cars used between the two grades – quite sudden and significant, especially when you remember that most police officers only drove at work; there are stories right up until the start of the 1960s of senior officers telling new officers that a constable who had a car must be corrupt, as police pay wouldn't stretch to that luxury. This wasn't nationally accepted, an indication of the variations that existed – although the content of training was very largely consistent across the country.

Police forces have always been autonomous bodies – often fiercely so – and differing practices and approaches in many activities are common. Driver training was conducted at force level, but there were regional driving schools established across the country, dealing especially with the more advanced-level training, which was given to far fewer drivers in most forces. This gradually broke down, and by 2000 just about every force was doing its own training in England and Wales, with Scotland retaining the structure of advanced and specialised driver training, which was all being conducted centrally at the Scottish Police College at Tulliallan (not far from Stirling). There has long been a coordination of the training, with a National Police Schools Conference meeting regularly and discussing a wide variety of issues, both strategic and detailed, to deal with driver training across the UK. The conference has a regional structure, and includes England, Wales, Scotland and Northern Ireland.

Various needs have generated specialised additions to the range of training offered by forces, and in more recent years some forces have been seen as having the lead in some specialisms. We have already touched on motorcycling, which has developed in style, content and amount enormously since the start in the 1930s, but there have been other areas to add as well; there is a need for some officers to be able to drive a full-sized truck, so lorry driver training started, and bus driving has also been a small but present issue in some forces. Changes in the law during the 1990s mean that officers who acquired their civilian licences from 1997 onwards cannot drive larger vans or mini-buses without additional training and testing, so that challenge was taken on as well. All this means that some police forces have appointed driving examiners, trained by the DVSA, who conduct the relevant driving tests along with their other skills.

Police driving courses have traditionally been very observant of the speed limits, using the areas that until the mid-1960s were de-restricted (had no speed limit) for higher-speed training. Following the introduction of the national speed limit in 1967, training continued to obey all other speed limits, but roads subject to the national speed limit continued to be used for higher-speed training, using the exemption that the emergency services have. In more recent years training using blue lights and sirens also exceeds the speed limit in other areas, when appropriate. Trainees are still expected to observe the limits punctiliously otherwise, though, which is a strong test of their ability to maintain control of their actions.

The challenge changes

Over the decades the big issues have changed. Those early drivers were dealing with unmade roads in cars with poor brakes and tyres, primitive suspension and gears that needed double de-clutching. But there was almost no traffic by modern standards on most roads . . . Now the car is faster, can achieve amazing levels of braking performance and is easier to handle and operate, but traffic systems and rules are far more complex, the roads are crowded, and the challenge of coping with the traffic has largely replaced that of managing the machine. Police cars have gone from being 'just a car' to having blue lights that can vary their pattern of display, variable sirens, headlamp flashers – all to help work their way through modern traffic conditions.

The IRA's terrorist bombings in the 1970s, subsequent terrorist attacks by other groups and the growth of serious organised crime has also had an impact on driver training. Some anti-hijack training was developed and operated, concentrating on identifying potential problems and avoiding them, but by the late 1980s this was regarded as inadequate. There was also some concern about differences in VIP escort practices on long journeys across force boundaries that seemed to make little sense, and a desire for national practices emerged. A longer and far more intense and thorough course was developed, including some extreme car-handling exercises as well as training drivers how to work with the protection officers and others involved in this area of activity. This was adopted nationally with some further refinement, and includes, where necessary, the techniques involved in secure escorts of high-risk prisoners, which are a different security challenge. This was one of the first courses where the use of a few forces having a regional 'lead' was developed – the numbers of officers involved being relatively low, and the skills involved being relatively high. There was also a strong desire to use the training to create standardisation in practice.

We saw in chapter 3 that a constable first commandeered a car to pursue a drunken horseman in 1900. This caused a great deal of what would then have been called 'fuss' in the press at the time, but by the 1980s the media were focusing on the negative aspects of police driving, especially on the issues arising from pursuits, which were becoming faster as car performance rose. There was also some attention from non-police-driving quarters on certain specific aspects of police driving practice, and a major review was completed in 1989. The national standardisation of anti-hijack training mentioned above has its origins in a recommendation of the review, and a general move towards bringing greater standardisation can perhaps be seen as starting with it.

The cars

The cars used in training have tended to reflect the fleet of the police force concerned, but there have been variations. Cottenham left a Lagonda with Lancashire Police that survived there until early this century and is now in the hands of an enthusiast. The Met driving school had a left-hand-drive Austin Allegro (they had plenty of right-hand-drive ones in the fleet as well) in the 1980s, which was used to familiarise officers with driving from the left seat as they could be required to remove parked cars or drive cars whose drivers had been arrested, and found themselves dealing with left-hand-drive cars in the increasingly cosmopolitan London traffic. One strong characteristic has always been that the training fleet is mainly plain unmarked cars, officers being expected to be able to operate safely without the help of the police uniform worn by marked cars. Marked cars are used where they are needed, but the vast majority of training requires the driver to operate safely without the help of livery, 'blues and twos' or other markings . . . After all, surveillance drivers have to do all that for real!

One of the big problems in police pursuits is that once you are chasing the bad guys (and pursuits rarely have female drivers as their focus) they are really in control – the police essentially are almost stuck with just following along whatever route the bad guys decide to take.

In the 1990s ingenuity was applied to the problem of how to actually stop the fleeing car full of villains. The 'Stinger' device – and later 'Stopstick' – were developed as ways of puncturing the tyres of the fleeing car without causing a traumatic tyre deflation. Sounds simple, but nothing is as straightforward as it sounds – to put the bed of nails across the road in front of the car you are chasing means being there first, guessing where it's going to go and being ready to throw the stuff across the road in front of it. Then you have to get back quickly so you don't also puncture the police car. . . . There is some risk attached to this; officers have been killed by the car that they were chasing while they were trying to deploy one or other of these devices.

A much more complex set of manoeuvres was developed in the south of England, and Surrey Police became the lead force in providing the training for a suite of tactics known as TPAC, Tactical Pursuit and Containment. TPAC is essentially about trying not to start a 'chase' until you are in a position to end it – and being able to end it by blocking the path of the pursued car, using police cars. There is a range of tactics employed here, but the one that makes it easiest to understand is simply surrounding three sides (front, side and back) of the pursued cars with police cars, using a barrier (like the central reservation barrier of a motorway) as the fourth side, then slowing down until you come to a halt. Of course, this means being so close that the pursued driver has no choice but to conform – which at the speeds involved at the start of the process can involve an exercise of very substantial driving skill and can only work with good radio communication between those driving all the police cars and utter trust in your fellow police drivers. It's not for the faint-hearted, and it's certainly not how the Highway Code advises you to drive, but I'd be lying if I said I wasn't scared the first time I did it for real. Training is given to already qualified advanced police drivers over several days and covers radio protocols, the different roles involved in achieving this safely amongst other traffic, as well as 'just' how to do it!

The training for this does involve some exercises that can be practised on an airfield or similar but there is a real need for some of this to be done live, as it can cover some distance. There is no substitute for real traffic to make the practicalities and difficulties real to the officers learning to do it, as controlling the traffic and creating a safe space to actually stop the car is part of the teamwork involved; this can also involve a rolling roadblock in front or behind to give the police a relatively safe space to operate in and thus not endanger the public whilst carrying out this risky but very effective tactic.

Another tactic – the most risky manoeuvre – is known as the PIT (Precision Immobilization Technique), and anyone who grew up watching British stock-car racing will be instantly familiar with the concept, if not the detail of the execution, which is complex and difficult. It's only used in real extremis as doing it on the road is very different than on a stock-car track! The PIT manoeuvre was never nationally recognised as a tactic, and has largely gone out of fashion now because TPAC is safer for both the police and the pursued and PIT needs lots of space to be relatively safe – an unusual situation on British roads! It is also dangerous and unpredictable, depending on the desperation of the driver being chased. The PIT manoeuvre has been used in the UK, but only very rarely.

It wasn't until the 1980s that training involving the use of blue lights and sirens ('blues and twos' in those days) became at all common, and in some places it has only been adopted far more recently. Indeed, only in the last few years has there been a nationally adopted manual of the 'tactics' of how to use the equipment. Various forces did adopt various specific exercises in their training; Lancashire police in the 1980s taught their panda car drivers how to stop another car in practical exercises, and the Met dealt with a need to train drivers on medium-sized Transit vans. These variations reflected local operational needs and practices, as well as the variations in fleet – London has little need for off-road driving, whereas in Cumbria or the Highlands of Scotland things might look a little different. Training in this is now integrated into the standard level of qualification, but is sometimes developed additionally at advanced level. It is a benefit of the quality of the general advanced driver training that all police driving students are given that allows this to be done relatively easily and effectively, as well as safely.

So what do they learn?

The image of police driver training that is most common is similar to the picture above – a car on a traditional skidpan, often with a group of officers watching. In fact, skid training has always only been a small (if spectacular and enjoyable) part of the whole programme in the UK, which has always had a strong ethos of avoiding problems, rather than dealing with them. With all the clever modern technology that is now fitted into cars, skid training has virtually become redundant anyway in recent years, and most forces that had skidpans have closed them down. The police are now taught a style based on a set of principles and themes that inter-link, intended to allow them to drive safely, and when necessary quickly, but with the two always firmly in that order of priority. The detail of driving is in *Roadcraft*. The learning outcomes of the training are laid down by the College of Policing.

The Highway Code

First published in 1931, this is the basis of how all road users are supposed to behave, and is fundamental knowledge for police driving students. It has been regularly updated (there is a new edition released in most decades since). The first edition had just 18 pages, with 32 pages in 1959 – it is now best seen online, and despite being over 140 pages it is constantly under pressure to include more information yet remain digestible.

Attitude

That concept of prevention being better than cure has always under-pinned the way police drivers have been taught – the use of highly developed observation and interpretation skills has been a really distinctive feature of the training, and when matched with some very positive attitudes and a systematic approach to dealing with whatever threat or opportunity is identified, creates a generally calm and surprisingly undramatic style, when the speeds attained are taken into account, and when it's done the way the instructors teach it! In recent years this has become a more high-profile and overt issue, with specific training about attitudes and risk being an integral part of the instruction. The basic idea is to prevent deliberate risk-taking, and to avoid unintended risks being taken, drivers are being encouraged to be very self-aware and to manage their performance in a way that reduces danger wherever possible. One of the things that is important to understand here is that for most police officers driving is simply one more skill they need to learn in order to do their job; it isn't necessarily something they want to do, or particularly enjoy – indeed, some would prefer not to drive. Of course, there are also many who do enjoy this aspect of their work – but it is important to recognise that for very many officers it is not at the core of their work, it's a tool.

Skidding

Police officers were taught for decades how to manage a sliding car on ice or other slippery stuff. The old adage 'steer into the skid' was drummed into them, and they *all* knew exactly what it meant! However, at that time cars were almost all rear-wheel-drive. Now most cars are front-wheel-drive, which changes that dynamic some-what, and all come with ABS and ESC systems of one kind or another, so this approach has become largely redundant. Even the rear-wheel-drive cars, and the police still use many BMWs in particular that are rear-wheel-drive, handle more like front-wheel-drive cars because the electronic traction and stability systems just don't let drivers have long over-steering moments in a Mk2 Escort rally style anymore, even if they had the skill to do it. Likewise, training to pump the brakes to retain directional control is now history due to the car's ABS technology, and police drivers are taught to operate with the new kit.

In fact, I can tell you a little TV secret here. When you see presenters hooning cars around sideways on an airfield doing long drifts in new cars loaned to them by manufacturers (meaning the tyres are free), they usually do it by disabling the electronic systems, a practice the manufacturers dislike because reinstating it is not as easy as you would perhaps think.

Observation, anticipation and planning

Over 40 pages of the current *Roadcraft* manual are devoted to observation and anticipating hazards, as well as specific advice throughout many other sections of it on what to look for in relation to specific aspects of driving. That's a lot about what seems like the obvious! Human nature being what it is, though, this training works – I can attest to that. It emphasises how much the police are trained to look far ahead (as well as behind and to the sides . . .) and to be prepared for whatever develops. A very distinct feature is how far ahead officers are constantly being pushed to look, combined with detailed priming of their minds about what to look for – where there might be ice, crosswinds or other hazards. Officers are trained to question what the implications of a road sign, a parked vehicle, muddy tyre marks or the fact that there's a school might be. It is hard to over-stress how important this underpinning skill and preparation is to the training. It's also true that training cops to be highly observant has other advantages for them when patrolling and looking for 'things that aren't right', the biggest key to spotting people doing what they shouldn't . . .

So if you want to drive really well, look a long way ahead, and learn to understand what you are looking at – what is likely to be there, what hints you are being given by what you can (or can't) see, what conclusions you can draw, or things you can be ready for, simply by interpreting what you can observe. Observation includes hearing things, feeling things and even smelling them (the smell of diesel on a wet day is a strong clue of a very slippery surface . . . essential to motorcyclists).

> ## 'System'
>
> All the time you are driving you absorb information from what you see/hear/feel/smell/etc. You also display information for others – deciding to signal, for example. When approaching a hazard you recognise it by doing this, then decide on the position to take through the hazard, such as the line you will drive along. Having done that, you decide what speed you will be doing along it, and then sort out the gears to suit that. Gears are only used to slow a car down when driving on very slippery roads; on all other occasions brakes will do that, not gears, and this has developed as brakes have improved of course. Then you simply accelerate away to the next hazard . . .

'Weather conditions such as fog, mist, heavy rain or snow, the fading daylight at dusk and the dazzling brilliance of the setting sun reduce visibility considerably. To meet these conditions speed must be reduced so that objects in the immediate foreground may be seen in time to take evasive action if necessary.'

Contrast this with today's plain English wording:

'Bright sun low in the sky can cause serious dazzle, especially on east/west sections of road: use your visors to reduce dazzle. If the sun is shining in your mirrors, adjust them to give you the best visibility with minimum glare. If you are dazzled by bright sun, other drivers may be too, so allow for this when overtaking.'

Roadcraft is now also available as an electronic book, with the possibilities of electronic learning being explored and introduced. It is perhaps less artful and elegant in its language, but it is full of information in every sentence, and still very well worth reading carefully.

A systematic approach

The second 'theme' that runs through the training is adherence to 'the system' developed by Cottenham in the 1930s and rewritten, but saying essentially the same, more recently. The objective is stated as placing the car in the right place at the right speed, with the right gear engaged . . . and the basics of the system, whilst it is about car driving, would not be unfamiliar to a racing jockey – sort out where you are going, decide how fast you can go there, then decide how to do it (horses don't have gears, but that's the next decision in a car . . .). It might seem odd that, rather than react by braking, the first thing to do should be to think about steering, but used in a positive and developed way, the 'system of car control', as the manual calls it, has been an enduring and strong feature of police driving. Most police driving instructors would describe it as the single biggest difference between police and other forms of driver training.

Attitude again

Motorcycle training

Training for motorcyclists follows the same general pattern as for car drivers. It started pre World War II, and developed further in the post-war years. Numbers are far smaller since the 1960s, when the 'Noddy bike' gave way to the panda car, and some forces make much more use of motorcycles than others.

Of course, driving a police car can often be about getting somewhere as fast as you can – safely. There are two big mental things going on in that sentence, and the training has to balance them against each other. Being safe can mean being slow, and getting there as quickly as possible sounds like the exact opposite. The training always starts with the 'safe and slow' bit and then develops – using those observation skills and applying that cold, logical approach to decisions built into the system. So police drivers are trained to drive working on the basis that their driving must first and foremost be safe, but within that to find ways to find opportunities, and develop understanding of what they can observe, so that they can take advantage of all the chances there are (or that they can create using the equipment of the car they are driving). That sounds simple enough, but it affects everything you are thinking about when you drive . . . it makes you look at the road itself, at the people around, at the other vehicles – and their drivers – and affects decisions about where across the width of the road to be, where and when to slow down or speed up, what gear to be in . . . There is also an acceptance that this means the driver needs to do more mental than physical work – identifying the dangers, finding the opportunities, working out the quickest way through a problem, keeping the average speed of the journey up, rather than concentrating on purely how fast to make the car go – a subtly but very different thing that goes to the heart of the way the best police drivers get around. And all done with safety as a prerequisite. For some years a book called *The Human Aspects of Police Driving* was used to deal with the attitudinal and physical health factors of the training. It acknowledged that mental fatigue is as big an enemy to driving quality as any other single factor.

Positioning

Where to position the car is also a distinctive feature of police driver training. One of the biggest visible differences to other driver training is the approach of police drivers to positioning for bends, especially left-handers. Given the policing need to be as quick as can safely be achieved at times, police drivers are taught specifically to position themselves so that they can see further (so have more time to react to things, be they problems or chances) and be seen earlier. This can show in a number of ways, but the most distinctive is the approach the police take to going around a left-hand bend, which places the car towards the centre of the road, rather than the more obvious safety option towards the kerb, giving them the slightly extended view ahead gained by a different angle in the line of sight through the corner. This needs to be learnt carefully, and not just taken up without care – there is a potential clash with oncoming traffic, as it will also be close to the centre of the road, and police officers are taught that if there is traffic coming towards them, they need to adjust themselves to the left and adjust their speed downwards as the result of the loss in how far they can see. Police driving works on the basis that you must be able to stop within the distance you can see to be clear, on your own side of the road – a concept traceable right back to Cottenham's influence, so the further you can see, the quicker you can drive. That means there is an advantage in being able to see further . . .

'Limit Points'

Another distinctive aspect of police driver training is the use of limit points to assess the safe speed to negotiate a bend that cannot be seen around. In very simple terms, the driver assesses how fast or slow the car needs to be driven in order to stop on the road surface visible from the driving seat, and adjusts the speed of the car to match that assessment. It is possible to develop this to some degree, and there is some finesse in using the technique, but essentially it gives the driver a way of making consistent, safe, and where necessary quick judgements and progress along a bending road.

Looking after the crew and the kit

Of course, the police have the job of making sure that the rest of us behave ourselves and obey the law. So they have to learn some of that themselves, and also learn what they are allowed to do themselves – when they are allowed to break speed limits and go through red traffic lights, or use sirens and blue lights . . . there's a whole bunch of extra rules they have to manage and understand, as well as the ones they supervise the rest of us about. Knowing how to make sure the car is fit for the road – for what they might use it for – and that it has all the equipment it should have is also a part of the daily routine that has to be learnt.

The Ten Commandments

Roadcraft for many editions contained 'The Ten Commandments', sometimes rumoured to originate with Cottenham himself, and certainly based on his analyses and teaching conclusions. Each has a short paragraph of explanation and a closing bullet point to go with these headings.

1. Perfect your roadcraft.
2. Drive with deliberation and overtake as quickly as possible.
3. Develop car sense and know the capabilities of your vehicle.
4. Give proper signals, use the horn and headlights thoughtfully.
5. Concentrate all the time to avoid accidents.
6. Think before acting.
7. Exercise restraint and hold back when necessary.
8. Corner with safety.
9. Use speed intelligently and drive fast only in the right places.
10. Know the Highway Code and put it into practice.

Whilst this style is a tad dated, there is little wrong with the content and ideas, although more explanation is perhaps needed with some of them now.

Roadcraft no longer has ten commandments; it is a style of presentation that does not fit with modern police culture. If it did, a modern version might read something like:

1. Know yourself, monitor yourself as you drive and be aware of your constraints.
2. Prioritise safety and never take a chance when driving – be certain before acting.
3. Read the road a long way ahead, and to the side (and behind). Understand what your observation is telling you.
4. Drive systematically.
5. Maintain your attention on the driving task – control the distractions around you.

6. Without ever compromising safety, position yourself for the best view ahead, on your own side of the road.
7. Always drive so you can stop in the distance that you can see is clear.
8. Know your vehicle, what it can and can't do – and how to control and use it safely and effectively.
9. Be deliberate and make decisions proactively – plan so you never need to react.
10. Know how to assess bends and corners and use the techniques of the manual properly.

Use of lights and sirens, exceeding speed limits . . . etc.

For the last few years there has been a standardised set of tactics for driving when using blue lights and sirens to deal with traffic agreed by the College of Policing. Circumstances vary infinitely and the advanced level of driving to which officers are trained before they engage in this aspect of driving is an important foundation in allowing them to deal with that variety. Their understanding of driving itself is important here – even if they do not consciously think about it in that way. It is this high level of development of advanced driving skill that allows them to drive above the speed limit, and for those trained to the higher, advanced level sometimes at very high speeds – safely. The police do not see a 'short cut' to driving on 'blues and twos' but regard the solid driving training as a fundamental preparation for that.

All is calm?

The drive will be, perhaps surprisingly until you really think about it and realise it has to be thus, wholly undramatic in character.

- because of all that early observation, and therefore early action to deal with whatever is there
- because of the training to be systematic and smooth, and to use the car sensibly and with some sympathy for it
- because of the attitudinal inputs
- because of the way all these things are brought together.

Calmly executed actions follow very quickly taken but rationally calcu-lated decisions, which are then acted on positively.

A drive executed the way it is taught is safe, systematic, smooth *and* quick.

The legacy

Many trained police drivers have become involved in road safety issues both during their police careers and after them, using the skills and logic of their training (and experience) to inform their activities. Police driver training has also informed some other forms of driver training, bringing the skills and methods into the wider community – mention is made of IAM Roadsmart and RoADAR, but there have been other examples over the decades. Whilst there are, of course, many other factors, it seems likely that the attitudes and approaches that were developed and tested in the stress of operational police driving have been a positive element in helping Britain be a safe place to use the roads . . . Britain is usually one of the top three countries in the world for road safety – statistics published by the World Economic Forum from the WHO in its Global Road Safety report in 2015 show Britain third, with only 2.9 road deaths per 100,000 people (Micronesia had 1.9, Sweden 2.8, and Libya, at the other end of the scale, had 73.4).

What next?

Quite where things will develop in the future is always a dodgy thing to try to predict. Where does police driving fit in with autonomous cars? How much will training techniques change? Who knows? We do know that the government changed the law about training for emer-gency service drivers in 2006 but has yet to bring that piece of legislation into effect. The police should be well placed to cope, however, when they do – since the 1930s they have been training widely and have developed a positive culture around doing so.

Like the roads, the traffic system and the cars themselves, training continues to develop and adapt, and seems likely to continue to be a source of support for road safety for some time yet. How things will develop in a world of autonomous vehicles remains to be seen – will there be autonomous emergency vehicles going faster than ordinary ones, or will police drivers still have to be trained to do that bit of the driving?

CHAPTER FIFTEEN

POLICING ON THE SCREEN

Growing up watching TV cop shows and old movies was a major influence on the young me. I loved the re-runs of *The Professionals* and was hooked on *Spender*, mainly because of that Cossie. The cars were cool and they were the draw that got me interested in policing as a career. We can't do a book about British police cars without looking at their portrayal on screen – sometimes brilliant, sometimes not and sometimes just plain mad. (I'm glad I never had to pursue an E-type with a Wolseley!)

The first crime film ever made was also one of the first British films; the plot of the 32-second, 1895 silent movie *The Arrest of a Pickpocket* concerned a sailor assisting the police with the apprehension of a rogue. Before World War II, murder-mysteries were a very popular form of British film; the 1929 Alfred Hitchcock classic *Blackmail* – the first British 'talkie' feature film – opens with a Met Police Crossley 20/25hp wireless lorry speeding through the capital. We believe they used a real police vehicle for this, not a fake stunt car as would become common once the industry grew and films became more numerous. The Crossleys could receive wireless messages in Morse and it is nearly impossible for a modern audience to imagine the impact of seeing police vehicles in action during the first reel of *Blackmail*.

For many, one of the best police moments in 1930s British films was in *Ask a Policeman* (1939) – who could forget Will Hay, Moore Marriott and Graham Moffatt fleeing the 'Headless Horseman'* in their 1930s Morris?

After World War II, British film policemen (rarely policewomen) were often depicted as CID officers in pursuit of spivs and wide boys, but the breakthrough picture in its depiction of the uniformed branch and police car chases was *The Blue Lamp*. The film sets out its stall early on by opening with a brilliantly choreographed car chase running under the credits with a Met Humber radio car in pursuit of an elegant 6-cylinder MG SA, which gets crashed (painfully for MG enthusiasts like myself; how rare are SAs today!) into some railings. The opening scene was inspired by the real-life shooting of Alec de Antiquis; in 1947 this brave man had tried to foil the getaway of three gunmen who had robbed a London jeweller.

In the late 1940s and early 1950s, Ealing Studios was world famous for its comedies and dramas alike, and this drama, which was shot in and around Paddington in 1949 and released in 1950, set future stand-ards for British cinema and television. Anyone who has seen *The Blue Lamp* will remember the shocking moment when Dirk Bogarde shoots Jack Warner's PC George Dixon† and the chase between the Met's Humber Super Snipe Mk1 and the villains' pre-war Buick Coupe. Before *The Blue Lamp* Warner was not always associated with screen law enforcement; one of his greatest performances is a chilling portrayal of a highly intelligent psychopath in *My Brother's Keeper*, in which he is chased through the countryside by Humbers and Wolseleys. Incidentally, his fellow escaping felon in that picture was played by an up-and-coming young actor named George Cole.

The film was regarded by *The Police Review* as 'the first time the

---

* Played by a pre-Q Desmond Llewellyn!

† There was a real George Dixon, a politician and reformer who pioneered schooling in the Midlands and lived from 1820–1898. *The Blue Lamp*'s producer, Michael Balcon, had won a scholarship to the George Dixon Grammar School in 1907 and it's widely believed the George Dixon character was named in honour of this life-changing oppor-tunity.

police of this country have been adequately presented on screen'*. *The Blue Lamp* was made with the full cooperation of the London Metropolitan Police; the force lent Ealing several of their Humber and Wolseley patrol cars and even permitted a select few officers to play small speaking roles. Some of the 1946 Super Snipe footage from *The Blue Lamp* was later used in the famous Ealing Comedy *The Lavender Hill Mob* and used in the scene in which Stanley Holloway and Alec Guinness 'borrow' a police Wolseley 18/85 and issue fake wireless messages: one of the great classic scenes of British cinema.

In fact, if you are a fan of British cinema of the 1950s and 1960s, it is almost certain that you will become a Wolseley fan. *Seven Days to Noon*, one of the best thrillers of its day, is dominated by 12/48s, 14/60s and 18/85s, while in *The Long Arm*, an unmarked 6/80 is on the trail of a ruthless safe-breaker in a Ford V8 Pilot; PC Nicholas Parsons helps to solve the case and doesn't give the crook benefit of the doubt.†

Town on Trial begins with a pair of mighty Wolseley 6/80s zooming towards the camera. Until the late 1960s police car enthusiasts might have two opportunities to see their favourite vehicles on a visit to their local picture house. At that time an evening at the cinema would last for about four hours; as well as the main feature, there would be the newsreel, the travelogue and – best of all for any police car spotter – the B-movie. This would usually be no more than 70 minutes long and over 75 per cent of them were crime films. With an average budget of between £10,000 and £20,000, not a large sum of money even then, and a shooting schedule of about two weeks, reliable and easy to source police transport was vital.

One of the best-remembered B-film outfits is the Danzigers; two US businessmen, brothers Edward J and Harry Lee Danziger, who churned out second features and TV series from the tiny New Elstree Studios between 1956 and 1962. One of their writers was a young chap named Brian Clemens, who would subsequently create *The Avengers* and *The Professionals*, but who at that time was at New Elstree producing scripts

---

*   Quoted on p. 15 of *Policing Images*, by Rob Mawby.
†   As he does regularly in *Just a Minute*.

based around props that the brothers Danziger had acquired. He once reflected that 'I'd receive a phone call on the Monday telling me to write a script based on 12 nuns' outfits, a nuclear submarine and a double-decker bus by Friday.' One such epic was 1959's *The Great Van Robbery*, starring the Danzigers' regular heavy, Denis Shaw, as 'Caesar Smith of Interpol' – plus a Vauxhall Victor F-Type that seemed to appear in quite a few of their productions.

Another studio associated with British crime B-films was Merton Park, in South London. Between 1953 and 1961 they churned out 39 editions of the *Scotland Yard* films, the early ones commencing with a Wolseley 6/80 racing along the Chelsea Embankment. The narrator of these epic dramas was always Edgar Lustgarten, a barrister and criminologist who looked about 150 years old, and the main actor was Russell Napier, as 'Duggan of the Yard', whose main job was to bark orders into the Wolseley's microphone. After 1958, the series began to use the 6/90, most brilliantly in 1959's *The Dover Road Mystery*. 'I'm flat out, chum!' exclaims the police driver as the Wolseley is in hot pursuit of a tuned-up Ford Zephyr Mk2. This chase was so exciting, I nearly dropped my cake as the 6/90 lurched all over the road, bell clanging; you can't beat a black and white movie on a cold wet Sunday afternoon.

In 1962 *Scotland Yard* was revamped as *Scales of Justice*, with a groovy new theme tune by The Tornados, and ran until 1967. One of the best episodes, from our point of view as police vehicle fans – and as drama actually – was *The Material Witness*, a surprisingly hard-hitting story about drink-driving, with the police's 'Big Farina' Wolseley pursuing a Rover P6 2000. Probably the best-remembered of the Merton Park crime B-films was their *Edgar Wallace* series, which ran from 1960 to 1965. You would be guaranteed an appearance of at least one Wolseley – 6/90 in the early editions, and a 6/99 or 6/110 in the later ones – plus some entertaining screen villains. *Five to One*, made in 1963, has a twenty-one-year-old pre-*Sweeney* John Thaw as a conman in a Vauxhall Victor FB Estate and 1962's *Solo for Sparrow* has a gang of hoods in a Vanden Plas Princess 3-litre, including Michael Caine as an Irish (!) gangster.

Of course, not every crime film, be it a main or a B-film, made in the UK between 1950 and the late 1960s featured a police Wolseley, although it certainly felt like it. A 6/90 turns up at the end of the brilliant second feature *The Man in the Back Seat* to investigate the Austin A125 Sheerline crash caused by the crime victim of the title, and another famously chases Peter Sellers' Aston Martin DB4 GT in the wonderful comedy *The Wrong Arm of the Law* – and appears to keep up! At one point 41 DPX, chassis 0157/R, suddenly becomes a standard, non-cowled, headlight DB4 during the bridge jump at Dolphin Road in Uxbridge, and a third car, 40MT, chassis no. 0167/R, was used when the GT developed serious engine problems. The last-mentioned Aston Martin was a works' experimental car that achieved a top speed of 152.5mph in a 1961 *Autosport* road test and was used by Aston works driver Reg Parnell to go from 0–100mph and back to 0 again in 24 seconds, which was then a record. One feels that even the best-driven of Wolseleys might have found it challenging keeping pace!

If you want to see some equally unfairly matched pursuits, how about the Wolseley 6/110 v Maserati 3500 GT by Touring in *He Who Rides a Tiger*, or even the Jaguar E-Type 4.2 Series One Roadster being pursued through Harwich by a 6/99 in the fab pirate-radio/pop-industry/diamond-smuggling flick *Dateline Diamonds*? And, yes, that neatly dressed, short-haired DJ is a young Kenny Everett.

Then there is the wonderfully bad horror film *Konga*, which features an overacting Michael Gough and a rather cross giant mutant gorilla; from our perspective most of the human cast is outperformed by a fleet of 6/90s, but oddly the film was later satirised in the TV comedy *The Big Bang Theory*. Two of the best police Wolseley-related comedies of the 1960s were directed by a young Michael Winner, long before he became associated with bad insurance adverts. *You Must Be Joking!* and *The Jokers* had plenty of tail-finned Wolseleys fitted with the new klaxon horns that the Met were then starting to use to augment the bells on their patrol cars. And, as proof that a low budget need not mean low quality, there is the brilliant 1967 Michael Reeves-directed horror film *The Sorcerers*. The moment when the 6/99 spins through 360 degrees as it chases a Jaguar Mk VIII was not in the script; apparently the stunt driver had lost control but the director thought the moment looked so good that he kept it in the final print.

However, there really are many non-Wolseley A and B films of the 1950s and 1960s. Humbers were regularly seen, from the Snipe in *The Long Memory* to the Hawk Mk III in *The Ladykillers* and *The Man in the Road*. *The Clouded Yellow* features several genuine police Austin A70 Hampshires patrolling the north of England, for when filmmakers left the capital this often gave filmgoers the chance to see a varied selection of police transportation. *Whistle Down the Wind* was mainly shot around Burnley and Clitheroe with genuine Lancashire Constabulary Ford Zephyr Mk2s and MGA 1600 Mk1s. *Violent Playground* was partially shot in Liverpool in 1957 and stars a CID Austin A90 Westminster, a 'Hi-Line' Ford Zephyr Mk2 and an E-Type Vauxhall Velox. There was also an exciting last-reel chase through the city – although the poor continuity means a bizarre mix of Ford and Austin exteriors and interiors. (Yes, the Teddy Boy who deliberately throws his bike in front of the Zephyr is a young Freddie Starr.)

The classic heist drama *Payroll* was largely filmed in Tyneside with Ford Consul Mk II police cars, although how they could hope to catch Michael Craig's mob of wage snatchers in their Jaguar Mk1 3.4 is anyone's guess. The writer-director Val Guest was renowned for using the British landscape in his pictures, and his fantastic thriller *Hell Is a City* uses real Oldham Borough and Manchester City Wolseley 6/90s and Austin A95 Westminsters. Even the unmarked 6/90 driven by Stanley Baker's Inspector Harry Martineau was a genuine police car. In Guest's East Sussex murder drama *Jigsaw*, the Consul Mk2 Deluxe and the Anglia 105E were Ford publicity vehicles but the A99 Westminsters and Morris Oxford Travellers apparently belonged to Brighton Borough Police (they also seem to share the same number plate!), just as the Wolseley in 1963's *80,000 Suspects* was from Bath City Police. Of the B-films that ventured out of London, *The Hi-Jackers* ends with unmarked Humber Hawks and various Vauxhalls – FB Victors and the PB Velox – arriving to arrest the gangsters who mug lorry drivers on

Chobham Common. A Hawk also appears, together with another police Oxford Traveller, in *Smokescreen*, one of the most fondly regarded of all British B-films. Meanwhile, back in the big city, the police use an Austin Se7en (the Austin-badged Mini's original name, as I'm sure most of you know) in the fast-paced thriller *Seven Keys*.

Robbery, which was released in 1967, is rightly regarded as one the finest ever British police films, one to rank alongside *The Blue Lamp*. It was made during the initial stages of the change to Regional Crime Squads,* when provincial forces would still 'call in the Yard' for serious crimes, and when senior officers still had a quasi-military bearing. It is hard to believe that Glynn Edwards was only 36 when he played the 'Squad Chief'. But now the grammar of film had moved on because the narrative's emphasis is on Stanley Baker's crime boss rather than James Booth's Detective-Inspector. *Robbery* featured one of the best car chases of all time. When shooting commenced, the Met were starting to use Jaguar S-types, supplied for the film by Nine-Nine Cars. The first reel is dominated by the pursuit of a gang of hoods in a Jaguar Mk2 3.8, with the director Peter Yates and his crew using handheld cameras. There is no distracting incidental music, just the accompaniment of two-note sirens and protesting Moss gearboxes. Yates had previously been the assistant works manager at HW Motors, in Surrey, and he was a filmmaker who really understood cars. When you see the film on DVD, just marvel at the sheer professionalism of the on-screen driving; at one moment the Jaguars narrowly avoid a Vauxhall Viva HB that has accidentally strayed into the shot and which was owned by a member of the public going about their day . . .

Robbery was made by Oakhurst Productions, which was co-owned by its leading man Stanley Baker and, as the title would suggest, was about the 1963 Great Train Robbery. For legal reasons, the real names of the villains couldn't be used in the film and the recently disused Rugby to Market Harborough line was used for the heist sequence. Some of *Robbery* was made in Ireland, to accommodate the leading man's tax status in the UK. One such was the roadblock scene with

---

* They were formed in 1964.

police Ford Zephyr 6 MkIVs. The film does have its faults – bad over-dubbing and the crashed Vanden Plas Princess 4-litre R clearly lacks an engine – but it is still essential viewing for anyone who appreciates British crime dramas and fine British cars. And the brief scene of that police S-Type trying to avoid children crossing a road is more dramatic than many a Hollywood thriller you could mention.

Other great police S-type moments of British cinema include *Frenzy*, Alfred Hitchcock's last ever film to be shot in the UK, and the incredible zombie-biker in Walton-on-Thames horror B-feature *Psychomania*. Another horror cult classic, *Scream & Scream Again*, features a quite amazing chase between the law in an S-type and a Humber Hawk Series IV and a 'space vampire' dressed like Scott Walker in an Austin-Healey 100/4. The conclusion of the chase defies all attempts at description – let us just say that it involves the Humber's starting handle bracket and a literally unhanded villain. *Universal Soldier* (a 1971 adventure film, as opposed to the Jean-Claude Van Damme vehicle) has the equally wonderful scene of a long-haired George Lazenby in a Reliant Scimitar GTE trying to outrun the Met – sample dialogue: 'We're chasing a mad bastard in a red Scimitar'. There is also *Loot*, *Sitting Target* – one of the few pictures where gangsters drive a Hillman Avenger – and *Villain*, which is also an early example of a movie using the S-Type as a getaway car. It must be said that the poor Jag does suffer from having to carry six gangsters with a flat tyre. In 1974 came *Callan*, the big-screen version of the 1967–1972 ITV series of the same name, and further proof that one brand of film that the British do very well is the seedy spy drama. It doesn't need pyrotechnics or Bruce Willis in a vest; instead it relies on a black Jaguar S-type, a white Range Rover and some level-crossing barriers in Kent, and is much better for it. Definitely one I'd recommend seeking out.

Callan was directed by Don Sharp, who had worked on the second unit of *The Fast Lady*, and the plot requires that Edward Woodward's David Callan, a reluctant hitman for The Section, unsettle the arms dealer Schneider using a Range Rover fitted 'with black glass all around'. Nine-Nine Cars were responsible for the stunts, and the darkly tinted panes both looked highly menacing and hid the fact that the stunt driver looked nothing like the lead actor (thereby avoiding a problem that always seemed to afflict the later Roger Moore 007 films). As any filmgoer in the 1970s could have told you, if a millionaire master criminal arms dealer favoured a battered Jaguar 3.8 S-type with a fake 'G' registration suffix (it's actually a 1964 car) rather than a new Daimler Double Six LWB it will eventually suffer a terrible fate, as it does in *Callan*. This was dictated by the economics of filmmaking; crashing new cars costs money unless the maker gives them to you, and it is a practice that continues today. The white Range Rover is another British film regular in this role because it still looked impressive even when old and thus cheap to buy. In *Callan*, Solihull's finest plays cat and mouse with the S-Type, and we then see a level crossing where the keeper is about to lower the barrier. All Jaguar enthusiasts might want to look away at this point.

A year later, another wrong-doer used an S-type to avoid the forces of law and order, although in *Brannigan* the hero was John Wayne's 'Det-Lt Brannigan' on attachment to Scotland Yard. The sequence in which Wayne 'borrows' a Ford Capri 3000 GT Mk2 required a double; the actor could not cope with a manual floor gear lever. The Jag genuinely leapt over an open Tower Bridge; the preparation for the stunt took the film company months of negotiations with the London Met and the port authorities until it was agreed that the Jaguar would not actually go over the bridge. However, the stunt maestro was the late great Peter Brayham, who was at the wheel of the S-type and saw this as a once-in-a-lifetime chance – he went over the gap! The 1966 Jag was nearly totally wrecked in the process and Brayham was not popular with the law for 'driving without due care and attention', so Wayne's chasing Capri jumping the bridge was done in post-production by driving the car up the ramp but stopping near the top – very obvious when you watch this pre-CGI-era movie. The Capri was lent to the production courtesy of Dagenham, in addition to 14 other cars, and Brayham had enthusiastic respect for Ford. Prior to *Brannigan*, he had been working on an ITV *Armchair Theatre* play called *Regan* – but more of that later.

By the end of the 1970s, Dave King's deeply corrupt Inspector Parky used a plain finish 'wedge' Princess 1700 HL in *The Long Good Friday*, playing against villain Hoskins' Se2 XJ6 (and glamorous moll Helen Mirren).

However, British police cinema wasn't all gritty realism. *Venom* is an everyday story of police Rover SD1s and an escaped deadly snake that bites Oliver Reed in the unmentionables; it's a very strange film . . . However, it was the SD1 that was to become to the 1980s and early 1990s British films what a Wolseley was to the 1950s and 1960s – from *The Great Muppet Caper* to the not very good big-screen version of *George and Mildred*, and from *The Tall Guy* to *Clockwise*, the Rover was always a welcome member of the four-wheeled cast. Even if the picture was of the standard of *Bullseye!*, a Michael Winner 'comedy' with all the humour value of a public information film about nuclear war, at least the police SD1 served as a point of distraction.

As a child, it was watching cop shows made for TV that got me interested in the police, though, so we certainly can't neglect the small screen here. The first British TV series to heavily feature police cars was aired in November 1954: the BBC's *Fabian of the Yard*. This opened, like *The Blue Lamp*, with a Humber Hawk speeding (well, trying to anyway) through the capital as Bruce Seton made the famous announce-ment – 'This is Fabian of Scotland Yard!' – and ran for 36 episodes until 1956. The first of many what are now rather unflatteringly called 'Police Procedurals', it was actually loosely based on the memoirs of a real-life detective, former Superintendent Robert Fabian.

ITV began transmitting in 1955 and one of its best-remembered early crime shows was *No Hiding Place*. From 1959 to 1967 Raymond Francis' Detective Superintendent – later Chief Superintendent – Tom Lockhart would arrive at the scene in a Wolseley 6/90, a Humber Super Snipe Series III and, in later years, a Wolseley 6/110. Sadly, unlike reality, crime never paid and Francis later observed that 'I used to plead; why can't the villain get away this week, just this once – I'll nick a couple next time to make up for it?'. He later, satirically, semi-reprised the role by appearing as the Inspector at the wheel of a 6/90 in *The Comic Strip Presents: Five Go Mad in Dorset*, lashings of ginger beer! And some cake. Sadly, although *No Hiding Place* was one of the most popular crime shows on British TV in the 1960s, only 25 editions of 236 programmes survive.

Wolseleys were also one of the major stars of *Gideon's Way* (1965), which was made on film by ITC. Almost every week Commander Gideon (John Gregson) would pick up a black Bakelite phone and order a horde of Wolseleys to descend on East London to stamp out crime. The episode entitled *The Reluctant Witness* offers especially good entertainment value, as the head gangster is played by Mike 'Jeff Randall' Pratt and one of his strong-arm men is Peter 'Here's one I made earlier' Purves!

For those who preferred French police cars there was, of course, the 1960–1963 BBC Television series of the *Maigret* novels, and any reader of a certain age will have an instant memory of Ron Grainer's theme tune, of the lead actor Rupert Davies striking a match on a wall in the opening credits and of his Citroën Traction Avant. A good deal of the *Maigret* series was shot in the studio, but there was a limited amount of location shooting in France with the very rare 1954 15/6, the late 6-cylinder model with hydro-pneumatic rear suspension and a 46-foot (yes, really, you read that right!) turning circle. The Citroën was provided by a French leasing company, and after the last of the 51 episodes was completed, Davies acquired the car and brought it to the UK. Ironically, Jules Maigret was not seen driving the Traction on duty; an officer with the rank of Commissaire was allotted a police driver.

Then there was *Z Cars*, the big daddy of car-based police series and, to put a myth straight to bed, it would have been called *Z Cars* whatever vehicles were featured; the fact that it became synonymous with Ford Zephyrs, which happen to begin with Z, was sheer coincidence. It was a controversial series that dominated the ratings during the 1960s. At the beginning of the decade Colonel Sir Eric St Johnston, who was then the Chief Constable of Lancashire Constabulary, had introduced the 'Crime Patrol' system, with unmarked cars patrolling a wide beat. This was to be the basis of the new BBC series and the Corporation's researchers noted that 'the Seaforth Crime Patrol use a Ford Zephyr and the Kirby Crime Patrol a Morris Oxford. Both cars are distinguished by twin aerials on the boot'.* It was the former that became a star of the TV screen when the Colonel gave his support for the new weekly police drama series that was to become famous as *Z Cars*. The show's joint-first scriptwriter and a key figure in its creation, Troy Kennedy Martin, also wrote *The Italian Job* (come on, you know it's really your favourite film, despite that intellectual-sounding French one you trot out at dinner parties. It's every car man's favourite film; no British police cars in the main car chase, though!) and the wonderful thriller *Edge of Darkness*. He was inspired by listening into police VHF broadcasts on his home radio while recovering from mumps and heard the voices of real, fallible human beings dealing with tragic, complex and sometimes even funny situations; things to which there were sometimes no right answers. He realised the police were not the plaster saints they were sometimes presented as, and he wanted to portray that realism.

However, after the pilot episode was first aired on 2 January 1962, the Colonel was not happy with the show's approach; he was apparently expecting the straightforward approach of the *Scotland Yard* B-films, whereas *Z Cars* was intended to be, and was from day one, a modern, gritty, realistic drama that would embrace the often-tragic nature of policing as so often experienced by young PCs. He visited Stuart Hood, the then Controller of BBC Television, and demanded that the series be withdrawn from the air, although what eventually happened was that the Lancashire Police withdrew their support. This meant that the Zephyrs and MGA Roadsters of the first season were subsequently replaced by Dagenham PR Zephyr 6 Mk3s, the car with which the programme is still indelibly associated. These were painted in Daffodil Yellow, which made them show up well on black-and-white videotape. Brian Blessed, who played PC 'Fancy' Smith, did not hold a licence

---

* Quoted on p. 173 in *Beyond Dixon of Dock Green: Early British Police Series* by Susan Sydney-Smith.

prior to the series and recalled in his memoirs that 'before I started to take lessons, someone at the BBC leaked to the press that I couldn't drive'. He also observed that 'as wonderful as the Ford Zephyr looked on TV, it was absolutely awful in the snow' – the straight-six engine made for a nose-heavy car with a light rear end.

Z Cars ran for 801 episodes over 16 years and used a wide gamut of Fords – Escort Mk1 panda cars seemed to dominate the colour shows of the 1970s – and the spin-off, *Softly, Softly*, made extensive use of Cortina Mk2s and Zephyr 6 Mk3s. The final edition, *Pressure*, was broadcast on 20 September 1978.

Many of the episodes of both shows sadly fell foul of the BBC's programme-wiping policy – most shows were recorded, but early video-tapes were very expensive and so the Corporation's accountants decreed that it be reused rather than store low-brow culture like *Z Cars* or *Doctor Who*; proof positive if any were needed that accountants should never be allowed any power! Looking at the website www.thiswaydown. org/missing-episodes/index.html is a very depressing experience; for example, every *Z Cars* episode transmitted between 19 October 1970 and 19 June 1972 is believed missing, as is every edition aired in 1968.

The early editions of those shows that survive are notable for being largely studio-based. Blessed noted that 'for front-on driving scenes, we'd have a camera on the bonnet and a backdrop at the reverse of the car, which would show some country lane or wherever it was we were supposed to be driving'.* Peter Lewis, a member of the production team, noted in 1962 that 'it's done live. Only the scenes showing open streets are filmed outside of the studio prior to live transmission/ recording. On the night the Z Car stands on wooden rollers for easy manoeuvring in front of back projection screens that produce 60mph illusions.'† Some episodes used more film work than others, and *People's Property*, from Series One, has Fancy and his partner Jock Weir (Joseph Brady) chasing two runaways to North Wales, their patrol car encountering a Mersey Tunnel Police Land Rover and an Austin A95 Westminster

---

* p.142, *Absolute Pandemonium*, Brian Blessed.
† Quoted on p.62, *British Television Drama; A History*, by Lez Cooke.

squad car en route. Of the many *Z Cars* storylines, one that really abides in the memory is the first series episode *Friday Night*, written by Troy Kennedy Martin. Jeremy Kemp's PC Bob Steele has to cope, single-handed, with the aftermath of a road traffic accident where a drunk motorist has crashed into a motorcycle. The ambulance is delayed, the pillion passenger has been killed and the teenage biker is dying, crying out for his father. There is no extraneous music; just a single set, a Ford Zephyr, and Kemp demonstrating why he was regarded as one of the finest character actors of his generation. *Z Cars* became twice weekly for a while and changed greatly over the years, as much due to broadcast techniques changing as politics and fashion, but at heart it remained true to its original ethos.

Other high-profile law-enforcement Ford users of the 1960s and 1970s included the disaffected spy David Callan, who drove a two-door Cortina Mk2 saloon when he was not being chauffeured in Lovely's Austin FX4 taxi. Then there is *Special Branch*, more of which later, and the ATV sitcom *Spooner's Patch*. The finest motoring moment of the latter was the car of Peter Cleall's DC Bolsover, who desperately wants a *Starsky and Hutch*-style Torino and so paints his 105E Anglia in red with a white stripe. However, just in case the reader thinks I am a Dagenham-obsessive, I'll mention the Vauxhall Viva HB in Yorkshire TV drama *Parkin's Patch* and the Jaguar S-type in the LWT drama *The Gold Robbers* before we start with what seemed to be the definitive big- and small-screen police car of the 1970s – the Rover P6B, although the smaller 4-cylinder P6s were used as well.

Be it BBC or ITV, the Rover would dash to the assistance of the ITC heroes of *The Persuaders!*, *The Protectors* or the incredibly bad *The Adventurer*, and regularly starred in the 1972–1974 ITV series *New Scotland Yard*. A P6 assisted with the hunt for Richard Burton's Ronnie Kray-like thug in *Villain* and (in very badly mocked-up form) arrested the gang boss Cyril Kinnear towards the end of *Get Carter*. Modern drama again took a cue from the past when, in 2007, the opening sequence of *Life on Mars* had lead character Sam Tyler (actor John Simm) apparently being run over by a late-model facelift Mk3 Cavalier (another example of using a cheap car to damage) and waking up in 1973 next to a Rover P6B. Amazingly, Tyler even uses the term Rover P6 when trying to explain his situation, which I remember thinking was odd when I watched it because although the car is now universally known as P6, like other Rovers the SD1, P4 or P5, hat slang has really come into use in the classic car world as a way of making clear which model you are talking about. Rover never sold the car as a P6; it was always Rover 2000, 2200 or 3500, depending on which engine it was fitted with, and I don't remember people calling them that when I was young. It's a retrospective nickname that's made its way into a (sort of) historical drama. We'll hear more about the car inconsistencies of my good mate Phil's two time-travelling detective series later.*

The Rover's main in-house rival, the Triumph 2000/2500 saloon, looked even more imposing when in 2.5-litre fuel-injected Mk2 form and was seen conveying Leonard Rossiter in *The Pink Panther Strikes Again* (1976) and in the controversial BBC series *Law & Order.* Meanwhile, ADO16 (an Austin or Morris 1100/1300, for you non-petrol heads) panda cars were often seen making guest appearances in such sitcoms as *Some Mothers Do 'Ave 'Em* and *The Good Life,* and some readers may have fond memories of sitcom *Rosie,* in which a Ford Cortina 1600L Mk4 patrol car was driven in a decidedly non-*Sweeneyesque* manner.

From 1973 onwards, the police cars from the British Leyland stable faced a new rival in the Ford Granadas used in *Special Branch.* This began in 1969 as a black-and-white drama shot on videotape, but after two seasons the Euston Films division of Thames Television took over the series. Seasons three and four were shot on celluloid, which gave viewers more of an opportunity to appreciate the Ford PR fleet Cortina Mk3 GT and Granada Mk1 against the familiar Rover P6s. There was also an excellent credit sequence of a Hillman Hunter being driven into

---

* *Rover did make a fantastic documentary film about the development of the car called* Project P6, *which the British Motor Industry Heritage Trust at the Gaydon Museum sells on DVD. It's well worth watching, one of the best period car industry development films you will ever see.*

a pile of cardboard boxes – this seemed to happen a lot on television during the 1970s and was easily replicable in your back garden with a pushbike when 10 years old . . .

After *Special Branch* concluded, *The Sweeney* was conceived as a replacement and the pilot *Regan* – with Mk4 Zodiacs amongst the squad's Granadas and the great character actor Lee Montague as the chief heavy – seemed so promising that a full series was commissioned even before it aired. Working on *Regan* was the aforementioned *Brannigan* stunt expert Peter Brayham, who knew that the show was 'going places'. (Readers will probably recognise him from the pre-credits scene of the fourth and final season – he is the hood driving the Escort Estate.) Another stunt expert much associated with the series is Rocky Taylor, who recalled in an interview with *Classic Car Weekly* that the Ford Consul GTs and Granadas were 'fast, comfortable and definitely looked the business as a police car'.

You can probably name so many of the main Fords used in the four series of *The Sweeney*, for they are the stuff of legend to us classic enthusiasts; Consul GT, Granada 3.0S Mk1, Granada Ghia 3.0 Mk1, Granada 2.8iS Mk2, Cortina 2000XL Mk3 and Cortina GT Mk3. These fine cars were all provided by Dagenham, whose press fleet all used Essex plates, and Peter Brayham once noted that if Regan and Carter are seen using a non-Essex registered vehicle – especially one several years out of date – it would probably be wrecked by the closing titles. 'Whenever the squad starts using a clapped-out old banger for no apparent reason, odds are that it is going to be totalled by the end of the episode'. Detective Inspector Regan's driver 'Bill' was played by Tony Allen, who was John Thaw's stand-in at one point, and one trick used during filming was the fitting of the police vehicles with laminated glass so as not to reflect the camera lights.

Rocky also recalls that the gangster's Jaguar S-type engines 'were always blowing up!', and some of the drivers would moan about their soggy suspension. Stunt teams were even advised to try to use no more than a gallon of petrol per episode, and although at that time the Flying Squad worked predominantly at night, costs ruled out extensive after-dark shooting! The tight schedule meant that the filming of each episode normally took ten working days, but the artistry on screen was unrivalled. Just look at the episode entitled *Stoppo Driver*, which for many is *The Sweeney* at its best, with Mr Taylor at the wheel of the Jag. Brayham planned the driving scenes so that 'no editor could ruin it with the cutting', and Regan and Carter were in a Granada Ghia Mk1 rather than their more familiar Consul GT; Euston did not want to risk damaging the latter. There were also two spin-off films, the first one *Sweeney!*, notable for the villain's fake Austin 1800 'Wedge' squad car, and *Sweeney 2*, featuring a liveried 3.0Si saloon. By the mid-1970s Thames Valley Police was using the big BMW for traffic work and the film car may have been the exact model that the company used for their successful print ad campaign: 'It takes one to catch one.'

By 1977 the BBC had their own alternative to Regan and Carter in the form of *Target*, which had Patrick Mower's Detective Superintendent Steve Hackett investigating crime in Southampton Docks, his trousers and nostrils equally flared. The show was once described as '*The Sweeney* as re-imagined by Alan Partridge', but it is often undeniably entertaining. It was shot on film, rather than the usual BBC video/film mix, Peter Brayham contributed to the stunt work and the theme tune is fantastic. As for the plots, they variously have Hackett 'borrowing' Ford lorries, staging arrests in deserted airfields with a Consul saloon and finding dead informers in CID Cortina Mk4s, which tends to make him irate. Our hero also likes to shout a lot when he is not crashing into – surprise, surprise – piles of cardboard boxes and/or tyres.

Meanwhile, a highly popular police show from 1955 to as recently as 1976 was *Dixon of Dock Green*, with Jack Warner reprising the character of George Dixon from *The Blue Lamp*. The programme developed a myth for being overly cosy, which is not quite fair as it's far harder-hitting than its reputation suggests. As with *Z Cars*, *Dixon* suffered very badly from tape wiping, and the earliest surviving colour episode, 1970's *Waste Land*, focuses on a Morris Minor 1000 panda car found abandoned in a derelict wharf. The driver cannot be traced, and, as the plot develops, the DCI, the Chief Superintendent – who arrives in a Rover 3500 P6 squad car – and Sergeant Dixon himself all fear the worst. *Dixon of Dock Green* was rarely a show that indulged in car chases, but Warner's performance and that shot of the PC's wife waiting in the back of an S-type area car for the news she secretly dreads make you realised just why the series remained on the air for so long – it had integrity. In a much lighter vein is the 1975 episode *Baubles, Bangles and Beads*, with an opening that is almost a parody of *The Sweeney*, with three hoods in a Ford Granada Mk1.

Incredibly, less than two years separate *Reunion*, the last edition of *Dixon*, which aired on 1 May 1976, and the first screening of *The Professionals* on 30 December 1977. Bodie and Doyle's early outings used British Leyland vehicles as the show's production company Mark One had recently produced *The New Avengers*, which used a fleet of BL PR cars. When *The Professionals* started filming, the executive producer Brian Clemens was already in negotiation with Ford. Until then George Cowley used a Turmeric Yellow Rover 3500 SD1 in three adventures, while Doyle drove a Pageant Blue Triumph TR7 and Bodie was mainly seen in a white Dolomite Sprint. Before long, though, the programme had settled down to its now-familiar remembered form. There was a new title sequence, with Peter Brayham crashing a Ford Granada Mk1 through plate glass at Wembley Stadium, replacing the frankly weird early one in which the team hoon a Rolls-Royce Silver Shadow around an abondoned military base while training. Cowley now had a Granada Ghia Mk1 in Jupiter Red as his official transport. This was replaced by a fuel-injected Mk2 Ghia in the same paint finish and, in the last seasons, by a pair of Midnight Blue injected Granada Ghias on TRX alloys.

As for Cowley's junior officers, Ray Doyle and his amazing permed hair (which should have been the name of a prog-rock band) were seen in an Escort RS2000 Mk2, which, despite its 'T' registration, was a 1976 model; a sunroof was later fitted during production to make it easier for the crew to light the interior. Martin Shaw was subsequently associated with, of course, Capris – initially a Strato Silver Capri Mk2 3.0S wearing an XPack body kit and rare Ronal alloys. He then used a Mk3 3000S finished in Solar Gold. His ever-smirking colleague favoured an Arizona Gold Capri Ghia Mk2 Automatic before William Andrew Bodie graduated to the most famous *Professionals'* car of all and the one that inspired a brilliant Corgi die-cast model, the Strato Silver 3000S Mk3s. Brayham recalled that Martin Shaw and Lewis Collins were '. . . both excellent drivers but I taught them a few tricks for the camera'.

It is sometimes forgotten that *The Professionals* used quite a wide range of CI5 vehicles, from Vauxhall Carlton Mk1s to Suzuki SJ40s. In *When The Heat Cools Off*, which was largely set in 1971, we see the then PC Doyle in an Austin 1100 panda car and, just like *The Sweeney*, whenever Bodie or Doyle was seen at the wheel of an old car, there would be automotive carnage to follow. Any fan of the show would have taken one look at that Morris 1800 Mk1 in *Blind Run* and shaken their heads. But it is the silver Capri 3000S Mk3 that will be forever associated with *The Professionals*, chasing a Series One XJ6 or a battered Vauxhall Ventora FD through Middlesex.

In contrast to Bodie and Doyle keeping law and order in the finest late-1970s leisure wear, there were those detectives with quirky cars. While the Americans had Detective Lieutenant Columbo arriving in his Peugeot 403 Cabriolet, British television in recent years has the Bristol 410 of *Inspector Lynley*, which always looked almost too magnificent to be used, unlike Bernard Cribbins' 1939 Austin Eight Military Tourer in *Dangerous Davies: The Last Detective*. On Sunday 18 October 1981, the BBC debuted *Bergerac*, which although filmed in Jersey and France had a production office at Pebble Mill in Birmingham, The stories of Detective Sergeant Jim Bergerac of Le Bureau des Étrangers was created by Robert Stewart Banks, who had previously devised the Bristol-based series *Shoestring*, where the uniformed police used P6s and the private-eye hero had a Ford Cortina Mk3 Estate. But in *Bergerac*, John Nettles would have to fight crime and keep law and order in a Triumph Roadster, which, with a heady 63bhp, could accelerate from 0–60 in approximately 25 seconds and take at least as long to stop again with its drum brakes. Villains could probably have saved their petrol monies and strolled away from that week's heist fast enough to evade the flat-out Triumph.

In a 2014 interview, Nettles remarked that he found the Triumph to be 'probably the worst sports car ever built. We filmed *Bergerac* on Jersey and the whole island is covered in narrow roads surrounded by hedges. It is difficult enough to drive a normal car out there but that thing had a 10-foot-long bonnet which made it impossible to drive'. Given such performance, a Reliant Robin might have proved an effective getaway car, and Jim's set of wheels was apparently decided upon at a fairly late stage in the show's development. The first Roadster was an 1800 model that was found in Yorkshire, and there was also a second white 2-litre engine model that had to be re-sprayed. By 1986 another 2000 Roadster was found in the West Midlands, and after the end of *Bergerac*, this Triumph was sold for £34,000 for the BBC's Children in Need appeal.

Between 1980 and 1985, viewers who preferred more down-to-earth police shows often tuned into early Saturday evening show *Juliet Bravo*, a programme dominated by Escort Mk2s and Cortina Mk2s in its early years, progressing to Austin Maestros – and as it won't get remarked upon anywhere else in this book I'm going to say here that the Maestro was a very clever bit of vehicle packaging; it's just a shame it looked so, well, Maestro-ish. It would also be unfair to judge the series on the strength of a clip from the 1980 episode *Fraudulently Uttered* that is doing the rounds on YouTube – a gripping chase involving a CID Ford Escort Mk2, a vicar in an Austin Allegro (and oddly, in more recent times one of the best Allegros in the world has been restored by the Vicar of Longbridge, Colin Corke, so maybe *Juliet Bravo* had some foresight there), a villain using a Mini Clubman and, best of all, a moped to effect his getaway. Another chase scene remembered by many is in *Runner*, where a nine-year-old boy steals an AC Invacar and is pursued by a police Range Rover; reversing the Wolseley V Aston Martin performance deficit seen in 1960s films.

Naturally, there are some SD1s, too, as throughout the 1980s Rover dominated small-screen crime drama. Just think of *The Gentle Touch*,

Widows and *Dempsey and Makepeace*, to name just three series. The car mainly associated with the last-named show may be the Ford Escort Mk3 1.6i Cabriolet in either 'A' or 'C' registration guise, but the mighty Rovers were always on backup. D&M, as us cop-show aficionados call it, marks the last time Ford PR loaned a Capri (a 2.8i in black) to a police show for 1986's episode *Jericho Scam*, in which it is used as a getaway car. The Capri's first high-profile role was as transport for 'Greene' (Patrick Mower), a crook in the last episode of the ITC drama *Department S*. Called *The Soup of the Day*, it was released on 4 March 1970, proving Ford's product placement team worked quickly, and they continued to promote the Capri in this way throughout its production life, as we shall see. The Capri's first really high-profile role was in the 1971 heist film *Villain*, which starred Richard Burton as Vic Dakin. It's now considered a cult classic, despite its lack of commercial success, and a gold Capri (wearing Chelmsford plate FPU 333H, a giveaway that it was a Ford PR fleet car) was perfect as the transport of Ian McShane's character Wolfe Lissner, an ambitious and ruthless young man involved in a complex relationship with the sadistic Dakin.

The SDI was in the opening credits of *Rockliffe's Babies*, together with a Mini Metro panda car which appeared in the quite entertainingly awful Cannon and Ball film vehicle, *Boys in Blue*, a 1982 remake of *Ask a Policeman*, and, of course, in *Minder*. It is quite a surprise to see the early shows as their look is far closer to *The Sweeney* than the comedy show they became; the first episode was made so long ago that the police also use a Mk1 Escort panda car. Of course, the police vehicles most associated with that fine show were the unmarked CID cars used by Detective Sergeant Chisholm and Detective Sergeant Rycott; forever arriving at the lock-up in a Cavalier, Montego, Sierra or even a Talbot Solara in their ever-futile pursuit of 'Arfur' Daley.

For those viewers who wished to see a Jaguar owned by a motorist on the right side of the law, there was *Inspector Morse*. The hero of Colin Dexter's novels originally drove a Lancia, but John Thaw apparently requested a British car for the series, even if he did once refer to the Jaguar 2.4 Mk2 as being 'a beggar to drive'. 248 RPA was a manual overdrive model that was registered in July 1960 and was acquired for *Morse* by Carlton TV. It appeared in all 33 episodes between 1987 and 2000 and was raffled in 2001. Since its TV stardom, the Jaguar had been extensively restored.

Meanwhile, some other British police TV shows of the 1980s and 1990s featured more up-to-date machinery that was not prone to off-screen breakdowns (although Morse waiting impatiently for the RAC or Jim Bergerac calling the AA might have proved entertaining). By 1991 Jimmy Nail's Detective Sergeant Freddie Spender favoured a Sierra Sapphire RS Cosworth to patrol Tyneside – 'It's a beast'. Ford loaned a new model for each of the three series, commencing with a G-reg model before the hero progressed to the four-wheel-drive 'Cossie' for the final two seasons. Ironically, the first Spender car was stolen and destroyed by the thieves; a £100-per-day security guard was appointed to guard the subsequent Sierras. This was a reflection of the theft problem that later led to Ford dropping the Cosworth, as sales plummeted because owners could not afford the insurance.

As so many of these programmes lasted for at least a decade, watching the repeats gives quite an insight into changes in police cars. The first episode of *The Bill* was broadcast in October 1984, and by the second series the opening credits featured an SD1 overtaking a Ford Cortina Mk5. *The Bill* lasted for 16 years, and for us fans of 1980s classics its first years are essential viewing. Over time, the unmarked Sierra GL Mk1s, Mk2 Transits, Leyland Sherpa and the Metro City X with its recessed headlamps would be retired and by the time of the final episode in August 2010 we are in a London of silver Astras. Incidentally, when filming the chase sequences, it was widespread practice to over-dub the police sirens afterwards. Nearly every reader will have their own idea of the most memorable storyline, but 1997's *Humpty Dumpty*, with plenty of shots of PC Stamp at the wheel of a five-door Metro and a straight and very sinister performance from Rik Mayall, is always worth a watch.

Similarly, *A Touch of Frost* began its 18-year run in 1992, with the Ford Sierra Mk2 and the liveried Vauxhall Senator and Cavalier IIIs of the first shows reminding anyone watching of John Major-era Britain. *Taggart* was made over an incredible 27 years; the original programme from 1983 has wonderful period detail of the police driving a Morris Ital. Even the Rover 214s and Peugeot 406s that appeared in the post-Mark McManus Taggart ('There's been a murder!') now look almost antique, and much the same applies to many crime series of the 1990s.

Because some programmes are familiar from constant re-screening on satellite television, the traffic can make the viewer double-take and then suddenly feel that they are very old. Just look at *Prime Suspect III*, with its third-generation Cavaliers and Astras, or *Between the Lines* (which was a superb precursor to the current *Line of Duty*, in that it dealt with the anti-corruption unit) with its Metros and Peugeot 405s.

The Customs and Excise drama *The Knock* seems to be awash with Vauxhall Senators and Rover 827s, while the Fiesta Mk4s and Vectra Mk1s of the early editions of *Midsomer Murders* are reminders that the series began in 1997. Even the traffic in *Waking the Dead*, with its Vauxhall Omega MV6s, seems to come from another world – all those Montegos in the background, while *Hamish Macbeth* saw an incredibly young-looking Robert Carlyle with a Ford Maverick. In *The Chief*, some of the early 1990s patrol cars now appear rarer than the 1956 Riley Pathfinder driven by the central character. In the Cornish-set *Wycliffe*, everything seemed to happen slowly; it might just be my memory, but gentle non-chase chases in a Land Rover Defender 90, Land Cruiser, Vauxhall Cavalier Series III or unmarked Vauxhall Astra 1.4 LS seemed to be regular occurrences.

And so we are nearly up to date (unless you are reading this book in 2037) with the Ford Mondeo Titanium X Mk4, Citroen C5s, Kia Sorento KX-2s, Volvo S60 D2 Gen2 and a rather elderly Ford Focus Mk1 Estate in the excellent *Line of Duty* (the best police procedural of recent years), whereas *Broadchurch* must surely be the first UK cop show to feature an electric police car, a marked Vauxhall Ampera.

One of the most off-beat TV police cars of recent years has been the SWB Land Rover Defender in *Vera*. Meanwhile, for those who want to escape into the past, there have been the Morris Minor panda cars of *The Indian Doctor*, the Riley RMA and Wolseley 4/44 of *Father Brown* and, from April 1992, the long-running *Heartbeat*. This last show was initially set in 1964, and the principal on-screen police car was a black Ford Anglia 100E, which was subsequently replaced by an Anglia 105E. There were also cameos from a P4 Rover and the Ford Zephyr 6 Mk3 of Inspector 'Smiler' Hackett, but over the years what had begun as a well-observed drama about policing a 1960s North Country where the 1950s had not quite ended turned into an unabashed nostalgia fest. A famous 'guest star' seemed to appear every week, the time period became vaguer and vaguer and car enthusiasts often became progressively more irate as cars from the early 1970s crept into a drama set in the 1960s.

However, it is important to bear in mind that such programmes are not primarily aimed at us enthusiasts. We all know when we see *George Gently* that Ro-Style wheels look (and are!) wrong on his 1966 Rover P5 3-litre Coupe, that the P6 2000 is not the 1963 model as stated by its registration number (the reversing lamps are wrong), and that Detective Sergeants in the late 1960s probably didn't share a hairstylist with the look of the lead guitarist of The Kinks. That said, two of the most entertaining British films set in period do have their errors – *Quadrophenia* has Austin Maxis, Citroen GSs and Renault 12s appearing in the 'mid-1960s', and the 1969-set *Withnail & I* has the same motorway traffic travelling to and from the north. But what does that matter when you can see a Morris Oxford Series VI in pursuit of Mods and Rockers, or that scene with Richard E. Grant and the BMC J4 Black Maria? All together now – 'Get in the back of the van!'

And this brings us to *Life on Mars* and the Ford Cortina Mk3 so beloved of Detective Chief Inspector Gene Hunt that he threatens a group of 'yobs' (to use a popular 1970s phrase) with deep unpleasantness should any harm come to it. A close look at the Cortina, especially of shots of the dashboard, reveals it is not quite what it seems, and, quite amazingly, you can blame Gordon Brown for that! One of his first acts as Chancellor in 1997 was to remove the rolling vehicle tax exemption on historic vehicles – which was then defined as anything over 25 years old – and create a hard date cut-off for VED exemption of January 1973. At that time Brown stated that this would not be altered (although subsequently it's been very sensibly changed back to a rolling exemption), and the very predictable result was a lot of nice 1973/74/75 cars losing their identities and reappearing as 'restored' earlier cars on 1971/72 identities taken from scrap or damaged cars so their owners could get free road tax. This practice was illegal, but sadly very common. The production company making *Life on Mars*, Kudos Films, literally just looked in the classifieds for a suitable 1972 car for Gene Hunt a few weeks before filming started and spotted the car for sale nearby. The props buyer went to look at it, thought it seemed okay and bought it. However, immediately after the screening of episode one the internet started to buzz that the car seemed wrong; it had a 2000E crest on the C pillar and a later-type 2000E dashboard, but an earlier GXL grille and twin round headlamps in place of the E's single rectangular units.

The following morning the BBC's *Top Gear* office research team talked about the car being incorrect (John Lakey, who is working on this book with me, was there), and as one of the *TG* team knew one of the props guys on the programme, he called him. They were amazed and had no idea they had bought a doctored car! Photographs were taken and emailed across that conclusively proved the Cortina was a 1974 or 1975 2000E, wearing the front panel and ID of a 1972 GXL. The 2000E was announced in September 1973 for the 1974 model year when the whole Cortina range was facelifted and replaced the GXL. The car was Ford Copper-Bronze, a colour announced with the facelift, while earlier Mk3s were a paler shade called Amber Gold. *Life on Mars* was set in early 1973, before the car was made! This information enabled Clarkson to good humouredly mug my mate Phil when he came on *Top Gear* to do his celebrity timed lap later that week. The car still wears this ID and is owned by a collector after being sold by the BBC for Comic Relief, for which it raised almost £13,000, making it, surely, the most expensive Mk3 Cortina on the planet!

You could also spend your days spotting other automotive anachronisms in *Life on Mars* – the Series 2 Austin Allegro panda car or a

Ford Zephyr 6 Mk3 with a G-registration suffix, to name but two examples – but 'Shut your gob! Or I'll come round your houses and stamp on all your toys!', as Gene might have elegantly put it. It is never a clever idea to make Mr Hunt cross, so we will also refrain from noting that, in 1981, few DCIs were much given to driving Audi quattros! However, the final episode of *Ashes to Ashes* explains all of these seeming lapses in detail. We won't reveal the plot (it is well worth watching from scratch if you've never seen it), but all we will say is beware any senior officer who wears glasses.

Possibly the most important aspect of *Life on Mars* or *Ashes to Ashes* is that they remind older viewers of the days when a Ford Cortina Mk3 with a vinyl roof meant that its owner had truly 'arrived', and they also inspire a future generation of car enthusiasts. In twenty years' time, will people watch programmes such as *Broadchurch* and marvel at the 'classic' fossil-fuel-driven police cars on screen? Let's hope so.

NINE-NINE CARS

When you watch a police drama of the 1950s, 1960s, 1970s and 1980s, it is impossible not to marvel at the skill and dedication of the stunt professionals. So many of the great police moments mentioned in this chapter are the work of Nine-Nine Cars, the forerunner of today's Action 99 Cars firm. According to the company's leading light and stunt ace, Richard Hammatt, 'I think it was formed around 1951 or 1952. At that time, there was no organised stunt fraternity per se in the UK – just "action men" who could also act a bit and drive a car.' The name behind Nine-Nine Cars was Joe Wadham, who many readers will recognise as the moustached police S-type driver in *Robbery* and from behind the wheel of countless Wolseley squad cars, not to mention the Consul GT in the opening credits of *The Sweeney*. Richard explains that 'Nine-Nine Cars was the only company registered at Scotland Yard for such work – Joe had the right contacts.'

Nine-Nine's roster of drivers included such names as Wadham himself, Jack and Jeff Silk, Alan Stuart – who began his career as Tommy Steele's saxophonist – and Rocky Taylor, to mention just a few often-unsung heroes of the screen. There was a fleet of cars, and converting a black 6/110 to resemble a London Met vehicle was a straightforward process – 'the bell could be quickly removed, the "police" signs were detachable and the blue lamp was magnetic. And in those days people did not recognise the same number plates appearing on police vehicles on screen.' Richard also points out the skills needed for a film car chase of bygone days – 'you were driving big heavy cars with "interesting" brakes!'

One famous comedy associated with Nine-Nine was the 1962 classic *The Fast Lady*, for which they supplied the police 6/99, with Wadham driving the gangsters' Jaguar Mk VIIM – in Richard's words, 'drivers would be on both sides of the law on screen'. The chase sequence took around a fortnight to shoot and Jack Silk had to put on a blond wig to double for Julie Christie'. Four years later, Stuart featured in the 1966 B-film *Circus of Fear*, with two 6/110s chasing the hoods' BMC J4 van. A 1965 *Autocar* article explained that, when planning a crash scene, the Nine-Nine team searched for 'not a wreck but a roadworthy and bodily sound vehicle selling at £100–£150. Particular attention is paid to the engine, and an on-the-spot compression test is given before the decision to purchase. Larger cars are preferred, such as Austin A95s and 2.4 Jaguars' – this was because they were stronger and less likely to injure the drivers. Richard himself notes that the cars often had to be 'available, cheap and with plenty of oomph'.

Richard trained as a helicopter engineer and his father was a well-known supplier of cars to the film industry. He began his own screen career at Elstree Studios; 'Around 1967, 1968, I worked on *Randall and Hopkirk (Deceased)* and *Department S*. I joined Nine-Nine in about 1970 and I was originally a stunt technician, doing the preparation. This might involve altering the handbrake ratchet system so that the driver could really yank it up hard without it locking on.' A crash sequence might involve 'removing the car's front seats, installing hand rails in the roof, and devising a centre punch hole in the windshield – this was in the days of "Zone Toughened" screens. There would be a button inside of the car which the driver would press to blow the glass out.' He eventually moved to driving work and recalls, 'One of the worst cars to spin was the FX4 taxi; the clutch used to collapse. The Sierra was really good for throwing about, whether it was a police car or not.'

Over the years, Richard has been knocked over in the *Think Bike!* public information films, doubled for Christopher Reeves in the junkyard crusher scene in *Superman III* and driven a Vauxhall Chevette down a railway line towards a train. And without his skills and dedication – and those of so many other great stunt professionals – many of our favourite films and TV shows would be unrecognisable.

GOING ABROAD

The BBC adaptation of *Maigret* is a prime example of how some British programmes and films were partially shot overseas to provide some prime car-spotting opportunities. How many of us have spent a Bank Holiday Monday watching the police Peugeot 403 in the 1966 Morecambe & Wise film *That Riviera Touch* or noted the black Ford Consul Mk2 squad car in the Jamaica locations of *Dr. No*? However, one of the oddest films in Sir Sean Connery's extensive career is 1974's *Ransom*, shot in Norway with plenty of black-and-white Volvo 145s. Sean plays Colonel Nils Tahlvik, possibly the only Scandinavian police chief with an Edinburgh accent. Then there is the charming but less than sporting DAF 66 driven by Barry Foster's off-duty hero in the Dutch-set *Van der Valk*, and the Polizei BMW 5 Series E12 in *Auf Wiedersehen, Pet*. The later episodes of *Bergerac* were increasingly made in France, which meant for police cars Renault 18 Breaks and Peugeot J9s.

Spanish-set adventures of *The Protectors* meant a Guardia SEAT 1500Ls appearing before the closing titles, and these seemed exotic when we were kids.

In Malta, the police used a Land Rover 109 in *Eyewitness*, and an East German police EMW 340 arresting Michael Caine in *Funeral in Berlin*. The 2016 Anglo-German thriller *Paranoid*, which was filmed around Cologne and Düsseldorf in Germany and Cheshire in the UK, showed that, sadly, livery apart, all European police cars are now quite similar, with BMWs of various types being featured prominently.

THE CI5 MOTOR POOL

I've mentioned quite a few productions where Ford provided the cars, and one of the best examples is *The Professionals*. The series expert and author of the guide *The Professionals*, Bob Rocca, explains how this worked, and how British Leyland missed out on a vast marketing opportunity. He thinks that the programme might have worked just as well had CI5 used BL products; 'had their fleet been bigger it may have proved fruitful for them. The cars loaned to that first series *of The Professionals* were on a left-over agreement with Mark One Productions for *The New Avengers*. When the second series of that didn't materialise, it was decided to finish that agreement on *The Professionals*. I think the contract eventually fizzled out when Ford, who'd just had great success on *The Sweeney*, put their cars forward, and British Leyland just couldn't compete with the amount of cars offered on a permanent basis, most importantly, for free! BL's Harry Donovan and his team were asked about the supply of further police vehicles but couldn't provide them from BL's fleet so they suggested Kingsbury Motors would be their best bet. The rest, as they say, is history for poor old BL.'

Peter Brayham suggested the Capri would be ideal for Bodie and Doyle – 'the producer Sid Hayers said if it was good enough for John

Wayne, then it was good enough for us.' In addition to the Capris, Escorts and Granadas used by The Big A, there was also 'as a courtesy vehicle, a beige Ford Cortina Mk4 2.0GL, VHK 536S. This car was driven by unit driver Alan Lind, whose job entailed collecting the two leads from their homes every morning and conveying them to each location. It did, however, also get coverage as an on-screen car in several episodes.'

Some other on-screen Cortinas were provided by Ford, and, 'as on the first series, Kingsbury Motors and Nine-Nine Cars continued to supply the other "Action" vehicles. When Ford took over the contract the cars were kept at Mark One Productions in Wembley for the duration of each series (including unit and producers' cars), and then returned during the hiatus of each; at which point the cars could be "kept on ice" for Mark One and returned the following season, as was the case with the RS2000; and indeed, the first Capri (VHK495S) – although this was replaced with COO 251T after a few episodes'. After that point, Mark One were unsure whether any more series would be produced, 'although in fact another two were made in 1980 and 1981', so 'new sets of cars would be supplied when ordered by the production company.'

In terms of the driving, the dangerous sequences were 'usually undertaken by the stunt team; obviously, when close-ups were called for, the actors would be involved; but not as often as one would think'. For Bob, it is the Capri that best encapsulates the appeal of *The Professionals* – 'the series increased sales of it; and Ford kept the model alive that much longer as a result. The Mk3 had only just been introduced in March 1978 and then promoted in the show within a few months . . . superb publicity and a great move on their part to supply that car. The RS2000 was nearing its end-of-shelf-life too, and the result of putting that car in the show didn't really have the same impact that Ford would have hoped.'

NEWSREELS

These were once as an essential a part of your local picture house as Kia-Oras and Pearl & Dean adverts for the local car dealer.

Both British Pathé (www.youtube.com/user/britishpathe) and British Movietone (www.youtube.com/channel/UCHq777_waKMJw6SZdAB-myaA) have some wonderful free online archives, so here are just a few choice police car moments.

Pathé

Women Best Drivers? – WPCs in Austin A95 Westminster patrol cars in Southend-on-Sea, flagging down dangerous cyclists.

Look Out Bandits! – Introducing the Daimler SP250 'Dart'.

Police Cadets 1 – With a Wolseley 6/110 Mk2.

A Job for The Police – Featuring a Humber Super Snipe Series III Estate.

New Police Equipment – Starring a Sussex Constabulary Ford Cortina Lotus Mk2.

Movietone

Skid Pan – How to drive a Ford V8 Pilot under very demanding conditions.

Dial 999 – A Wolseley 6/80 chases a Sunbeam-Talbot 90 accompanied by the *Dick Barton* theme tune.

A Fair Cop – WPCs at the wheel of London Met's MGB Roadsters and Liverpool City Police Riley One-Point-Fives.

Question of Adhesion – Demonstrating the Ferguson converted police Ford Zephyr 6 MkIV.

Vascar Speed Recorder – An Essex Constabulary Triumph 2000 MkII uses the new VASCAR – note the string-operated 'Police Stop' roller blind in the rear window!

THE SWEENEY

The creator of this seminal police series was Ian Kennedy-Martin, the younger brother of Troy, and readers should look at his fascinating webpage, www.iankennedymartin.com. In an interview for the book *Talk of Drama*, by S. Day-Lewis (ULP Publications, 1998), Kennedy-Martin recalled:

I was looking at *Softly, Softly,* which by then was a very poor extension of Troy's original, and around the same time I was meeting this guy from the Met's Flying Squad and finding out some of the police realities of the day. While the Home Secretary was protesting 'We'll never arm the police', there were at that moment 200 armed policemen in London. They were called the Embassy Protection Squad. What an extraordinary coincidence that there was a bank robbery in South Kensington, and the Embassy Protection Squad happened to turn up outside, fully armed and ready to shoot the people as they came out of the bank. So I knew what was happening and talked about it to George Markstein, who was Head of Script Development at Thames-Euston.

He invited me to write an episode for their *Special Branch* (1969–74) police series. I said I'd seen it and it was terrible. He agreed and suggested I should try and think of a replacement. So I went to Euston on that basis and talked to Lloyd Shirley, the Thames drama boss, about the changes in policing signalled by the arrival of Commander Robert Mark as the new chief at Scotland Yard. I said there was a series to be made about the new-style Flying Squad. The *Regan* pilot was commissioned for the *Armchair Theatre* slot and Euston committed to a series even before the single went into production.

ACKNOWLEDGEMENTS

Thanks to:

John Lakey: well-known automotive encyclopaedia whose credits stretch across TV, film and numerous publications. Without whom this book would simply not have happened.

Steve Woodward: Former Traffic Officer and Britain's most knowledge-able police car historian who was a fantastic co-writer and a font of great stories.

Bedina Steatham: Researcher who trawled through legions of old car magazines, transcribed endless taped interviews, wrote sections of text and now really wants an MGA police car . . .

Andrew Roberts: Film historian and police car enthusiast who was a joy to work with as co-author on the film and TV chapter.

Peter Rodger: Former Chief Inspector and driving instructor at Hendon Police Collage, now Head of Driving Advice for IAMRoadSmart

(Institute of Advanced Motorists), who was the perfect co-author for the police driving chapter.

Richard French: Former Principal Engineer of the Metropolitan Police whose knowledge of how to create a police car was invaluable.

Former Police Traffic Officers who all gave freely of their time:
Roger Parker
Chris Morgan
Robert Marshall
Brian Homans
Alan Matthews

The many others who contributed stories and information, thank you. The members of the Police Car UK Club for bringing their cars to events and in some cases even allowing me the pleasure of driving them.

The men and women of the British Police, who keep our country safe and, sadly, very occasionally pay the ultimate price. It was an honour to serve with you and a pleasure to learn something about the service's motoring history.

PICTURE CREDITS

All illustrations and photographs are courtesy of the author or in the public domain, except for the following:

Pages 66, 73, 76, 80, 81, 100, 117, 181, 200, 260, 294, 340: John Lakey; page 25: Courtesy of Daimler-Benz AG; pages 28, 29, 50, 62, 157, 251: Beaulieu/National Motor Museum; pages 32, 208: Metropolitan Police; page 40: British Motor Industry Heritage Trust; pages 41, 181, 313: Carrie Wilson; page 49: Vauxhall Motors; page 67: Bryan Goodman; page 68: PC John Oliver; pages 13, 69, 70, 72: Brian Homans; pages 71, 96, 304: Hampshire Police; page 74: West Yorkshire Police/John Berry; page 75: Tom Caron Collection; page 77: City of London Police; pages 78, 98: Rootes Archive Centre Trust; pages 82, 157, 182, 236, 241, 311, 315, 330, 331: Steve Woodward; page 91, 121: Ford Motor Company; page 95: Julius Thurgood; page 96: Bedford Constabulary; page 102: Vauxhall Motors; page 103: The British Motor Museum; page 133: Jaguar Land Rover; page 248: Subaru UK; page 251: Cranhill Arts / glasgowfamilyalbum; page 305: Christopher Davies; page 309: PM Photography

Plate section credits: Plate 1: top image: John Lakey (with thanks to the British Motor Museum for the location); middle image: Beaulieu/

National Motor Museum/Ford Motor Company; bottom image: Ford Motor Company; plate 2: top image: Ford Motor Company; middle image: British Motor Industry Heritage Trust; plate 3: top image: John Lakey; middle image: British Motor Industry Heritage Trust; bottom image: John Lakey; plate 4: top image: John Lakey; middle image: Vauxhall; bottom image: Beaulieu/National Motor Museum; plate 5: all images: Ford Motor Company; plate 6: top image: Tom Caron Collection; middle image: Clifford Gray; bottom left image: Beaulieu/National Motor Museum; bottom image: John Lakey; plate 7: top image: Humberside Police; middle left image: Vic Lakey; middle right image: John Lakey; bottom left image: Vic Lakey; plate 8: middle image: John Lakey; bottom left image: Carrie Wilson; bottom right (top) image: John Lakey; bottom right image: Briggs Automotive Company (BAC); plate 9: top image: Roy Childs/Alamy; middle right image: Vic Lakey; middle right image: Thames Valley Police; bottom middle image: Steve Pearson; bottom image: John Lakey; plate 10: top image: John Lakey; middle right image: John Lakey; middle left image: BMW; bottom image (top): John Lakey; bottom image: Steve Pearson; plate 11: top two images: Steve Pearson; middle image: The Rootes Archive Centre Trust; plate 12: top image: Chief Inspector Phil Vickers/Lincolnshire Police; middle image: Ford Motor Company; bottom image: Steve Pearson; plate 13: top two images: John Lakey; bottom image: Metropolitan Police; plate 14: British Motor Museum; plate 15: top image: British Lion Film Corporation Ltd; middle image (top): John Lakey; middle image (bottom): Chris Morgan; bottom image: Cambridgeshire Constabulary; plate 16: top image: Vauxhall; middle image: John Lakey.